Reaping Digital Dividends

Reaping Digital Dividends

Leveraging the Internet for Development in Europe and Central Asia

Tim Kelly, Aleksandra Liaplina,
Shawn W. Tan, and Hernan Winkler

1818 H Street NW, Washington, DC 20433
Telephone: 202-473-1000; Internet: www.worldbank.org

1 2 3 4 20 19 18 17

ISBN (paper): 978-1-4648-1025-1
ISBN (electronic): 978-1-4648-1027-5
DOI: 10.1596/978-1-4648-1025-1

Cover photo: © Getty Images. Used with permission; further permission required for reuse.
Cover design: Debra Naylor, Naylor Design, Inc.

Library of Congress Cataloging-in-Publication Data has been requested.

Contents

Boxes

Figures

Maps

Tables

About the Authors

Tim Kelly has been a lead information and communication technology policy specialist at the World Bank Group since 2008 and is based in the Nairobi, Kenya, office. He wrote the policy chapter of the 2016 edition of the *World Development Report* on the theme "Digital Dividends." He previously managed World Bank reports on *Maximizing Mobile* and *eTransform Africa,* as well as the *Broadband Strategies Toolkit.* On the operational side, he is currently managing information and communication technology programs in Comoros, Malawi, Somalia, and Tanzania. He was formerly the head of the Strategy and Policy Unit of the International Telecommunication Union and worked with Logica and the Organisation for Economic Co-operation and Development. Over the past 25 years, Tim has specialized in the economics of information and communication technologies. He has written or coauthored more than 30 books on the subject, including the World Bank's *ICTs for Postconflict Reconstruction*, the International Telecommunication Union's *Internet Reports* and *World Telecommunication Development Report,* and the Organisation for Economic Co-operation and Development's *Communications Outlook.* He has an MA (Hons) degree in geography and a PhD in industrial economics from Cambridge University.

Aleksandra Liaplina has been an economist at the World Bank Group since 2013 and is based in Washington, DC. She specializes in infrastructure investment and finance and has contributed to a number of World Bank publications addressing the economics of infrastructure as well as the role of regulation and policies in attracting private and institutional capital to infrastructure investments. She has been managing a portfolio of public-private partnership projects under a program with the European Commission, providing policy and regulatory advisory services, participating in investment operations in Africa and Europe, and leveraging donor partnerships and client engagements. She was also involved in designing and launching the new Impact Evaluation program in the Transport and ICT Global Practice. Aleksandra was previously at Intel Corporation, working across sector development; the Russia's Ministry of Economic Development, where she was involved in Russia's World Trade Organization accession process; the United Nations; and other organizations. She is a Fulbright Scholar with a dual master's degree in international development policy from Georgetown University and in economics from St. Petersburg State University.

Shawn W. Tan is an economist in the World Bank's Trade and Competitiveness Global Practice, focusing on countries in Europe and Central Asia. His research interests are in international trade, the effects of institution, policy or regulation changes on firms, and economic geography. Before joining the World Bank, he served as a senior officer in the International Policy Division at the Singapore Economic Development Board, where he was involved in Singapore's trade agreement negotiations, in the Association of Southeast Asian Nations trade and investment forums, and in trade facilitation for multinational corporations in Singapore. He holds a PhD in economics from the University of Melbourne.

Hernan Winkler is a senior economist in the Jobs Group at the World Bank. During the completion of this report, he was an economist in the Office of the Chief Economist for the Europe and Central Asia region of the World Bank. He specializes in applied microeconomics, with a particular focus on issues related to labor markets, and the sources and consequences of poverty and inequality. His research has been published in peer-reviewed economics journals, including the *Review of Economics and Statistics* and the *Journal of Development Economics*. He was part of the core teams of several World Bank regional reports, including *Diversified Development, Golden Aging,* and *Risk and Returns*. He was previously a researcher at CEDLAS (University of La Plata, Argentina), where he conducted research on poverty and distributional issues affecting countries in Latin America and the Caribbean. He holds a master's degree in economics from the University of La Plata and a PhD in economics from the University of California at Los Angeles.

Foreword

When the Internet began spreading across the world in the late 1990s, it held out the promise to overcome impediments to economic development. If distance no longer mattered, people, governments, and firms would become increasingly interconnected, and the place where one lives and works would no longer be the main determinant of economic success. Online education could reach places where teachers and schools are scarce. Easier access to information would allow people to make better economic decisions and would enable revolutionary increases in productivity.

In the intervening years, the Internet delivered some of its promises. First, it has reached developing countries faster than other technological advances. Nowadays, citizens in the South Caucasus and in Central Asia are more likely to use the Internet and have a mobile phone subscription than to have a bank account or a credit card. Thousands of workers in Europe and Central Asia already use online platforms to offer their services to clients in high-income countries. Transparent and competitive platforms have the potential of transforming business environments that too often have been dominated by special interests or shielded from global markets.

Unfortunately, the Internet has fallen short of its promises in some other aspects, as its benefits have not fully reached people in the bottom of the income distribution. For example, skilled workers are better able to leverage the Internet to increase their earnings, whereas unskilled workers face a higher risk of having their jobs automated. High-productivity firms are more likely to use the Internet to grow their business, while mom-and-pop stores face the risk of being displaced. Rich countries, which tend to have stronger institutions than poorer countries, are better equipped to use the Internet to fight corruption and hold public officials accountable.

In other words, the past three decades have shown that the benefits of achieving universal Internet access or increasing the size of the information and communication technology sector will not fully materialize unless governments improve their business environment, invest in human capital, and enhance their institutions. This message cannot be emphasized enough, especially for countries in the ECA region with a long tradition of distortionary industrial policies aimed at cherry-picking winners and losers.

Reaping Digital Dividends: Leveraging the Internet for Development in Europe and Central Asia provides a framework for governments in the region to maximize the impact of the Internet on poverty reduction and shared prosperity. It highlights the diverse yet surmountable set of challenges. Although many people in the East remain unconnected, the experience of their neighbors to the West shows that achieving nearly universal Internet access does not guarantee success. For example, several factors hamper technology adoption among firms, while rigid regulations constrain the expansion of the sharing economy.

This report argues that reaping digital dividends requires policies focused not only on the telecommunications sector but also on the analog complements, such as skills and the business environment. It also highlights that governments should be prepared to address the disruptive effects of new technologies and facilitate the transition of displaced workers to new and more productive jobs. With the right policy mix and investments, countries in ECA have the exceptional opportunity to embrace the Internet to narrow the gap with richer economies, to level the playing field for all firms, and to make labor markets more inclusive. *Reaping Digital Dividends* aims to provide a blueprint to help governments make this a reality.

Cyril Muller
Vice President
Europe and Central Asia Region
World Bank Group

Acknowledgments

This report was prepared by a team led by Tim Kelly, Aleksandra Liaplina, Shawn W. Tan, and Hernan Winkler. The work was carried out under the direction of Hans Timmer, Chief Economist for the Europe and Central Asia region, and with the guidance of Cyril Muller, Vice President of the Europe and Central Asia region.

- The Overview was written by Hernan Winkler, with inputs from Tim Kelly, Aleksandra Liaplina, and Shawn W. Tan.
- Chapter 1 ("Internet Adoption") was written by Aleksandra Liaplina, with inputs from Tim Kelly, Aude Coponat Schoentgen, Ali Hassan, and Maria Emilia Cucagna.
- Chapter 2 ("Supply of Internet Networks and Services") was written by Aleksandra Liaplina, with inputs from Tim Kelly, Jan van Rees, and Aude Coponat Schoentgen.
- Chapter 3 ("Firm Productivity and Dynamics") was written by Shawn W. Tan, with inputs from Ali Hassan, Jelena Kmezic, and Ehui Adovor.
- Chapter 4 ("Digital Trade") was written by Shawn W. Tan, with inputs from Ali Hassan, Jelena Kmezic, and Ehui Adovor.
- Chapter 5 ("The Internet Is Changing the Demand for Skills") was written by Hernan Winkler, with inputs from Maria Emilia Cucagna, Maya Eden, Emmanuel Vazquez, Patrizia Luongo, Kirill Vasiliev, and Dmitry Chugunov.
- Chapter 6 ("The Internet Is Changing the Labor Market Arrangements") was written by Hernan Winkler, with inputs from Maria Emilia Cucagna, Emmanuel Vazquez, Georgi Panterov, Ievgeniia Viatchaninova, Christine Shepherd, Siddhartha Raja, and Natalija Gelvanovska.
- Chapter 7 ("Toward a Digital Single Market") was written by Tim Kelly, with inputs from Aude Coponat Schoentgen, Shawn W. Tan, and Hernan Winkler.

The team is grateful for the feedback and support provided by Anabel Gonzalez, Pierre Guislain, Boutheina Guermazi, Randeep Sudan, Paulo Guilherme Correa, and Javier Suarez. The team is grateful for the guidance, support, and technical inputs from the *World Development Report 2016: Digital Dividends* team, and peer reviewers Juan Navas-Sabater, Marc Tobias Schiffbauer, Mary Hallward-Driemeier, and Phillippa Biggs (Coordinator for Broadband Commission for Digital Development, International Telecommunication Union).

In addition, two workshops offered opportunities for different parts of the report to be presented and reviewed. In particular, a brainstorming workshop with senior management from the Europe and Central Asia region, and the Trade and Competitiveness and the Transport and ICT Global Practices, helped build the policy relevance of the report. In addition, an authors' workshop with attendance from World Bank staff and external reviewers contributed to strengthen the technical framework of the report.

Valuable comments and suggestions were also provided by Andras Horvai, Johannes Zutt, Carl Patrick Hanlon, Jakob Kopperud, Eskender Trushin, Roumeen Islam, John Mackedon, Wolfgang Fengler, Ganesh Rasagam, Andrew Kircher, Irene Wu (Federal Communications Commission), Elisabetta Cappanelli, Piotr Lewandowski (Institute for Structural Research), and Caroline Paunov (Organisation for Economic Co-operation and Development).

The team especially acknowledges the valuable contributions of Ekaterina Ushakova, who oversaw the production of this report; Rhodora Mendoza Paynor, who provided continuous support; and William Shaw and Colin Blackman, who provided insightful comments and changes to the report, and whose editing skills improved the report substantially.

Rumit Pancholi was the production editor for the report, working with acquisitions editor Jewel McFadden, executive editor Pat Katayama, and editorial and production manager Aziz Gokdemir in the World Bank's formal publishing program. Elizabeth Forsyth edited the report in its various iterations, Zuzana Johansen proofread the report, and Datapage International prepared the typeset pages. Debra Naylor designed the report and cover, using a photograph courtesy of Getty Images. The team is grateful for their professionalism and expertise.

Abbreviations

ASYCUDA	Automated System for Customs Data
B2C	business-to-consumer
BPO	business process outsourcing
CRM	customer relationship management
DAI	Digital Adoption Index
DSL	digital subscriber line
EAEU	Eurasian Economic Union
EAP	East Asia and Pacific
ECA	Europe and Central Asia
EGDI	e-government development index
ERP	enterprise resource planning
EU	European Union
FDI	foreign direct investment
GATS	General Agreement on Trade in Services
GDP	gross domestic product
GMM	generalized method of moments
GNI	gross national income
GSM	Global System for Mobile
HSDPA	High Speed Downlink Packet Access
ICT	information and communication technology
ID	identification
IoT	Internet of Things
IP	Internet Protocol
ISP	Internet service provider
IT	information technology
ITA	International Technology Agreement
ITU	International Telecommunication Union
IXP	Internet exchange point
KILM	Key Indicators of the Labour Market
LAC	Latin America and the Caribbean
LFS	labor force survey
LLU	local loop unbundling
LPI	Logistics Performance Indicators
LTE	Long-Term Evolution
M2M	machine-to-machine
MOOCs	massive online open courses

MTS	Mobile TeleSystems
MVNO	mobile virtual network operator
NGA	next-generation access
NGN	next-generation network
NTMs	nontariff measures
OECD	Organisation for Economic Co-operation and Development
OFDM	orthogonal frequency-division multiplexing
OTT	Over-the-Top
PIRLS	Progress in International Reading Literacy Study
PISA	Programme for International Student Assessment
PPP	purchasing power parity
R&D	research and development
RFID	radio frequency identification
SMP	significant market power
SPS	sanitary and phytosanitary
STEM	science, technology, engineering, and mathematics
TASİM	Trans-Eurasian Information Super Highway
TBTs	technical barriers to trade
TEIN	Trans-Eurasia Information Network
TFP	total factor productivity
VoIP	Voice over Internet Protocol
VtI	vehicle-to-infrastructure
W-CDMA	Wideband-Code Division Multiple Access
WiMAX	Worldwide Interoperability for Microwave Access
WITS	World Integrated Trade Solution
WoW	Women in Online Work
WTO	World Trade Organization

Country and Economy Codes

AFG	Afghanistan
AGO	Angola
ARE	United Arab Emirates
ARG	Argentina
ARM	Armenia
AZE	Azerbaijan
BDI	Burundi
BEL	Belgium
BEN	Benin
BFA	Burkina Faso
BGD	Bangladesh
BGR	Bulgaria
BIH	Bosnia and Herzegovina
BLR	Belarus
BLZ	Belize
BOL	Bolivia

BTN	Bhutan
CAF	Central African Republic
CHE	Switzerland
CHL	Chile
CHN	China
CIV	Côte d'Ivoire
CMR	Cameroon
COD	Congo, Dem. Rep.
COG	Congo, Rep.
CRI	Costa Rica
CZE	Czech Republic
DJI	Djibouti
DNK	Denmark
DOM	Dominican Republic
ECU	Ecuador
EGY	Egypt, Arab Rep.
ESP	Spain
EST	Estonia
ETH	Ethiopia
FRA	France
GBR	United Kingdom
GEO	Georgia
GHA	Ghana
GIN	Guinea
GRC	Greece
GRD	Grenada
GTM	Guatemala
HKG	Hong Kong SAR, China
HND	Honduras
HRV	Croatia
HTI	Haiti
HUN	Hungary
IND	India
IRN	Iran, Islamic Rep.
IRQ	Iraq
ISR	Israel
JAM	Jamaica
JOR	Jordan
JPN	Japan
KAZ	Kazakhstan
KEN	Kenya
KGZ	Kyrgyz Republic
KHM	Cambodia
KOR	Korea, Rep.
KSV	Kosovo
KWT	Kuwait
LAO	Lao PDR

LKA	Sri Lanka
LTU	Lithuania
LUX	Luxembourg
LVA	Latvia
MAR	Morocco
MDG	Madagascar
MEX	Mexico
MKD	Macedonia, FYR
MLT	Malta
MMR	Myanmar
MNE	Montenegro
MNG	Mongolia
MOZ	Mozambique
MRT	Mauritania
MWI	Malawi
MYS	Malaysia
NAM	Namibia
NER	Niger
NGA	Nigeria
NLD	Netherlands
NOR	Norway
NPL	Nepal
NZL	New Zealand
PAN	Panama
PHL	Philippines
POL	Poland
PRT	Portugal
PRY	Paraguay
ROU	Romania
RUS	Russian Federation
RWA	Rwanda
SDN	Sudan
SEN	Senegal
SGP	Singapore
SLE	Sierra Leone
SLV	El Salvador
SRB	Serbia
SVK	Slovak Republic
SVN	Slovenia
TGO	Togo
THA	Thailand
TJK	Tajikistan
TKM	Turkmenistan
TTO	Trinidad and Tobago
TUR	Turkey
TZA	Tanzania
UGA	Uganda

UKR	Ukraine
URY	Uruguay
USA	United States
VCT	St. Vincent and the Grenadines
VEN	Venezuela, RB
VNM	Vietnam
WSM	Samoa
YEM	Yemen, Rep.
ZAF	South Africa
ZMB	Zambia
ZWE	Zimbabwe

Regional Classifications Used in This Report

This report covers 47 countries referred to as Europe and Central Asia (ECA) countries. These are divided into 10 groups: Western Europe, Southern Europe, Central Europe, Northern Europe, Western Balkans, South Caucasus, Central Asia, Russia, Turkey, and Other Eastern Europe.

Western Europe	Southern Europe	Central Europe	Northern Europe	Western Balkans
Austria	Cyprus	Bulgaria	Denmark	Albania
Belgium	Greece	Croatia	Estonia	Bosnia and Herzegovina
France	Italy	Czech Republic	Finland	Kosovo
Germany	Malta	Hungary	Latvia	FYR Macedonia
Ireland	Portugal	Poland	Lithuania	Montenegro
Luxembourg	Spain	Romania	Sweden	Serbia
The Netherlands		Slovak Republic		
United Kingdom		Slovenia		
South Caucasus	**Central Asia**	**Russia**	**Turkey**	**Other Eastern Europe**
Armenia	Kazakhstan			Belarus
Azerbaijan	Kyrgyz Republic			Moldova
Georgia	Tajikistan			Ukraine
	Turkmenistan			
	Uzbekistan			

Overview

Information technologies have been transforming the world for centuries. The movable-type printing press, which originated in the Europe and Central Asia (ECA) region about 500 years ago, was a fundamental condition for the diffusion of literacy in Renaissance Europe. This invention gave birth to a new culture of information exchange, where rising professionals benefited from cheaper books to conduct business. Moreover, cities with printing presses attracted scholars and universities as well as novel industries through backward links (such as paper mills). As a consequence, cities that were early adopters of this technology witnessed more rapid and dramatic improvements in their living standards. Others lagged behind (Dittmar 2011).

Even earlier, about 1,000 years ago, a more basic technology helped to create a culture of knowledge and information exchange in Central Asia. The large-scale production of paper was a driving force behind the proliferation of books—and of reading and writing—in this region. While libraries existed in Europe during this time, their number and wealth of classical texts were profoundly weaker. Moreover, during this Enlightenment, the development of scientific knowledge in Central Asia flourished. Not only were they advanced in astronomy, geology, and medicine, but they also wrote the first book of algebra, which gave its name to the field (Starr 2013).

Like the printing press and the large-scale production of paper, the Internet is creating new opportunities for individuals and businesses across the globe. However, not everyone is benefiting equally. The economic gains—or digital dividends—associated with the Internet often accrue to groups with higher incomes, with the right set of skills, or in a suitable enabling environment.

As a result, growing inequalities between and within countries may follow. As this book argues, this need not be the case. If the right set of policies is established, digital dividends have the potential to be the driving force behind poverty reduction and shared prosperity in the ECA region.[1]

How Is ECA Different?

From east to west, the role of the Internet among the economies of ECA countries could be characterized in five words: strong governments, risk-averse private sectors. However, this statement carries vastly different meanings across the region. In the richer economies of the west, government strength is associated with both fear of change and strong regulations that address many of the market failures of the telecommunications sector. Among the developing economies of the east, government strength is associated mostly with Internet providers that are state-owned monopolies. Private sector risk aversion means that vested interests as well as concerns about privacy and cybersecurity have kept the depth and complexity of Internet use at very low levels.

The experience of the European Union (EU) shows both the benefits of strong government involvement in solving the market failures that plague Internet provision and the negative consequences of government aversion to the changes introduced by new technologies. As chapter 2 shows, most countries in the EU even outperform the United States in affordability and quality of Internet access. This was achieved through a mix of policies intended to foster competition in the sector by promoting infrastructure sharing and the development of Internet exchange points, a factor to promote competition among providers. Moreover, the EU's regulatory framework has been a major driver of regulatory reform, not only in its members but also in countries seeking to join the EU.

The experience of the EU shows that affordable and nearly universal access does not necessarily imply that firms and individuals will make the most of the Internet. In fact, given their level of technological development, this group of countries lags in e-commerce (figure O.1), use of the Internet for sophisticated tasks beyond e-mail, and development of firms that are global leaders in the Internet market. So why is Europe still awaiting its Google or Facebook?

At the other extreme of ECA, countries in Central Asia and the South Caucasus face a different set of challenges. Their residents pay some of the highest prices in the world for Internet access. In return they get very poor service, with mobile speeds barely sufficient to send and receive text messages. While geography—being landlocked and having a mountainous terrain and low population density—plays an important role in driving up prices, other issues exacerbate these limitations. For example, among landlocked countries, good diplomatic relationships and connectivity with neighboring countries are crucial. But geopolitical issues in Central Asia and South Caucasus represent a major hurdle to making the Internet affordable for everyone. Additionally, the reluctance of some governments in the region to foster development of the broadband market makes achieving affordability nearly impossible. As a result, about 40 and

60 percent of the population in Central Asia and South Caucasus, respectively, remains unconnected (figure O.2). This implies that firms and individuals in this part of ECA will be excluded from the productivity and wage gains associated with digital technologies in the near future. What will it take to make this happen? This book argues that *embracing change* will be a key ingredient.

FIGURE O.1
ECA countries lag behind other countries in e-commerce sales

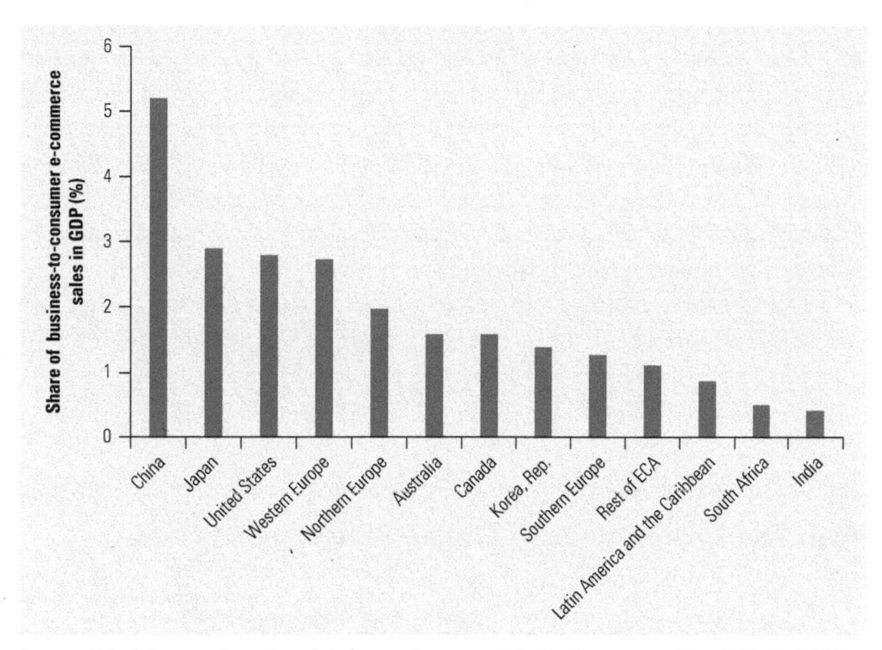

Source: Calculations are based on data from various reports by the Ecommerce Foundation for 2014.
Note: The averages are calculated for the regions. "Rest of ECA" countries are Belarus, Bulgaria, Croatia, the Czech Republic, Hungary, Poland, Romania, the Russian Federation, the Slovak Republic, Slovenia, Turkey, and Ukraine. GDP = gross domestic product.

FIGURE O.2
The digital divide is high in Eastern ECA

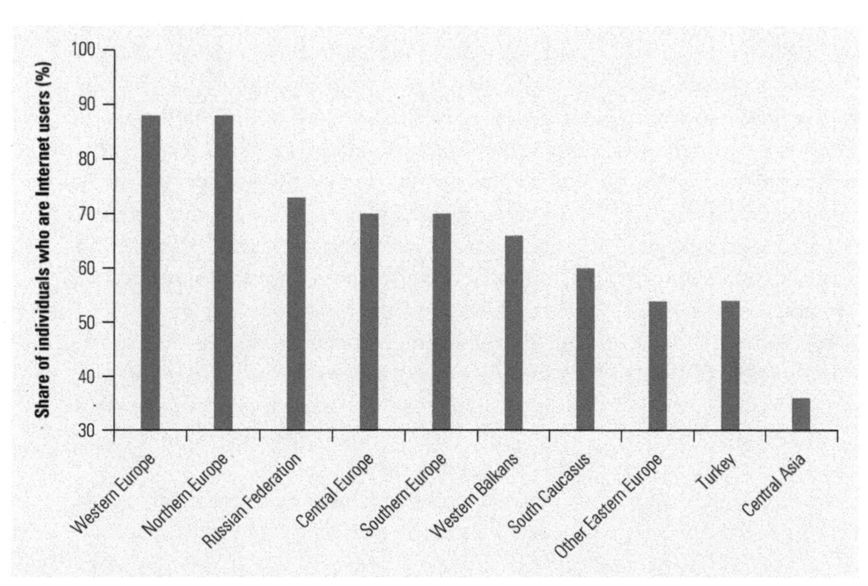

Source: Based on World Development Indicators. International Telecommunication Union, World Telecommunication/ICT Development Report and database, and World Bank estimates.

Embracing Change

The main challenge facing ECA countries is an aversion to the changes that new technologies may bring. This aversion is manifest in the backlash that sharing-economy platforms have suffered in the richer economies of the west (box O.1) and the efforts of monopolistic telecommunications providers to prevent the entry of new competitors in the economies of the east. While this book acknowledges that the Internet—as with most technological changes—may bring disruptions, its potential efficiency gains would outweigh those costs. Instead of protecting existing jobs by inhibiting the application of new technologies, policies could focus on assisting workers who lose their job in finding new job opportunities. Instead of a top-down attempt to control information, policies could focus on reaping benefits for societies through bottom-up innovation, unleashed by access to the Internet and big data (see chapter 6).

The trade-offs at the center of this argument are not new. They are just another version of the process of *creative destruction* outlined by

BOX O.1 Declared War: The Misfortunes of Online Platforms in the EU

Ever since Uber decided to jump-start its businesses in the old continent, governments across the region did not hesitate to block or curtail the expansion of this ride-sharing platform. With the support of taxi and chauffeured car trade associations, legislators engaged in legal battles with the platform. For instance, French lawmakers passed a bill banning Uber from using global positioning system technology to display the location of cars and potential customers and to require drivers to return to their place of dispatch if they were not immediately booked for another ride. Even more onerous, the government temporarily issued a decree imposing a 15-minute wait between the booking time and the time passengers could actually board the car (this decree was eventually struck down). The war against Uber has extended well beyond France, with UberPOP (the lowest-fare version of Uber) being banned in many countries and cities.

The controversy regarding online platforms did not end with Uber, as home-sharing platforms have also experienced turbulent times in Europe. For instance, Berlin imposed a de facto ban on short-term rentals targeted at people renting more than 50 percent of their apartment. Regulators defend this legislation on the grounds that Airbnb (a home-sharing platform) is increasing property prices and contributing to a housing shortage.[a]

Concerns about the rise of online platforms are justified, as they do generate disruptions. As is the case with international trade and technological change, they create adjustment costs, especially for displaced workers who do not have the skills to find a new job or who made significant investments in their previous occupation. Most important, they generate significant challenges to the existing social contract by creating new forms of work. The institutional arrangements originally designed to protect salaried workers are not easily available to the independent workers of online platforms. Those who lack access to unemployment and health insurance as well as to future pensions will face a higher risk of falling into poverty in old age or when facing negative shocks.

However, online platforms also create efficiency gains. For instance, Airbnb has brought significant

(Continued)

BOX O.1 *(continued)*

benefits for consumers, as the platform's additional competition has helped to bring down hotel rates (Zervas, Proserpio, and Byers 2016). Similarly, Uber is proving to be an important resource for reducing income volatility for many workers (Farrell and Greig 2016). A recent study shows that Uber brings huge efficiency gains to the transportation industry, as Uber drivers spend a much larger share of their driving time and drive more miles with a passenger in the car than do taxi drivers (Cramer and Krueger 2016). The higher the inefficiency of the incumbents, the more dramatic the efficiency gains from online platforms will tend to be. For example, the taxi industry is a textbook example of the distortionary effects of overwhelming government regulations. The technology and practices of ride-sharing platforms help to (1) avoid expensive licensing systems, (2) achieve better matching of drivers and passengers, (3) elude regulations such as allowing taxi drivers to drop off passengers outside their license area, but not allowing them to pick up another passenger there, and (4) match supply and demand better by having more flexible supply and prices.

The efforts of governments to stop the diffusion of the sharing economy stand in sharp contrast with the actions of consumers and providers of services in online platforms. The growth of this new business model has been outstanding. Already about a third of individuals age 25–39 have used online platforms in the European Union (see chapter 6). Accordingly, McKinsey Global Institute (2016) estimates that about 24 million independent workers in Europe and the United States have provided services in online platforms.

Should governments continue to discourage technological change? If housing shortages exist, eliminating distortionary regulations that constrain the supply could be a more effective way to promote affordable housing than artificially restricting demand by discouraging the entry of home-sharing platforms (see, for example, Cheshire and Hilber 2008). Analogously, policies designed to facilitate the transition of displaced workers toward new jobs and to adapt labor market institutions to new forms of work may prove to be more effective for economic development than regulatory measures to prevent inevitable changes.

a. See "Berlin Ban on Airbnb Short-Term Rentals Upheld by City Court," *The Guardian*, June 8, 2016 (https://www.theguardian.com/technology/2016/jun/08/berlin-ban-airbnb-short-term-rentals-upheld-city-court).

Schumpeter, in which innovative sectors expand and replace old and traditional ones, and creative destruction is an essential component of economic development.

Naturally, this change involves far more than simply increasing access to new technologies for everyone. It also entails creating an enabling environment for firms and individuals to make the most of them. This is not only a condition for *digital dividends* to improve household living standards in the long term, but also to mitigate the disruptive effects that may occur in the short term. As this book argues, efforts to avoid change will only delay—and certainly not prevent—change, as new technologies always find a way to breach barriers. As the history of the movable-type printing press showed, unintended consequences of myopic policies could last for centuries—too high a price to pay.

How should ECA countries embrace the changes that the Internet may bring? The next section describes the framework used in this book.

Reaping Digital Dividends in ECA: A Framework

This book follows the *World Development Report 2016: Digital Dividends* (WDR16) in arguing that, to reap the maximum digital dividends, the economies of ECA not only need to implement the reforms necessary to improve Internet access, but also need to focus on the "analog foundations" of the digital economy, specifically skills, institutions, and regulations (World Bank 2016b). Thus, developing a strategy to reap the digital dividends requires addressing the digital divide by removing the barriers to an Internet that is universal, affordable, open, and safe for firms, citizens, and governments. It also requires strengthening the analog "complements" to digital technologies to ensure that innovation, efficiency, and inclusion prosper and that the countervailing risks of concentration, inequality, and control are mitigated (figure O.3).

Throughout its seven chapters, this report provides many examples of bottlenecks that are affecting the improvement of Internet access and analog complements in the ECA region. The experience of the richer economies of the west illustrates that nearly universal and affordable Internet access is far from being the magic bullet for the emergence of a strong Internet economy. For instance, the developed economies of ECA have smaller shares of e-commerce sales in gross domestic product (GDP) than Japan and the United States. The causes of the low levels of e-commerce in the EU do not come from poor Internet provision. Instead, a host of complementary factors such as lack of online payment systems, poor logistics infrastructure, and stringent regulations could explain this result. More specifically, the development of e-commerce relies on the availability of instruments of electronic payment, such as debit or credit cards. With regard to credit card use, the economies of Western and Northern Europe are well behind the United States, while countries in Central and Eastern Europe underperform the economies of East Asia (figure O.4). In Central Asia, South Caucasus, and the Western Balkans, the use of credit cards is almost negligible. Complicating the issue, an overwhelming fraction of people in ECA report lack

FIGURE O.3
Transforming the digital divide into digital dividends

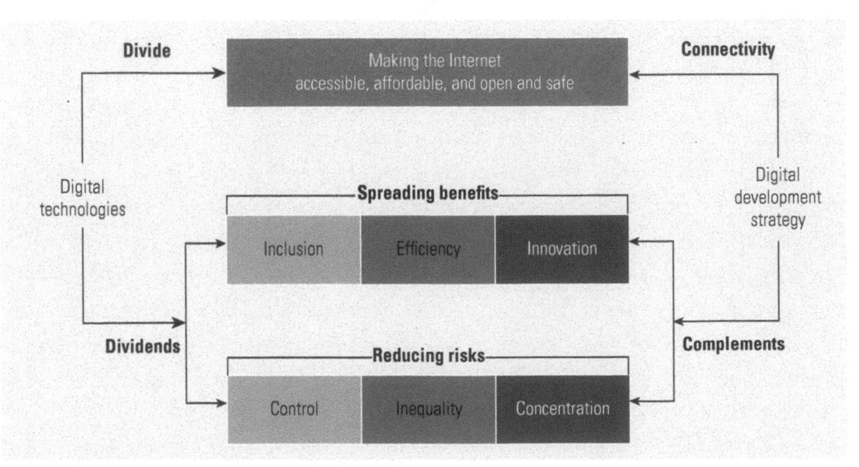

Source: WDR 2016 team.

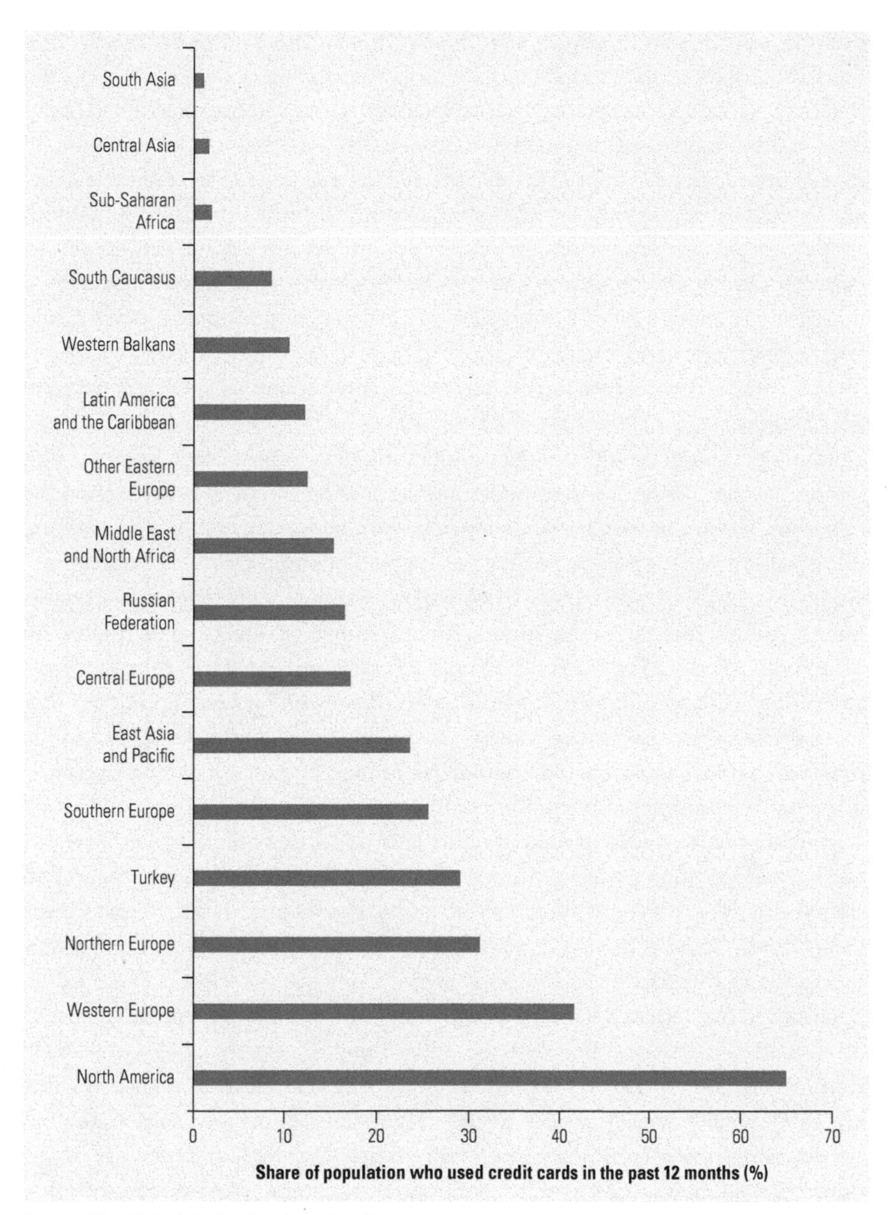

Source: The Global Findex Database 2014.

FIGURE O.4
Low use of digital payments is a barrier to e-commerce in ECA

of trust in financial institutions as the main reason for not having a bank account, which could be a difficult bottleneck to overcome. In addition, privacy and security are large concerns for both individuals and firms in Europe: many individuals do not engage in online shopping because of security concerns, and one out of every four firms in most European economies refuses to use cloud computing because of security concerns.

Furthermore, highly regulated markets inhibit the ability of firms to sell online. EU countries have attempted to ban the online sale of services from other EU countries: Germany tried to ban the online sale of over-the-counter pharmaceuticals, and Hungary tried to ban the online sale of contact lenses. These countries argued that the requirement was applied equally to all firms, but the bans provided extra advantages to local firms and went against the EU principle of free movement of goods and services. While these countries were ultimately unsuccessful in imposing the bans, it took a legal case to settle this issue, which creates uncertainty and costs for exporting firms. In summary, without these complementary factors—efficient regulations for a competitive market, strong trade infrastructure, and financial systems—the richer economies of ECA may not experience all of the positive effects that digital technologies could have on efficiency and innovation.

In the eastern part of the region, covering mainly countries in Central Asia and South Caucasus, a different set of bottlenecks is limiting the role of the Internet. Landlocked countries with sparse populations face higher costs to expanding Internet access. They also have poor international connectivity with neighboring countries. Countries that are relatively poorly interconnected across borders are doubly disadvantaged because they can be held hostage by Internet transit providers in neighboring countries. There are, for instance, only three fiber connections into Turkmenistan, while Armenia's borders with several of its neighbors are effectively closed for transit because of strained political relationships. Accordingly, the absence of a close relationship between Tajikistan and Uzbekistan hampers schemes to improve regional connectivity in Central Asia, resulting in prohibitive prices to access the Internet. More than 80 percent of the population in Armenia, Georgia, and the Kyrgyz Republic would have to spend at least 10 percent of their household expenditure to obtain a basic mobile plan. A vicious circle emerges: high prices and poor service quality for the Internet mean that demand is low, which, in turn, fails to generate incentives for infrastructure investment. Would achieving universal Internet access be enough to bring digital dividends to this part of ECA? The answer to this question is a definite no, as countries in this region are also lagging in the area of analog complements such as skills and institutions. For instance, more than 50 percent of 15-year-olds in Central Asia are functionally illiterate (that is, they know how to read and write, but cannot make inferences or understand forms of indirect meaning). Given this deficit, better Internet provision may not necessarily complement their skills and translate them into wage gains. Accordingly, poor competition policies and infrastructure mean that firms fail to have the incentives to incorporate the Internet into their production processes (figure O.5). Without skills or competitive markets, digital technologies may bring risks such as inequality and market concentration for this group of countries.

Other unique challenges exist for countries between the developed economies of Western Europe and the developing economies of Central Asia and the South Caucasus. These countries have achieved an adequate level of Internet access and quality of provision. As a result, firms and individuals are already taking advantage of digital dividends to exploit job and business opportunities. However, the diffusion of the Internet across economic sectors is already exposing the tensions that technological change typically brings about. For instance, telecommuting falls short of promoting the labor market inclusion of women and older workers in countries where these groups face barriers to labor market

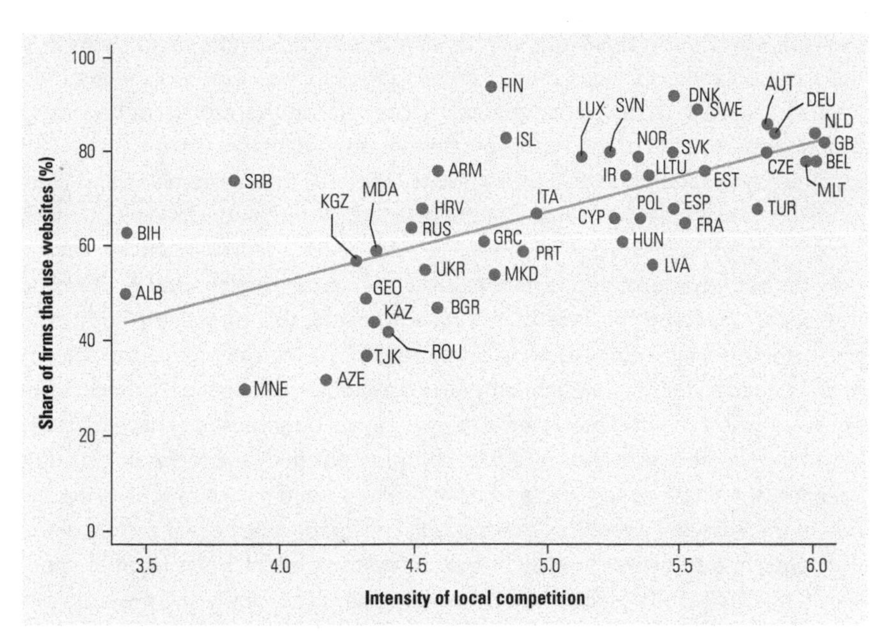

FIGURE O.5
Internet use by firms is positively correlated with the intensity of local competition

Source: Calculations based on using data from Enterprise Survey, Eurostat, and WEF Global Competitiveness Report.
Note: The 2013 data is used for most countries, except for Russia, which has Internet use data from 2012, and for Tajikistan, which has competition data from 2014. WEF = World Economic Forum.

FIGURE O.6 Older workers and women are more likely to take advantage of telecommuting in countries where they face lower participation barriers

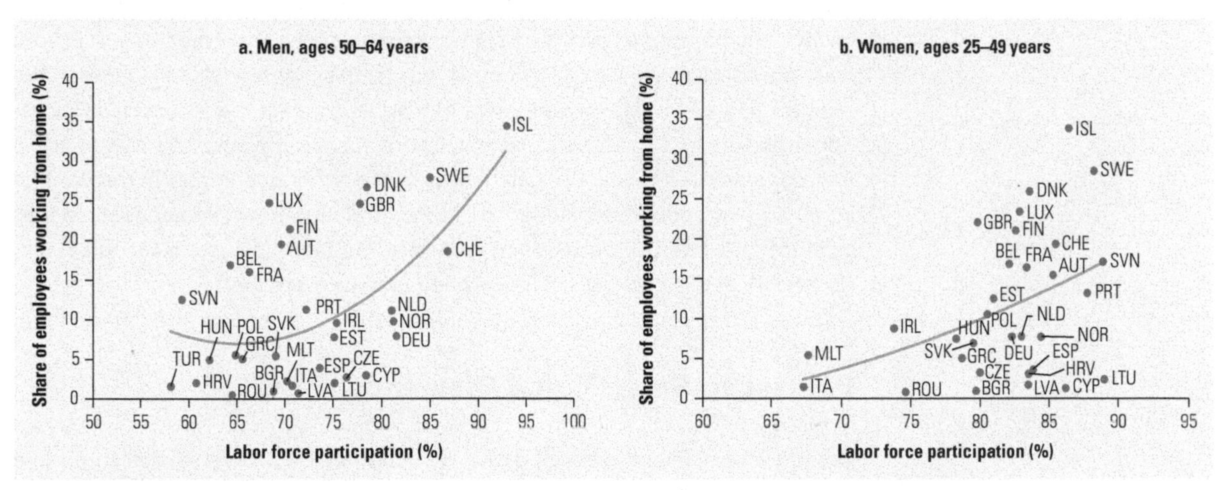

Source: Eurostat 2014.
Note: Turkey was removed from panel b, as per a reviewer's suggestion, because it is an outlier. LFS = labor force survey.

participation (figure O.6). By allowing traditional tasks to be broken into smaller and more specialized tasks, the Internet is also driving the rise of the "gig" economy, where traditional employment—defined as a full-time, permanent salaried job—shrinks, and alternative work arrangements grow. In fact, countries that were early adopters of telecommunications reforms aimed at improving the availability and affordability of the Internet experienced a more rapid increase in the share of workers telecommuting and engaging in part-time work in Internet-intensive sectors. At the same time, many countries in this group are becoming global leaders in the online supply of talent through digital platforms.

These new labor market developments create both opportunities and challenges. For example, while many of these countries are global leaders in online freelancing, most online freelancers are working in the shadow economy. That is, they do not have legal contracts that are enforceable (see chapter 6). If online freelancers become the face of the *new informal*, this may exert additional pressure on social security systems in the region. At the same time, the scarcity of skills needed for the new economy—such as socioemotional or high cognitive skills—means that firms interested in technological upgrades may have difficulty recruiting workers with the skills that complement the new technologies. For instance, the highly specialized and compartmentalized nature of tertiary education in the Russian Federation, often closely affiliated with specific sectors of the economy, fails to deliver the flexibility needed by workers in the Internet economy (OECD 2013). The incentives to be in the formal economy and the supply of skills required in the new economy are underdeveloped in these countries. Without these complements, digital technologies will not improve inclusion and might increase inequality.

In order to reap digital dividends, ECA countries need to implement not only reforms in the telecommunications sector, but also reforms affecting the development of skills, labor markets, and institutions. After all, the high correlation between Internet penetration and economic development may reflect both the impact of technologies on economic growth and the fact that richer economies with high levels of human capital and market-friendly regulations are more likely to absorb new technologies (figure O.7). In other words, this high correlation between the Internet and economic development highlights the fact that economic success depends on a kaleidoscope of factors. Consequently, it would be naïve to argue that countries in the east can catch up with the richer economies in the west by improving Internet access alone. The high correlation between economic development and Internet penetration provides the stepping-stone for the policy framework of this book, which is outlined in the next section.

Policies to Reap Digital Dividends in ECA

Countries at different levels of digital development face diverse challenges and hence a different prioritization of policies. This book tailors the policy framework of the WDR16 to the ECA region by splitting countries into three groups according to their level of digital development: emerging, transitioning, and transforming (figure O.8). Emerging digital countries are those where the level of Internet penetration is so low that they have barely experienced any of the transformational changes that this new technology brings. Transitioning digital countries are those with a decent level of Internet access, but where this new technology is generating disruptions due to the lack of certain analog complements. Finally, transforming digital countries are those with a well-developed telecommunications market,

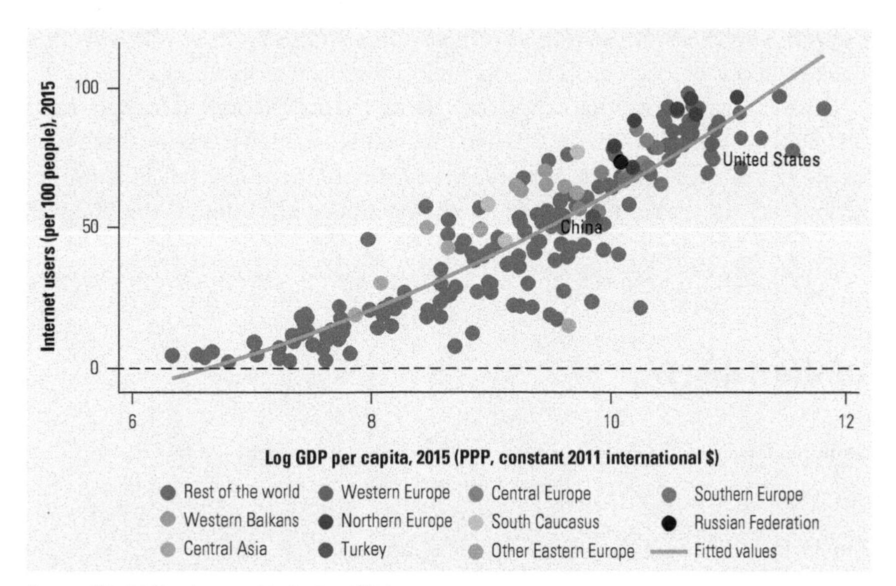

FIGURE O.7
Internet penetration and economic development are highly correlated

Source: World Development Indicators 2016.
Note: Regarding Internet use, Internet users are individuals who have used the Internet (from any location) in the past 12 months. Internet can be accessed using a computer, mobile phone, personal digital assistant, games machine, digital TV, and so forth. GDP = gross domestic product; PPP = purchasing power parity.

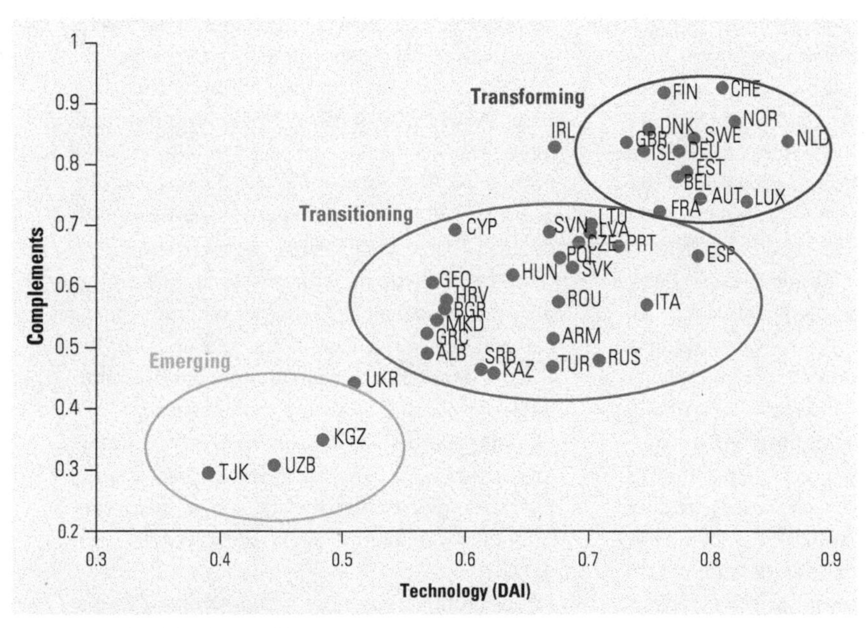

FIGURE O.8
Emerging, transitioning, and transforming digital economies in ECA

Source: Based on World Bank 2016.
Note: Data for Turkmenistan are not available. "Technology" (*x* axis) is measured by the Digital Adoption Index (DAI), a composite measure of progress in digital technologies. DAI is based on three sectoral subindexes covering businesses, people, and governments, with each subindex assigned an equal weight: DAI (Economy) = DAI (Businesses) + DAI (People) + DAI (Governments). Each subindex is the simple average of several normalized indicators measuring the adoption rate for the relevant groups. For further detail on DAI, see the WDR16 background papers at www.worldbank.org/wdr2016. Similarly, "Complements" (*y* axis) is the average of three subindicators: starting a business; years of education adjusted for skills; and quality of institutions.

but some insufficiently developed complementary factors, which will inhibit the positive effects of the Internet on economic growth and shared prosperity.

Each chapter of this book considers in depth the policy priorities for each topic and group of countries (box O.2). Given substantial variation in the level of development achieved in different areas, some countries may fall in different groups across chapters, depending on the policy priorities of the chapter.

BOX O.2 Structure of the Report

The rest of the book examines the following policy challenges facing countries:

- Chapter 1 summarizes the main facts about Internet adoption in ECA (demand side). It investigates how governments, firms, and individuals adopt the Internet, which countries are ahead of the curve, and which may be lagging. Overall, the region is not yet gaining the full anticipated benefits, or "dividends," from the investment in digital technologies.
- Chapter 2 investigates how the Internet is provided in the ECA region (supply side). It analyzes physical and nonphysical barriers as well as financial and nonfinancial measures to address these and other bottlenecks to provision of the Internet in the region. Binding constraints range from the lack of broadband capacity to insufficient competition among Internet providers and from inadequate regulation of markets for goods and services to a lack of skills. These binding constraints differ from country to country.
- Chapter 3 investigates how the Internet is improving the productivity of firms and encouraging the entry and growth of more productive firms, while encouraging the exit of less productive firms. It examines how a lack of competition, weak management capacity, and lack of skills can prevent firms from using the Internet more effectively, reducing its positive impact on firm productivity and entry.
- Chapter 4 examines the digital trade in ECA and how the Internet is increasing firms' potential to produce new goods and services for new markets, facilitating more firms to participate in international trade. It investigates how the enabling environment—logistics infrastructure, payment systems, a well-oiled trade system, and access to international markets—is critical to facilitating more digital trade and to attracting and retaining firms, particularly as the Internet expands firms' mobility.
- Chapter 5 analyzes how the Internet is changing the demand for labor within the economies of the ECA region and which demographic groups are more likely to benefit from or to lose out on greater use of the Internet by businesses in the region. It also envisions what types of skills will be in demand in the future and identifies how to cope with the likely shortages of such skills.
- Chapter 6 analyzes how the Internet is changing labor market arrangements by allowing new and more flexible forms of work, such as telework and freelancing. The chapter also investigates why not everyone in ECA is benefiting from these new forms of work and how this differs by age and gender.
- Chapter 7 offers a policy framework, built on the previous chapters, to guide government policy makers, sector regulators, and business strategists. It aims to promote the embrace of digital technologies and to secure the necessary foundations regarding analog complements to digital technologies. The net result should enable the ECA region to reap the anticipated dividends from the investment in digital technologies.

Although this book does not dedicate a whole chapter to e-governance issues, these issues are analyzed across the different chapters. For example, chapter 1 investigates governments' Internet adoption, chapter 3 analyzes the effects on firms of introducing more online government services, and chapter 6 discusses the importance of cooperation between governments and online platforms to facilitate the payment of social contributions and taxes through the Internet.

For example, Serbia and Turkey belong to the transitioning group of countries with regard to Internet access, but belong to the emerging group with regard to skills development. Figure O.9 provides a snapshot of the main policy priorities for each group of countries, as defined in figure O.8.

Among emerging economies, the bulk of the digital policy agenda is concentrated on making the Internet accessible and affordable. Chapter 2 discusses some policy reforms that may contribute to this goal. For instance, countries in Central Asia have yet to liberalize their international gateway, which remains with the incumbent operators, which tend to be state-owned. Moreover, their domestic markets—and even the mobile sector—are characterized by the dominant role of incumbents as well. Reforms to foster competition in the telecommunications sector are an important policy tool to increase Internet adoption (figure O.10). Needless to say, reforms aimed at fostering infrastructure sharing are basically nonexistent. This absence reduces the scope for competitive market entry, because it obliges potential market entrants to build their own networks from scratch, rather than offering services on lines rented from the incumbent. The poor connectivity of emerging economies could be a blessing in disguise, as the experience of the new EU members shows that laggards may have more options regarding which technology to adopt first (box O.3).

Emerging economies could also learn from the experiences of transforming economies such as the United Kingdom, whose government is setting deadlines for terminating service through copper lines in order to incentivize investment by operators in noncopper (and thus faster and higher-quality) networks. Finally, good relationships with neighboring countries are crucial to improving Internet access in landlocked countries. Unless governments move beyond outdated

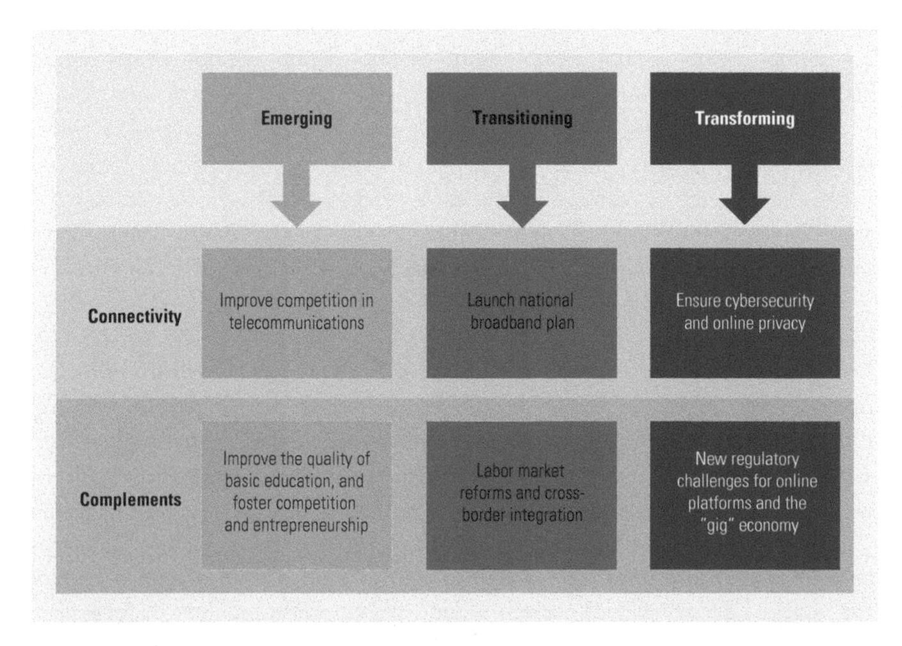

FIGURE O.9
Policy priorities for emerging, transitioning, and transforming digital economies in ECA

FIGURE O.10
Early reformers experienced larger increases in Internet penetration

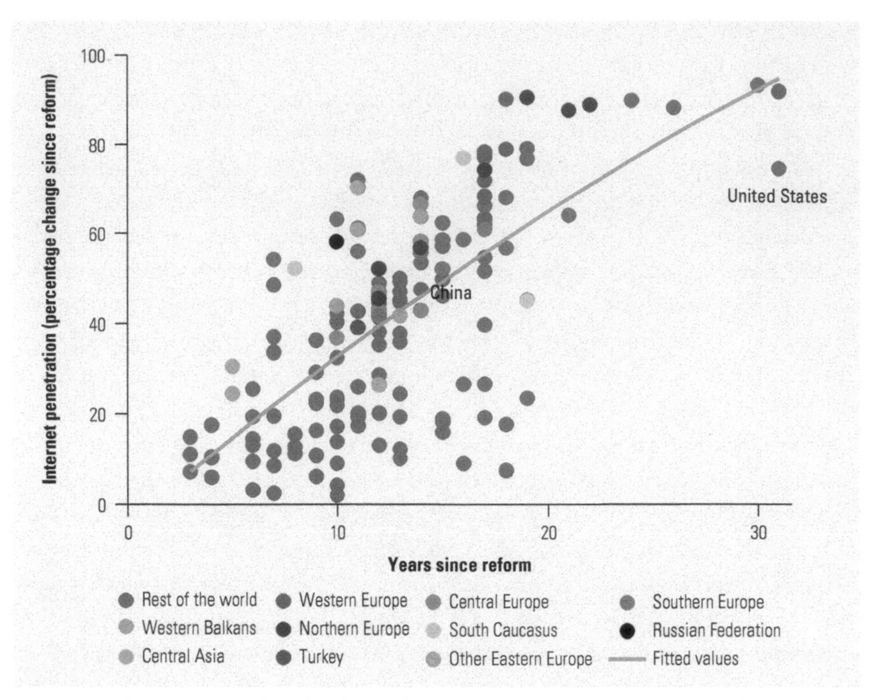

Source: Calculations are based on data from TeleGeography, Regulatory Status Dataset (http://www.telegeography.com) and World Development Indicators 2016, International Telecommunication Union, World Telecommunication/ICT Development Report and database, and World Bank estimates.
Note: Internet penetration is the number of Internet users per 100 people. The reform is the liberalization of the international segment of the telecommunications sector.

disputes, the citizens of some of the countries in the South Caucasus and Central Asia may have to wait decades before the digital dividends can translate into higher wages and employment opportunities.

However, even if emerging countries manage to improve their connectivity, high levels of functional illiteracy will constrain Internet adoption. As chapter 5 argues, measures to improve the quality of basic education (primary and high school) are an essential complement of the digital agenda in emerging countries.

In transitioning digital economies, where the Internet is already widespread, more focus is needed on the complements of the digital development agenda, one of them being the labor market. Creating online jobs in the shadow economy would place significant pressure on the social protection systems of ECA, which tend to be based on contributions. For example, if online freelancers do not make pension contributions, they are at risk of falling into poverty in old age and becoming eligible for social assistance. Estonia offers a good example of cooperation between the government and online platforms to improve the collection of contributions in ride-sharing platforms: transactions between customers and drivers in the online platform are registered, and the tax-relevant information is transmitted to authorities, who use it to prefill tax forms to facilitate future tax payments (European Commission 2016). Cooperation between governments and online platforms to facilitate tax and social contribution payments would nudge workers out of the shadow economy and provide them with some employment protection.

BOX O.3 Successful Policies for Digital Development: The Case of New EU Members

The European broadband landscape is diversified in terms of speed, and countries with the highest speeds are not always those one would expect. Indeed, some EU newcomers outperform Western European countries with regard to broadband speed. For instance, the Czech Republic, Latvia, and Romania each perform better than the EU average, with broadband speeds comparable to those in Scandinavia. The new EU members have built their national broadband networks more recently than most Western European countries. This has enabled them to benefit from more advanced technologies by jumping straight to a fiber network and to benefit directly from the advantages of speed and economies of scale. Furthermore, in several of them, new market entrants have built their own fiber-based networks rather than relying on the incumbent's copper local loops.

Estonia went one long step further. Since the early 1990s, Estonia has progressively become a digital economy that champions the use of information and communication technology (ICT) and digital services in both the public and the private sectors. Estonia's digital transformation was founded on a strong political will for action, notably to widen access to the Internet and promote inclusion. In 2002, some 460 public access points were established in municipal centers, public libraries, and schools. This has been reinforced through relevant legal and administrative reforms. The digital identity (ID) is one of the foundations of the Estonian e-government ecosystem. The Estonian system gives all citizens a personal identification number that enables them to sign legal documents or contracts remotely.

These decisions paved the way to the digital economy that Estonia has today. Since 2001, the Estonian government has been building a technology platform called X-Road, which provides a data-exchange layer between the public and private sectors. By facilitating open interactions among government agencies, between government and citizens, and between public and private sectors, it offers improved efficiency. For example, a law requires that no government agency can refuse a digital form from a citizen and ask for a paper copy instead. Also, thanks to integration of the digital ID with X-Road, citizens never have to provide the same information more than once.

Besides access to services, the e-government system has contributed to a move toward a more accountable and transparent government. Estonia rose from 78th (in 1996) to 40th (in 2010) on a ranking of 144 countries for its control of corruption. Moreover, the system facilitates democratic participation via easy and accessible Internet voting. Today, one-third of Estonians vote online.

The government also invests heavily in cybersecurity; indeed, the whole system of digital IDs, signatures, and X-Road is powered by secure 2048-bit encryption. Besides this, a specific department (called Critical Information Infrastructure Protection) evaluates the security of information systems and risks.

The experience of the Baltic and Central European economies suggests that governments could play a vital role in promoting digital development. Could it be repeated in the far east of ECA?

Source: Adapted from World Bank 2016b; see also Vassil 2015.

At the same time, the rise of the gig economy poses important challenges for social security systems that were originally designed for salaried forms of employment. Whether to treat individuals using online platforms to make a living as independent contractors or employees of the platform is yet to be decided; however, this new mode of employment will certainly require more portability of social security contributions, in the sense that the contributions become increasingly attached to the worker and not to the job.

Transitioning digital economies also face the challenge of making international trade easier and simpler for firms. Streamlining procedures is the trade facilitation measure that would have the largest impact on trade flows and could reduce trade costs between 2.2 and 2.8 percent. Trade facilitation improvements could also increase online trade. eBay introduced a global shipping program that handles shipping and customs clearance for its sellers. Sellers selected for the program had 2.7 percent more exports than those not selected. Countries can leverage ICT to modernize customs agencies and procedures. For example, Albania employed the Automated System for Customs Data to improve its risk management and inspection processes from 2007 to 2012. This program reduced the customs clearance time and increased trade significantly.

Transitioning digital countries could also focus on regulations that hinder e-commerce. Some e-commerce firms that sell services and products such as health services and cosmetics are regulated by government agencies for public health and safety reasons. Government agencies, however, may not be familiar with the operations of online companies. A government unfamiliar with the e-commerce business model may subject the company to many different regulations, imposing an unnecessary and possibly prohibitive administrative burden on the company. One of the first e-commerce companies in the former Yugoslav Republic of Macedonia, Grouper.mk, faced these issues when it started, as the legal framework was not ready for e-commerce activities. However, the company was able to resolve these issues through consultations with the government and support from aid agencies.

Finally, transitioning digital economies are also lagging in their connectivity agenda, as many have yet not launched any national broadband plans. Compounding this delay, many still lack strong representation in international organizations dealing with telecommunications policies. Membership is important because it confers greater visibility on the international stage as well as access to discussions on best practices.

Transforming digital economies, such as the Netherlands and the United Kingdom, have some of the most affordable and highest-quality Internet services in the world. Accordingly, their human capital stock and regulations are contributing to the rapid diffusion of the Internet across the economy. There are, nevertheless, important challenges ahead. Efforts to make the Internet safe are crucial to spreading the benefits of digital dividends. Concerns about cybersecurity are inhibiting the adoption of more sophisticated technologies among firms (figure O.11).

Compared with simpler uses of the Internet, firms may face larger risks of security breaches when adopting newer Internet technologies such as cloud computing and the Internet of Things (IoT), as many devices are connected over the Internet and data are shared across different business functions. Instead of creating barriers to cross-border data flows, governments can increase their efforts to educate firms on information technology (IT) security methods, create regulations to ensure that most firms have a minimum level of IT security, and ensure that the national Internet infrastructure is not

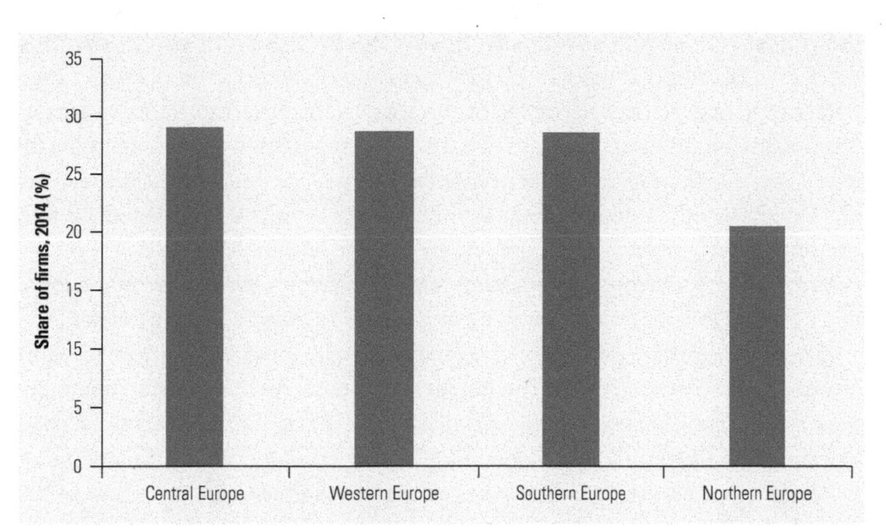

FIGURE O.11
At least one out of four firms in Europe is not using cloud computing because of cybersecurity concerns

Source: Eurostat Database (multiple years).

compromised. As another example of good practices in this area, the EU has recently concluded the new Privacy Shield Framework agreement with the United States to govern transatlantic data transfers that meet privacy concerns. Countries that impose data barriers to protect local industries may generate adverse effects on the economy, as data barriers reduce GDP and exports by 1.7 percent each (Bauer and others 2014).

Accordingly, transforming economies may also focus on one important complement of the digital strategy—namely, adapting competition policies to the new market structures that the Internet is creating, such as multisided markets. These markets or platforms are different from the usual markets, as there are often two (or more) distinct groups of customers, network effects within and across these groups, and an intermediary that brings these two customer groups together (such as eBay). In standard economic models, if prices are higher than marginal costs, firms have some form of market power. With a multisided market, prices do not necessarily equal marginal costs, as it may be efficient (to maximize the size of the market) for the platform to charge certain customer groups higher prices and others lower prices or even to offer the product for free. Thereby, the competition agency cannot consider price without considering the effects on the other parts of the market (OECD 2009). Transforming economies have the chance to be at the forefront of proposing policies to foster competition in this new environment to avoid the risks of increasing concentration.

Reaping Digital Dividends in ECA: Toward a New Enlightenment

The seven chapters of this book describe the multiple ways in which the Internet can bring enormous benefits to the ECA region. They also highlight the existence of important challenges. ECA countries are in a unique position to reap the digital dividends. For example, governments in the east could adapt the successful policies that the new EU members implemented to improve and even outperform the richer economies of the west with regard to Internet access. Moreover, Central Asian countries could benefit from their closeness to China for improving their connectivity and bolstering future e-commerce possibilities. The development of faster and cheaper mobile access to the Internet could also benefit sparsely populated countries, by making distance a less binding constraint. Finally, Russian-speaking countries would benefit from important economies of scale in content development and online freelancing opportunities.

Countries in the ECA region are facing the dilemma of whether to delay changes or to embrace them. The longer they postpone this decision, the more their residents will have to wait to enjoy the digital dividends. In the same way that the large-scale production of paper 1,000 years ago brought the Central Asian Enlightenment and the printing press brought the European Renaissance 500 years later, the future payoffs of information technologies could be unexpectedly high in ECA.

Note

1. This book follows the *World Development Report 2016* (WDR16) in its use of terminology (World Bank 2016a). WDR16 covers all digital technologies that facilitate the creation, storage, analysis, and sharing of data and information. These technologies include information and communication technologies, notably the Internet, mobile phones, and other electronic devices. Similarly, this book covers the developmental impact of digital technologies as a whole. These technologies include primarily mobile phones and the Internet, but also satellite technology as well as the applications and services delivered over them, sometimes referred to as "Over the Top" (OTT) applications, like social media.

References

Bauer, M., H. Lee-Makiyama, E. van der Marel, and B. Verschelde. 2014. "The Costs of Data Localisation: Friendly Fire on Economic Recovery." ECIPE Occasional Paper 3/2014, European Centre for International Political Economy, Brussels.

Cheshire, P. C., and C. A. L. Hilber. 2008. "Office Space Supply Restrictions in Britain: The Political Economy of Market Revenge." *Economic Journal* 118 (529): F185–221.

Cramer, J., and A. B. Krueger. 2016. "Disruptive Change in the Taxi Business: The Case of Uber." NBER Working Paper 22083, National Bureau of Economic Research, Cambridge, MA.

Demirguc-Kunt, A., L. Klapper, D. Singer, and P. Van Oudheusden. 2015. "The Global Findex Database 2014: Measuring Financial Inclusion around the World." Policy Research Working Paper 7255,

Dittmar, J. E. 2011. "Information Technology and Economic Change: The Impact of the Printing Press." *Quarterly Journal of Economics* 126 (3): 1133–72.

Ecommerce Foundation. 2014. BRICS B2C E-commerce Report 2014 (Light version).

———. 2015a. Central Europe B2C E-commerce Report 2015 (Light version).

———. 2015b. Eastern Europe B2C E-commerce Report 2015 (Light version).

———. 2015c. Latin America B2C E-commerce Report 2015 (Light version).

———. 2015d. North America B2C E-commerce Report 2015 (Light version).

———. 2015e. Northern Europe B2C E-commerce Report 2015 (Light version).

———. 2015f. Southern Europe B2C E-commerce Report 2015 (Light version).

———. 2015g. Western Europe B2C E-commerce Report 2015 (Light version).

European Commission. 2016. "Communication from the Commission to the European Parliament, the Council, the European Economic and Social Committee, and the Committee of the Regions, a European Agenda for the Collaborative Economy." European Commission, Brussels.

Farrell, D., and F. Greig. 2016. "Paychecks, Paydays, and the Online Platform Economy Big Data on Income Volatility." JP Morgan Chase Institute, San Francisco, February.

McKinsey Global Institute. 2016. "Independent Work: Choice, Necessity, and the Gig Economy."

OECD (Organisation for Economic Co-operation and Development). 2009. *Two-Sided Markets: Policy Roundtable.* Paris: OECD.

———. 2013. "Russia: Modernising the Economy." Better Policies Series, OECD, Paris, April.

Starr, S. F. 2013. *Lost Enlightenment: Central Asia's Golden Age from the Arab Conquest to Tamerlane.* Princeton, NJ: Princeton University Press.

Vassil, K. 2015. "Estonian e-Government Ecosystem: Foundation, Applications, Outcomes." Background Paper for the *World Development Report 2016*, World Bank, Washington, DC.

World Bank. 2016. *World Development Report 2016: Digital Dividends.* Washington, DC: World Bank.

World Economic Forum. 2013. *The Global Competitiveness Report 2013–2014.* World Economic Forum. Geneva.

———. 2015. *The Global Competitiveness Report 2015–2016.* World Economic Forum. Geneva.

Zervas, G., D. Proserpio, and J. Byers. 2016. "The Rise of the Sharing Economy: Estimating the Impact of Airbnb on the Hotel Industry." Research Paper 2013-16, Boston University, School of Management, Boston, MA.

1

Internet Adoption

Internet adoption is highly correlated with economic development. That is, while rich economies have achieved almost universal access, poor countries lag behind. Similar correlations exist within countries. For example, Internet adoption is substantially lower among individuals living in rural areas than among individuals living in urban areas. Younger and skilled workers are more likely to use the Internet than workers who are older and unskilled. The fact that higher Internet penetration is highly correlated with a host of measures of economic success suggests that achieving universal access requires not only reforms to the telecommunications sector, but also policies for helping individuals and firms to make the most out of the Internet. This chapter examines these gaps in Internet adoption across and within countries and sets the stage for the rest of the book.

Introduction and Main Messages

This chapter examines various aspects of Internet adoption in Europe and Central Asia (ECA).[1] It analyzes adoption with regards to the number or category of users (breadth of Internet adoption) and the different kinds of applications used (depth of Internet use) (table 1.1). About 69 percent of ECA's population has access to the Internet, although less than half of those who live in areas with mobile coverage use high-speed mobile Internet and only one-fifth of those with access to fixed Internet have high-speed connections. This chapter argues that differences in

TABLE 1.1 Measures of Internet Adoption in Europe and Central Asia

Breadth	Depth
Basic quantitative parameters of Internet adoption to measure Internet uptake and connectivity by: • Category of user, that is, firms, governments, and households (% of firms with broadband Internet) • Socioeconomic and demographic characteristics (Internet use by gender, education, and age) • Geographic dimension (Internet use in urban versus rural areas) • Internet use by different sectors of the economy (schools, health clinics)	Intensity and sophistication of Internet use by category of user: • Frequency and duration of use (% of daily active users, hours spent online) • How firms, governments, and households or people use the Internet (use of the Internet to perform more complicated transactions and operations as opposed to just basic search and download) • Internet as a tool to achieve specific goals (for example, to build a sustainable smart city, to leverage intelligent transport systems, or to monitor and reduce emissions and congestion)

economic development can explain a large extent of the sharp contrasts in Internet use within the region. Four out of five European Union (EU) citizens have used the Internet, but more than 60 percent of individuals in Central Asia are not connected. Few households in middle-income countries have access to speeds above 10 megabits per second (Mbit/s), which limits the user experience and the ability to take greater advantage of the Internet.

At the same time, Internet use varies significantly within countries: level of income is positively correlated with rates of broadband Internet penetration, urban areas tend to have greater Internet connectivity than rural areas, and gender differences are evident in Internet use (particularly in rural areas).

The intensity of Internet use—for example, as measured by the frequency of Internet use and hours spent online—varies considerably in ECA; the share of daily Internet users ranges from less than 50 percent in Bulgaria, Romania, and Turkey to almost 90 percent in Denmark, the Netherlands, and Sweden. Even in countries where frequency of use is high, significant differences may exist between socioeconomic groups. While the difference in frequency of use between men and women is less than 10 percent, level of education, employment status, or age can be an important determinant of differences in the frequency of use. Internet use varies from basic functions such as e-mail or web surfing to high-bandwidth applications such as high-definition streaming of video or gaming. The adoption of devices and the availability of adequate speed to use devices efficiently remain low in certain parts of ECA, which contributes to the inequitable distribution of benefits.

Many ECA firms that use the Internet do not take full advantage of the technology to sell their products or to perform complex tasks. For example, the share of firms with their own website varies a lot between ECA countries. The level of digital intensity in the EU is not closely correlated with income levels. The penetration of high-bandwidth applications—for example, video conferencing—is low, but growing rapidly.

ECA countries with very different levels of income have similar levels of Internet adoption by government, although the kinds of services offered vary considerably. Most lower-middle-income countries provide "emerging" types of information services, rather than being proactive in involving citizens in government activities. Only a few countries in ECA have developed "transactional" services or have introduced biometric identification. The take-up of services also varies considerably. Within the EU, Bulgaria, the Czech Republic, and Romania have the lowest levels of overall interaction with the government and online submission of completed forms.

In explaining these differences, this chapter prepares the ground for the rest of the book. The first section considers the breadth of Internet adoption (quantity and types of users), the second discusses the depth of Internet adoption (intensity of Internet use), and the third analyzes government use of the Internet to provide services. A final section presents conclusions.

Breadth of Internet Adoption in ECA: Uptake and Connectivity

The decision to use the Internet is a process that has several requirements (figure 1.1), including the availability of service in the area and the ability to use the service. Each of these factors is associated with particular policy challenges. For some uses, such as e-commerce, the range of factors stretches far beyond the digital ecosystem and involves transportation, logistics, and some other sectors.

Of the global population of about 7.4 billion in 2015, some 4.2 billion are offline, of which 0.4 billion live in areas with no mobile or fixed coverage (World Bank 2016 based on 2016 GSMA data; ITU 2015; Meeker 2015; United Nations data 2014; World Development Indicators 2014). About 250 million of the 4.2 billion people offline are in ECA.[2] To put it another way, within ECA's

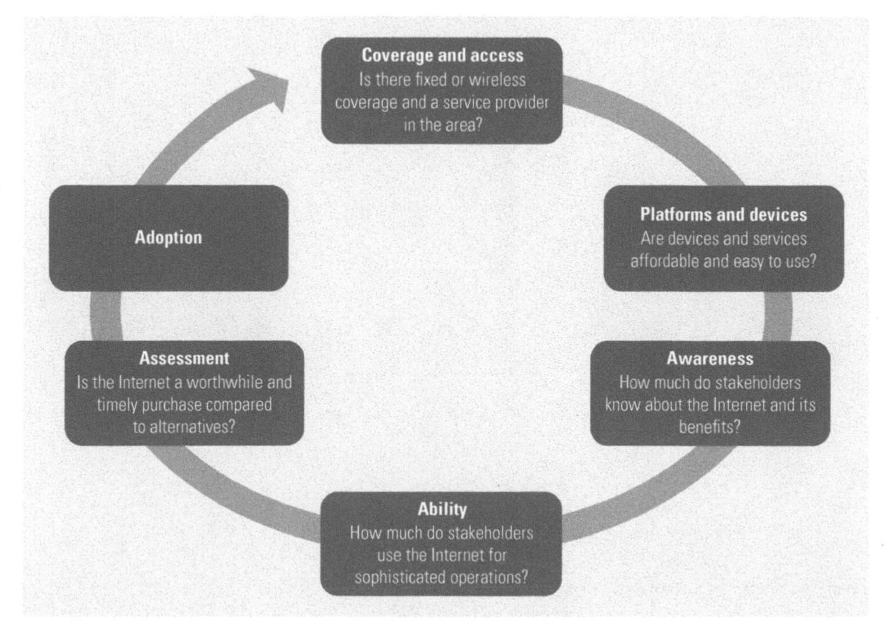

FIGURE 1.1
Prerequisites for Internet adoption: The user perspective

population of about 818 million, about 568 million (69 percent) are online. But the mere presence of access does not mean that the Internet is used or is used to its full extent. Less than half of persons living within areas of mobile coverage use high-speed mobile Internet, and only one-fifth of those having access to fixed Internet have high-speed connections (figure 1.2).

Internet access is distributed unevenly across the region (figure 1.3). In the European Union, four out of five citizens have used the Internet,[3] but elsewhere far fewer people are online. In Central Asia, more than 60 percent of the population is

FIGURE 1.2
Internet users in the ECA region

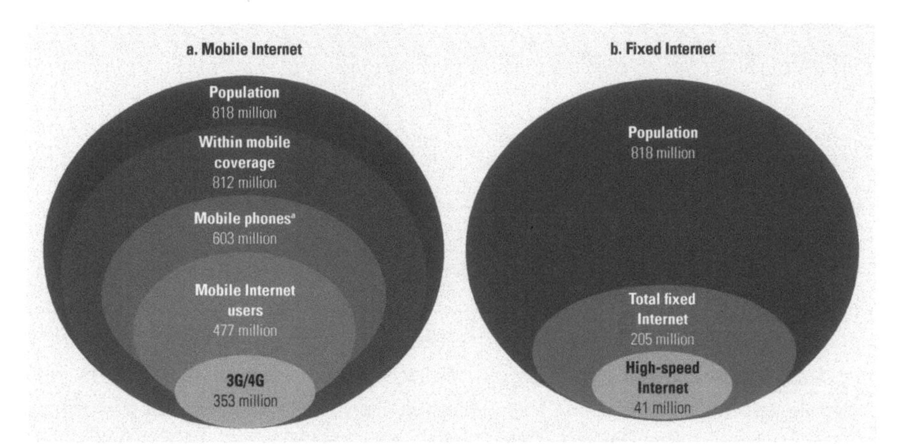

Sources: TeleGeography data 2016 (fixed Internet subscribers; https://www.telegeography.com/research-services/globalcomms-database-service/); GSMA data 2016 (mobile subscribers; https://www.gsmaintelligence.com/); International Telecommunication Union data 2014 (mobile coverage; http://www.itu.int/en/ITU-D/Statistics/Pages/default.aspx); World Bank data 2015 (population; http://databank.worldbank.org/data/reports.aspx?source=2&series=SP.POP.TOTL&country=).
a. Mobile phones here refer to all types of devices, smartphones, and non-Internet devices.

FIGURE 1.3 Who is online globally and in ECA, 2014

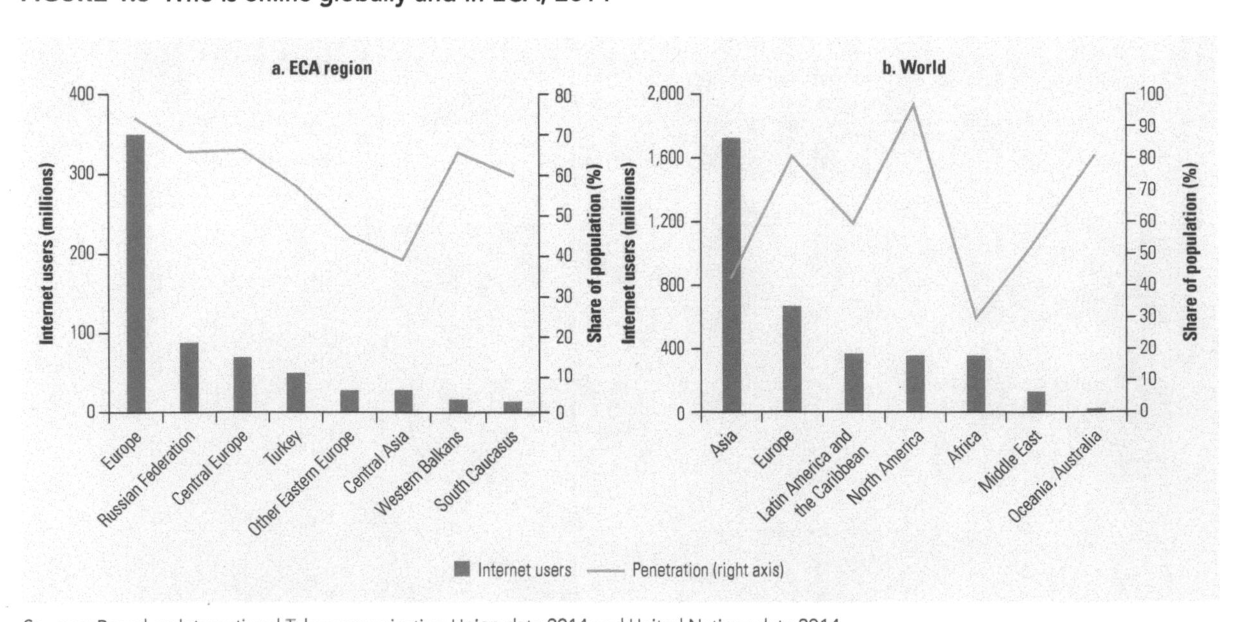

Sources: Based on International Telecommunication Union data 2014 and United Nations data 2014.
Note: Europe here refers to geographic Europe.

not connected; the proportion of individuals using the Internet (on any device) ranges from a high of 55 percent in Kazakhstan and 44 percent in Uzbekistan to 28 percent in the Kyrgyz Republic, 17 percent in Tajikistan, and just 12 percent in Turkmenistan.[4]

Many of these disparities may be explained by differences in economic development. Overall, high-income countries have higher household rates of fixed-broadband penetration, and lower-middle-income countries have lower penetration rates (figure 1.4). Nevertheless, disparities exist within income groups. The difference between Armenia and Ukraine, for instance, exceeded 20 percentage points in 2014 (and exceeded 30 percentage points in 2015), up from only 12 percentage points in 2009. Ukraine's gross national income (GNI) stagnated over 2009–15, including ups and downs—in part due to political turmoil—whereas Armenia's GNI rose. A similar disparity is evident between France and Italy and some other countries.

Central Asia and South Caucasus have the lowest rates of fixed broadband penetration in the region, while Western Europe and the Baltic States have the highest. However, there is significant variation within each subregion. In Central Asia, penetration ranges from less than 1 percent in Tajikistan and Turkmenistan to more than 30 percent in Kazakhstan. Albania and Greece fall well below the average for their subregion. At the same time, Europe is home to 8 of the 14 countries in the world where fixed Internet penetration exceeds 90 percent.[5] Significant regional

FIGURE 1.4
Is country income level an accurate predictor of household broadband penetration?

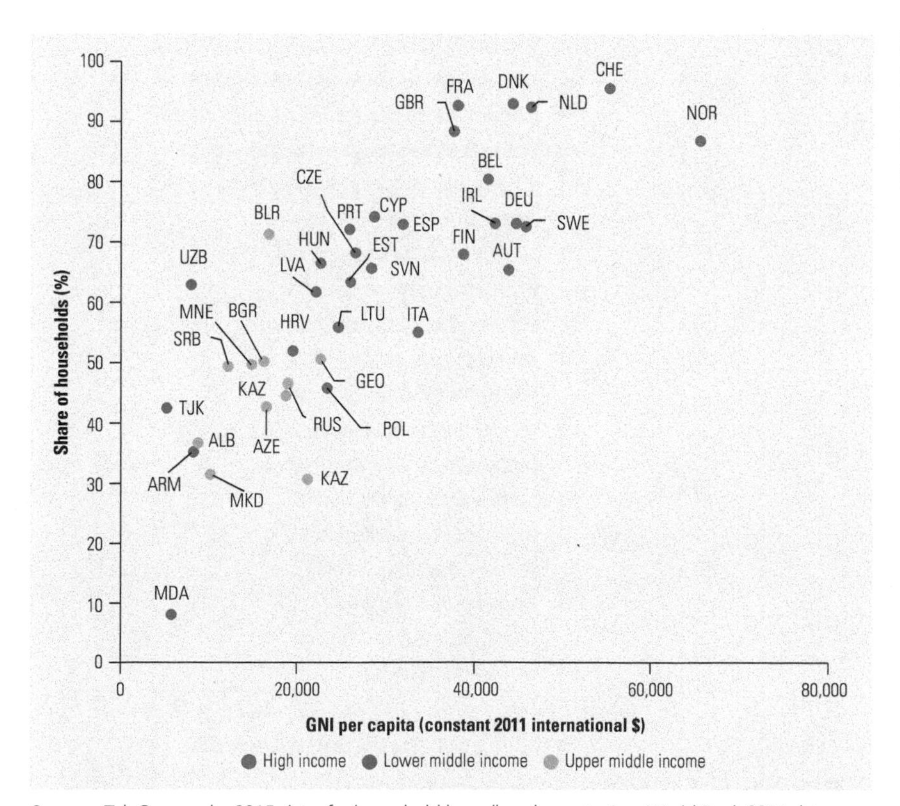

Sources: TeleGeography 2015 data, for household broadband penetration; World Bank 2014 data, for GNI. GNI = gross national income.

disparities exist within countries as well. In Moldova, for instance, the household penetration rate for fixed broadband reached 75.6 percent in the capital Chisinau in 2015, but was much lower in the rest of the country, for example, 21.1 percent in Soldanesti.

Penetration, however, has little to say about the quality of access. Few households in middle-income countries have access to speeds above 10 Mbit/s (figure 1.5), which limits the user experience and the ability to take greater advantage of the Internet. Even for "light service use," which involves basic functions such as e-mail, web surfing, and basic streaming of video at home (if four devices are connected), a speed of 6–15 Mbit/s is required. If several high-bandwidth applications are running at the same time (even with only one device connected), this is the minimum speed required (FCC 2014). According to recent studies (Ericsson, Arthur D. Little, and Chalmers University of Technology 2013; Kongut, Rohman, and Bohlin 2014), faster broadband speed can bring economic benefits (greater innovation and productivity in business, higher household income, consumer surplus), social benefits (improved access to services, more efficient health care,

FIGURE 1.5 Some countries in ECA still lag in broadband speed

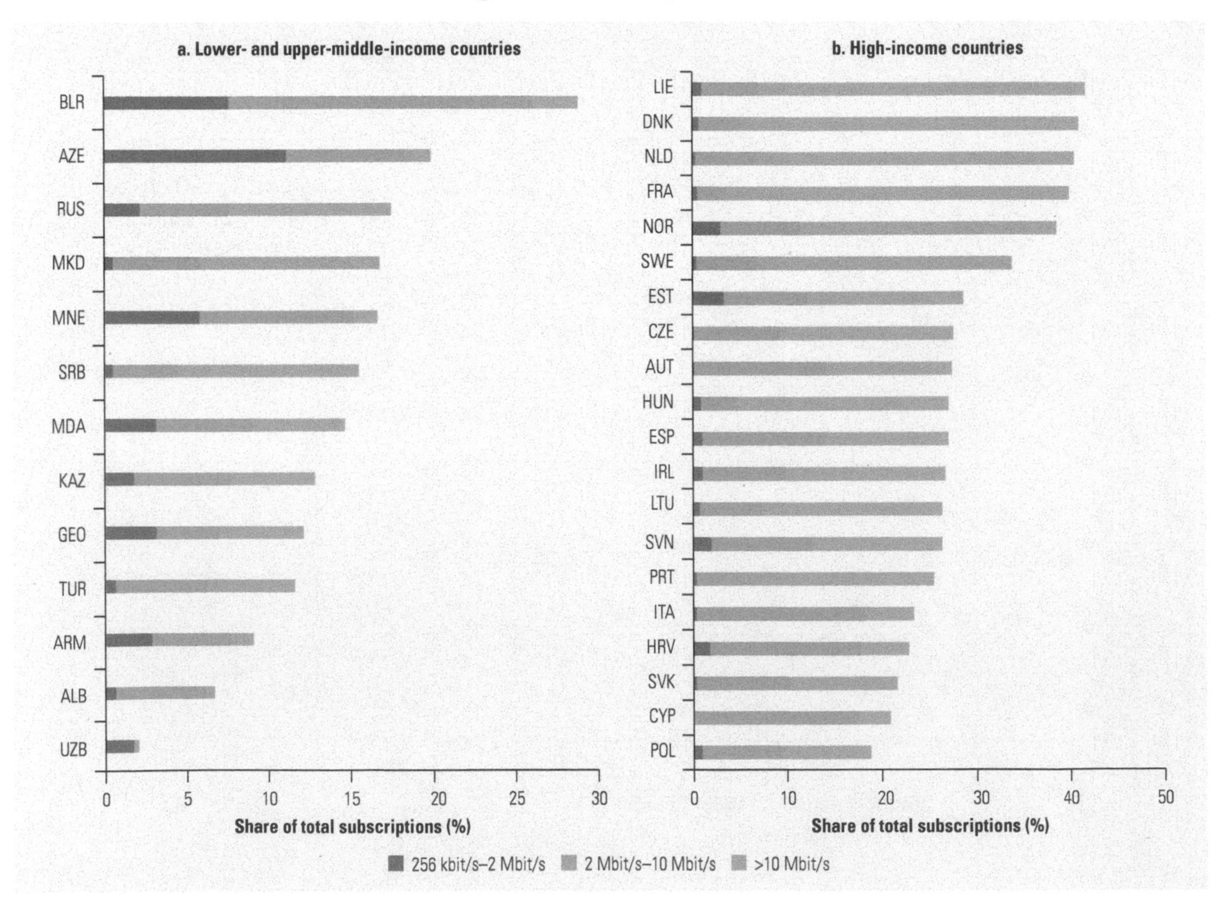

Source: International Telecommunication Union 2015.

education, banking services, entertainment), and environmental benefits (more efficient energy consumption).

Albania, Bosnia and Herzegovina, Hungary, and Montenegro have the lowest wireless broadband penetration rates in the region, while Finland, Denmark, and Sweden have the highest (ITU 2014). There is a significant difference between narrowband (second-generation—2G) and broadband (third- or fourth-generation—3G or 4G) mobile-cellular phone use, with 3G and 4G allowing high-speed Internet and a much greater consumption of content and services. In some countries, like Montenegro or Ukraine, fewer than one in three mobile subscribers has upgraded to a broadband connection. Overall, the uptake of 3G and 4G is below the world average in 15 ECA countries, most of them outside the EU (figure 1.6). All of the countries in the Western Balkans, except Serbia, are lagging behind, and Uzbekistan, with a penetration rate of 33 percent, has the highest penetration in Central Asia, where the adoption of narrowband is higher than broadband. Other Eastern European countries (Belarus, Moldova, and Ukraine) also have penetration rates below the global average, either because service is expensive or because the user base is equipped with former generations of mobile phones.

Disparities in network coverage, especially 4G, explain some of the differences in Internet uptake (4G coverage is addressed in more detail in chapter 2). 4G services are not yet offered in some countries, for example, Ukraine. Moreover, even 4G has significant limitations when it is used for video streaming or other bandwidth-intensive multimedia applications (for example, in the education or health sectors).

In Europe, MVNO (mobile virtual network operator) markets and regulations are a sign of liberalization, competitiveness, and maturity of markets. Europe has the largest MVNO market, with more than half of the world's 1,200 MVNOs.[6] Germany has the largest MVNO market in the world, with more than 100 MVNOs that have a market share above 50 percent. The market share of MVNOs exceeds 20 percent in at least eight Western European countries. The healthy MVNO market in Luxembourg contributes to its 153.28 percent market penetration (Q1 2016, GSMA) and targets mainly cross-border users and international travelers, with large price reductions on roaming charges. Within Eastern Europe, the Czech Republic and Kosovo lead the way, with MVNOs holding more than 10 percent of the market, even though the former was one of the last European markets to allow the entrance of MVNOs (at the end of 2012). This success may be due to the introduction of new price regulations for mobile services, which facilitated the entry of new players. Following an assessment in 2016, the regulator asked operators to lower the wholesale prices charged to MVNOs in order to maintain competition. Within Central Asia, in contrast, the MVNO market is almost nonexistent. In Tajikistan, Aiva Mobile was launched in October 2014, the first and only MVNO in the country to date.

Disparities in overall telecommunication infrastructure may also explain differences in Internet uptake across ECA. For example, fewer than 1 in 200 schools is connected in the Kyrgyz Republic. Even in Belarus, Poland, and the Russian Federation, the proportion of schools with Internet access was near or less than 80 percent, and with access to fixed-broadband less than 10 percent in Belarus and less than 50 percent in the Russian Federation (ITU 2014). As discussed in chapter 6, telecommunications reforms have had a major impact on increasing the use of the Internet. For example, in Europe employment creation was higher

FIGURE 1.6
The race to upgrade to mobile broadband (3G + 4G)

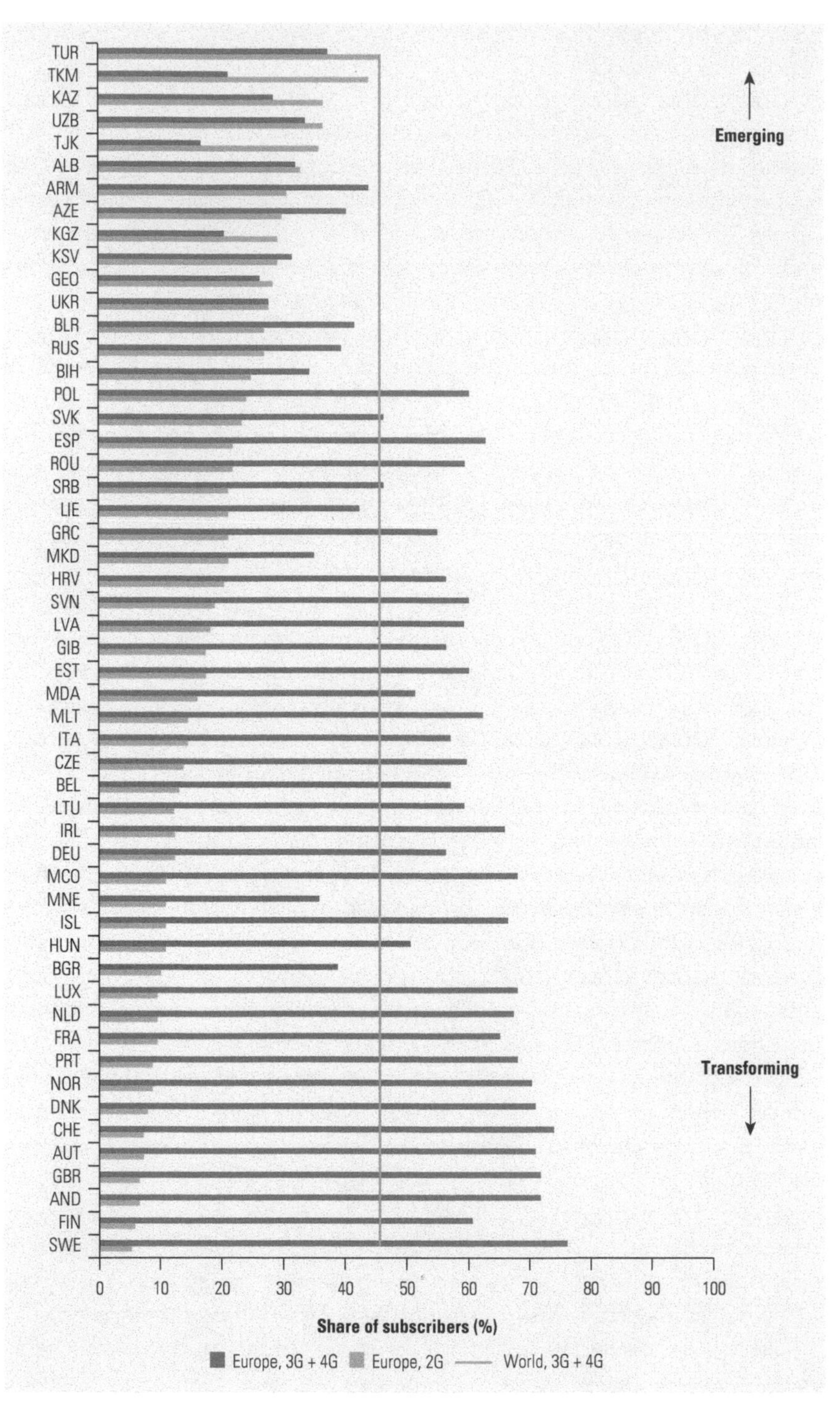

Source: GSMA data 2016 (https://www.gsmaintelligence.com), Q1 (first quarter).

in Internet-intensive sectors and there was a larger rise in teleworking in information and communication technology (ICT)–intensive sectors after telecommunications reforms.

Within countries, there are persistent digital divides across income, geography, and gender. The level of income is positively correlated with broadband Internet penetration rates (at entry-level speeds). As discussed in chapter 5, the Internet may also be increasing income inequality, as people in the bottom 40 percent of income distribution are least likely to possess the skills needed to participate in the new economy. In Central and Eastern Europe and, especially, in Bulgaria and Romania, wages for more highly skilled jobs have risen significantly in regions where the rate of Internet adoption has been the highest. The relative demand for skills seems to be higher in countries with lower Internet adoption (such as Central Asia and South Caucasus), and Internet-intensive sectors seem to be absorbing a larger share of skilled workers, suggesting that the lack of skills itself could be creating bottlenecks to expanding the Internet among the poorer economies of ECA.

Urban areas tend to have greater Internet connectivity than rural areas, although this relationship varies considerably across the region. In Bulgaria and Romania, for instance, the urban-rural divide in Internet access is greater than 30 percentage points (figure 1.7),[7] while in Belgium and Luxembourg penetration rates are higher outside cities. In Tajikistan, fixed-line Internet access remains limited to major urban areas, and the primary method of access is via dial-up or leased-line connections, while a handful of Internet service providers (ISPs) also provide satellite and fiber broadband services.

Internet use also differs considerably by gender in ECA. An estimated 1.7 billion women in low- and middle-income countries are not connected. ECA has the lowest number compared with other regions, but the share of unconnected women is still very high—at 42 percent (GSMA 2015). The gender gap in mobile phone ownership stands at 4 percent—second lowest after East Asia and Pacific at 3 percent (GSMA 2015). Countries with lower gross domestic product (GDP) per capita have wider gender gaps (GSMA 2015). Comparing ECA countries, the gender gap in Internet use is about 3.5 percent, on average, but in Western Balkans it is among the highest in the world (figure 1.8). Cost is reported as the main reason for not owning a device or not using the service. Network quality or coverage is the second reason, followed by security or harassment, technical literacy, and operator trust.

Some countries attempt to measure the various determinants of adoption on a regular basis, with a view to assisting groups that have limited access. For instance, as part of its Government Digital Inclusion Strategy, the United Kingdom developed a nine-level digital inclusion scale for individuals ranging from "never have, never will" to "expert" (United Kingdom Cabinet Office 2014). The study found that 14 percent of the U.K. population was offline, including those who were "willing but unable" to connect. For 62 percent of those offline, lack of interest was identified as the main reason, although answers varied within subgroups (table 1.2)—those who were "willing but unable," for instance, were identified as having problems with access, confidence, or skills. Of these people, 83 percent were age 45 years or older and 58 percent were in

FIGURE 1.7
Urban-rural divide in household fixed Internet access, 2014

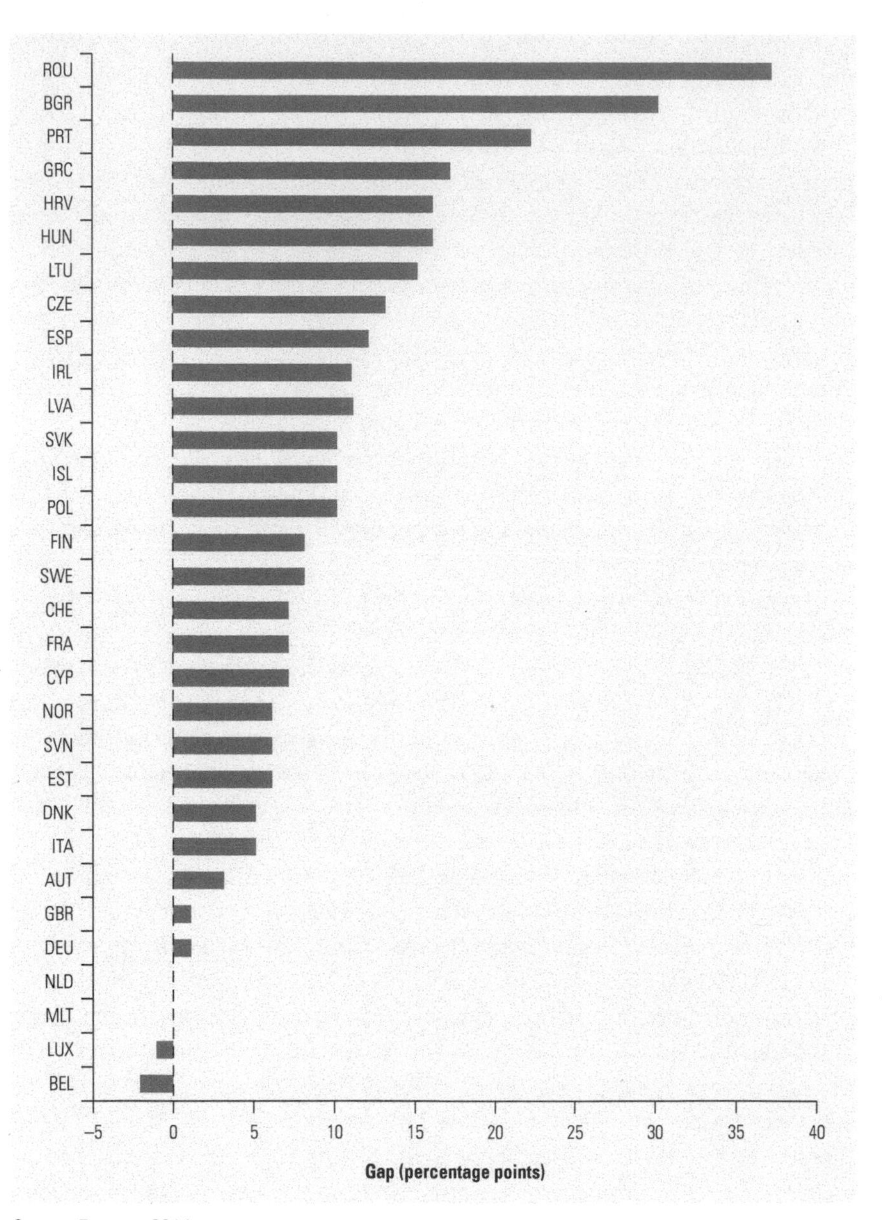

Source: Eurostat 2014.
Note: The number of households with broadband access in urban areas corresponds with the number of households in densely-populated areas (with at least 500 inhabitants/km^2). The number of households with broadband access in rural areas corresponds to the number of households having broadband access in sparsely populated areas (with less than 100 inhabitants/km^2). The area gap is calculated by taking the difference between urban and rural areas.

semiskilled or unskilled manual occupations or were unemployed. More generally, as shown in chapter 5, older workers are less likely to know and to learn how to use new technologies.

The Digital Adoption Index (DAI) provides an alternative presentation of differences in Internet adoption. The DAI, introduced in World Bank (2016), is a composite index measuring the extent of the spread of digital technologies within and across countries for all of the key agents in an economy: people, businesses,

FIGURE 1.8
Gender gap in Internet use, in ECA and globally, 2013

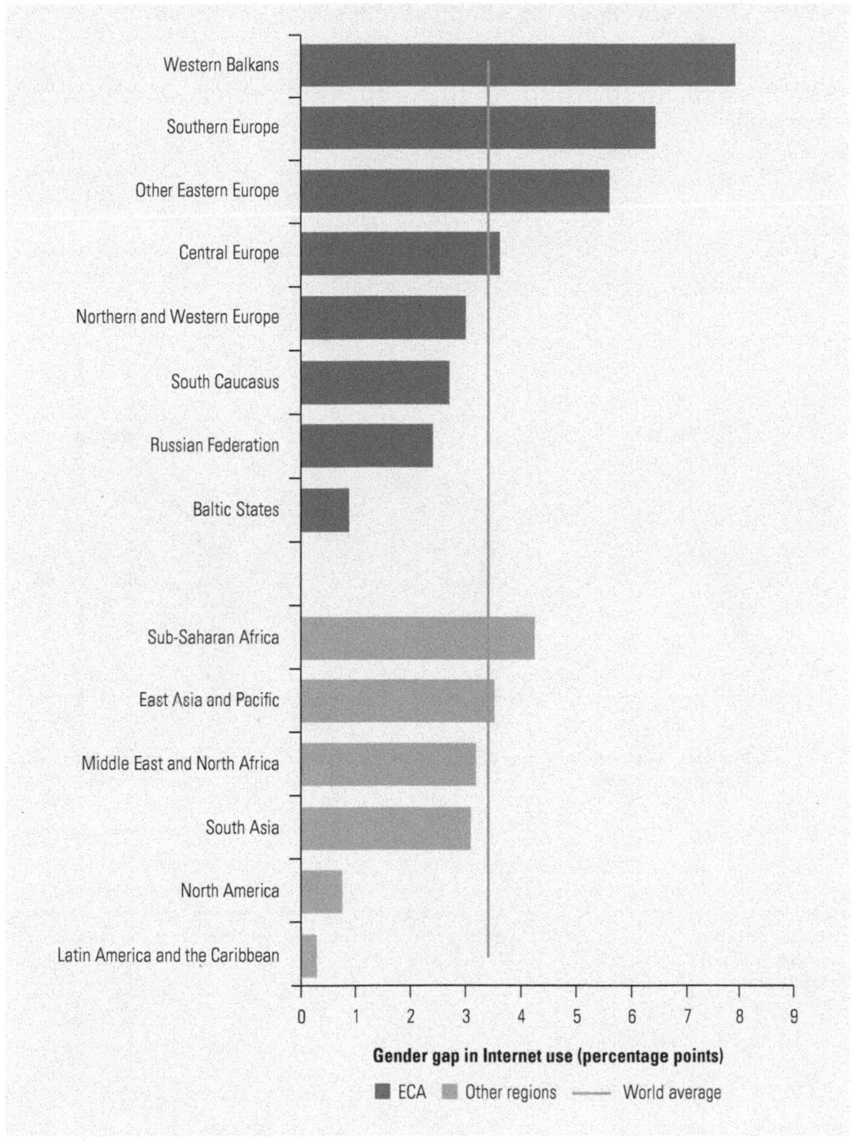

Source: Based on International Telecommunication Union 2013.
Note: The gender gap is calculated by taking the difference between the percentage of males and females using the Internet. Baltic States include Estonia, Latvia, and Lithuania. Central Asia is not shown in the figure because of unavailable data.

and governments.[8] Figure 1.9 illustrates the differences in adoption of the Internet by businesses, governments, and people between the ECA region and the average of countries in other parts of the world at a similar level of GDP per capita. Adoption of the Internet by firms and citizens is higher in ECA countries than in other countries at a similar level of development, while adoption by ECA governments is similar to the global average (as discussed later in this chapter). Within the ECA region, countries at a similar level of GDP per capita have significantly different levels of digital adoption.

TABLE 1.2 Factors Hindering Adoption of Internet Use in the United Kingdom

Lack of access	Lack of skills	Lack of motivation	Lack of trust
Accessibility	Literacy skills	Risks	Identity
Location	Digital skills	Necessity	Security
Cost	Security skills	Financial benefits	Standards
Technology	Confidence	Social benefits	Reputation
Infrastructure		Health and well-being benefits	
Language			

Source: Adapted from United Kingdom Cabinet Office 2014.

FIGURE 1.9
ECA Internet adoption by firms, people, and governments, relative to the rest of the world, 2014

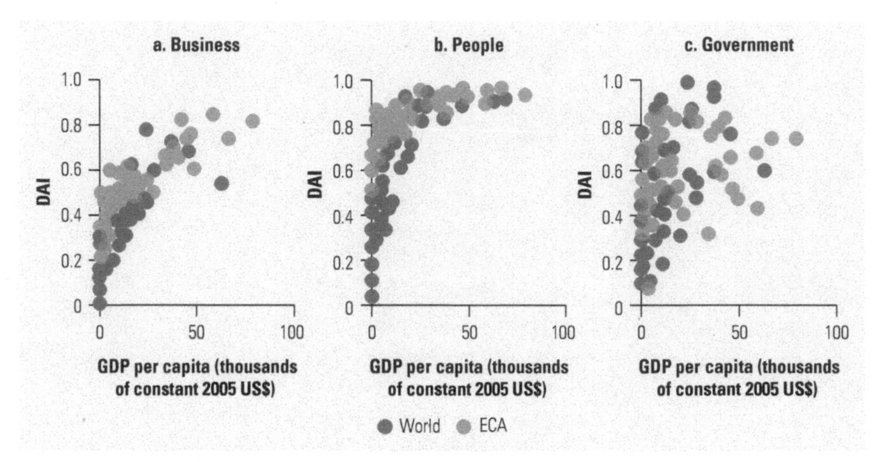

Source: World Bank, for GDP per capita (constant 2005 US$).
Note: The Digital Adoption Index is a composite index measuring the extent of spread of digital technologies within and across countries. It is based on three sectoral subindexes covering businesses, people, and governments. For a detailed description of the Digital Adoption Index, see http://pubdocs.worldbank.org/en/587221475074960682/WDR16-BP-DAI-methodology.pdf. DAI = Digital Adoption Index; ECA = Europe and Central Asia.

Considering digital adoption for each country category (see figure O.8 in the overview and annex 1A for more details on country categories), some ECA "transitioning" countries (for example, Italy and Russia) have a higher level of adoption than some "transforming" countries (for example, Hungary and Ireland). Moreover, the gap between "emerging" and "transitioning" is not as great as one might expect. Finally, the variation in the business and government subindexes is relatively high between countries (figure 1.10).

The overall level of mobile-cellular and Internet access at home (two indicators constituting the citizens cluster of the DAI index) is relatively high in ECA and is equally spread throughout the income groups. The highest variation is within the lower-middle-income group, with Armenia leading. Albania scores the lowest within the upper-middle-income group (figure 1.11).

Although firms have high access to broadband overall, they vary by other parameters. At least 80 percent of firms have access to broadband in almost all ECA countries. There are, however, some stark outliers: only 28 percent of firms in Uzbekistan and 45 percent in Turkey have access to broadband Internet.

FIGURE 1.10
Digital Adoption Index, 2014

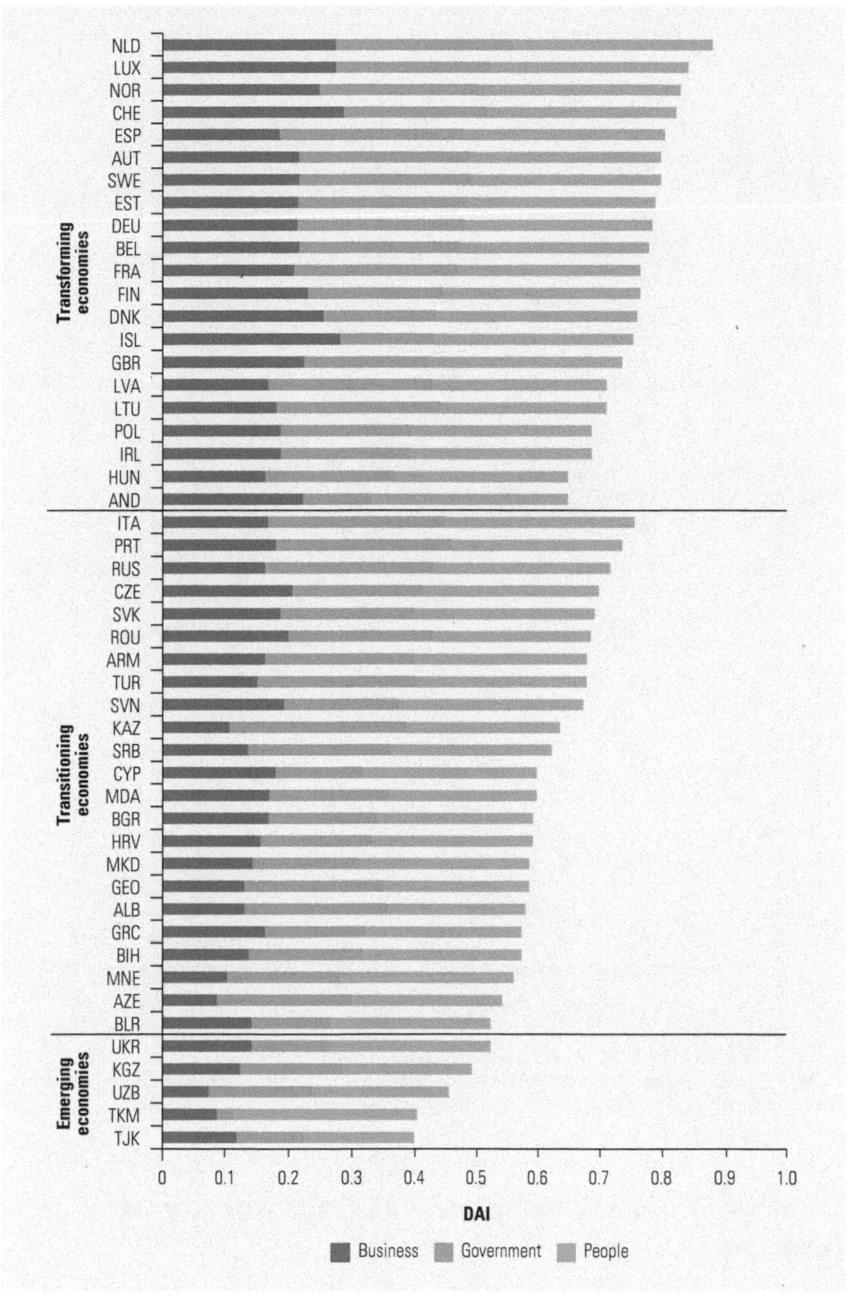

Source: Adapted from World Bank 2016 (http://wbgfiles.worldbank.org/documents/dec/digital-adoption-index.html).

Also, differences in firms' use of the Internet are significant even between upper-middle-income countries (for example, Kazakhstan and Romania) and high-income countries (for example, Austria and the Netherlands). The DAI for the business cluster[9] confirms that Central Asian firms are lagging behind in the use of the Internet (figure 1.12). Some countries of Central Asia (Kazakhstan) and South Caucasus (Azerbaijan) also score significantly lower on the subindexes for digital adoption by governments and people (figure 1.10).

FIGURE 1.11
Overall Internet access for people is high in ECA

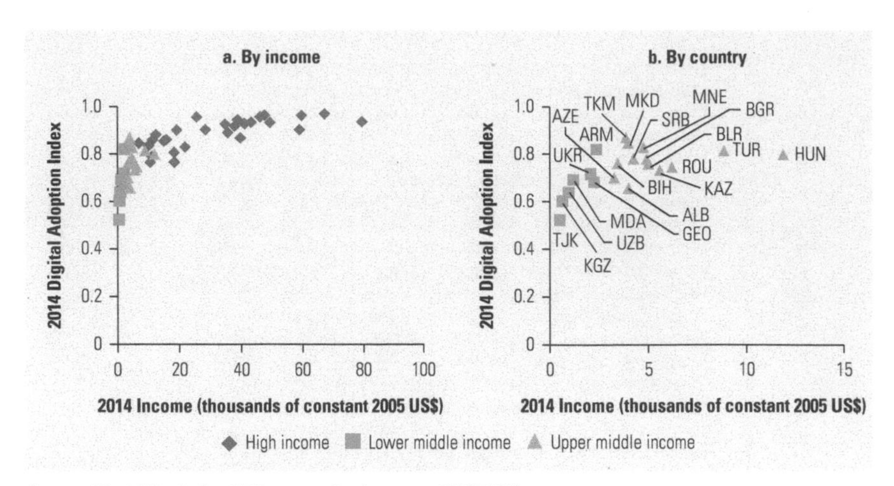

Source: World Bank, for GDP per capita (constant 2005 US$).
Note: For a detailed description of the 2014 Digital Adoption Index, see http://wbgfiles.worldbank.org/documents/dec/digital-adoption-index.html.

FIGURE 1.12
Digital Adoption Index for firms in ECA

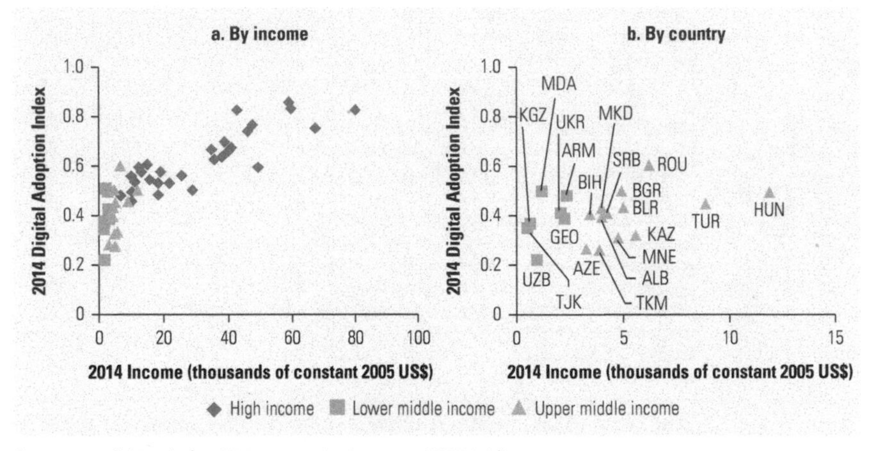

Source: World Bank, for GDP per capita (constant 2005 US$).
Note: For a detailed description of the 2014 Digital Adoption Index, see http://pubdocs.worldbank.org/en/587221475074960682/WDR16-BP-DAI-methodology.pdf.

Depth of Internet Adoption in ECA: Intensity of Internet Use

While the presence of fixed or mobile networks in the area (along with service providers) is a prerequisite for Internet access, it is only a starting point in terms of the ability to take full advantage of the digital economy. Better access does not necessarily mean more intensive use.

Intensity of Internet Use by Households and Individuals

While it is difficult to provide a comprehensive indicator that measures the intensity of Internet use, a rough approximation of intensity is based on the frequency of Internet use and hours spent online (figure 1.13). Globally, the frequency of use

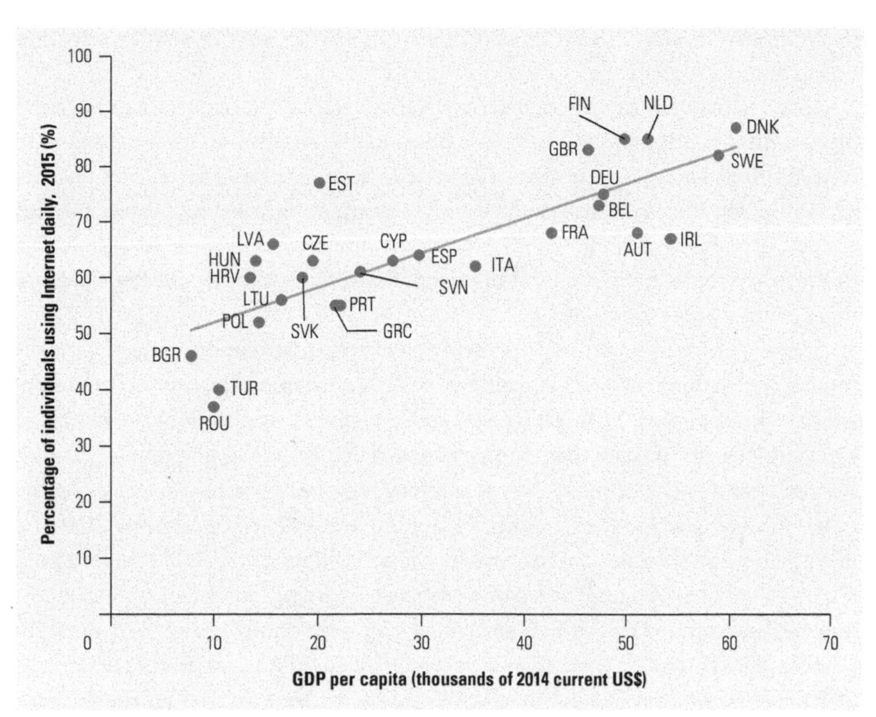

FIGURE 1.13
Frequency of Internet use varies significantly in ECA

Sources: Eurostat data (http://appsso.eurostat.ec.europa.eu/nui/show.do?dataset=isoc_ci_ifp_fu&lang=en) for frequency of use; World Bank data 2014, for GDP.

is highly correlated with GDP per capita. In ECA, the proportion of daily Internet users ranges from less than 50 percent in Bulgaria, Romania, and Turkey to almost 90 percent in Denmark, the Netherlands, and Sweden. Some countries have leaped ahead of their peers in the subregion. For example, in Estonia almost 80 percent of the population accesses the Internet daily, close to the levels in Denmark and Sweden, which have much higher levels of mobile and fixed Internet penetration and higher adoption of smartphones. In contrast, Greece, Poland, and Romania, for example, have considerable potential for greater Internet use, given their relatively high level of fixed and mobile Internet penetration.

Even in those countries where frequency of use is high, significant differences may exist between socioeconomic groups. While the difference in frequency of use between men and women is less than 10 percent, level of education, employment status, or age can be important determinants of differences in the frequency of use. The difference for all but five countries exceeds 10 percent, reaching 20 percent and more in Hungary, the Slovak Republic, and Slovenia, if one of three factors—low level of education, unemployed, or retired—is present (Eurostat 2016).

In addition to frequency and hours spent online, Internet use varies from basic functions such as e-mail or web surfing to high-bandwidth applications such as high-definition streaming of video or gaming. Video now accounts for more than half of total consumer traffic in ECA (Cisco 2016). The share of Internet Protocol (IP) video in total consumer traffic is expected to increase in Poland (from 56 percent in 2015 to 74 percent in 2020), in Spain (from 64 to 82 percent), and in Sweden (from 63 to 81 percent) (Cisco 2016). Central and Eastern Europe is expected to follow the global trend, with digital television (especially personal or digital video recording services) and social networking having the highest penetration rates and online gaming experiencing the largest growth.

As for mobile consumption, mobile social networking is expected to have the highest rates of adoption, while mobile banking and e-commerce are projected to have the highest rates of growth. The mobile segment has a large potential for growth, given that services are increasingly being consumed on mobile devices (Cisco 2016). For example, countries with a high rate of mobile penetration and a low level of banking (figure 1.14) offer significant opportunities for the development of m-services (box 1.1).

Some high-bandwidth applications, for example streaming video, typically require a minimum connection speed of 4 Mbit/s, compared with web browsing, which only requires up to 1 Mbit/s. Moreover, if several applications are launched simultaneously or if more than one device is in use (that is, if one family member uses several devices or more than one family member is connected at the same time), the demand for speed grows. In Central and Eastern Europe, by 2020 the average household is forecast to have four fixed devices or connections, including high-definition televisions, personal computers, gaming consoles, smartphones, and tablets, similar to the average American household today.[10] In Western Europe, the average household is expected to have 10 devices or connections by 2020.[11] Web-enabled televisions are likely to become the largest residential device or connection category in households and smartphones to become the largest consumer mobile device or connection category. The distribution of these devices, however, is expected to remain unequal. Adoption of smartphones remains low in certain parts of ECA, such as Azerbaijan or Bosnia and Herzegovina, where penetration is less than about 40 percent, mainly because of the high cost of smart devices and import duties relative to income levels (figure 1.15). Albania (not displayed in the figure) has the lowest penetration in the region, at slightly above 30 percent as of the second quarter of 2016.

FIGURE 1.14
M-money opportunities in countries with low numbers of banks and high mobile penetration

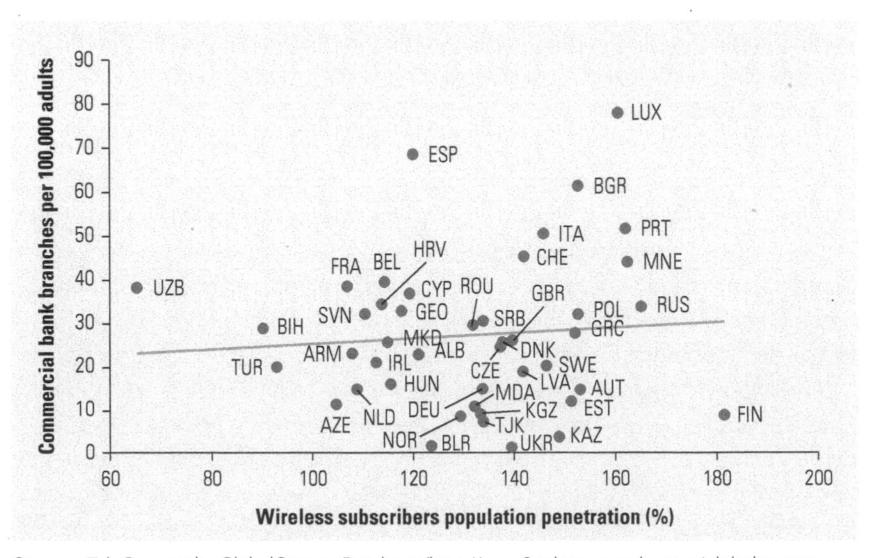

Sources: TeleGeography GlobalComms Database (http://www2.telegeography.com/globalcomms-database-service), for 2014 Wireless subscribers population penetration; International Monetary Fund, Financial Access Survey (http://data.worldbank.org/indicator/FB.CBK.BRCH.P5) for 2015 commercial bank branches, per 100,000 adults.

BOX 1.1 ECA Has a Great Potential for Mobile Banking and Commerce

Only 30 percent of ECA countries currently have access to m-money services (figure B1.1.1), totaling 1.7 million registered m-money accounts and 240,000 active (90 days) accounts. The region, however, has significant potential (alongside the Middle East and North Africa), as these services are expected to grow 50 percent in the next few years.

M-services in Central and Eastern Europe, where cash is still used for the majority of payments, have significant potential for growth. However, initiatives so far have been piecemeal, for example, transportation solutions in the Czech Republic and the Slovak Republic and m-payments in Poland. More recently, Vodafone introduced m-payments through M-Pesa in Romania, a service originally launched in Kenya in 2007 by Vodafone and Safaricom. M-Pesa enables users to pay their utility bill, buy a drink in a bar, or send cash to family and friends.

FIGURE B1.1.1 Share of developing markets with mobile money, by region, December 2015

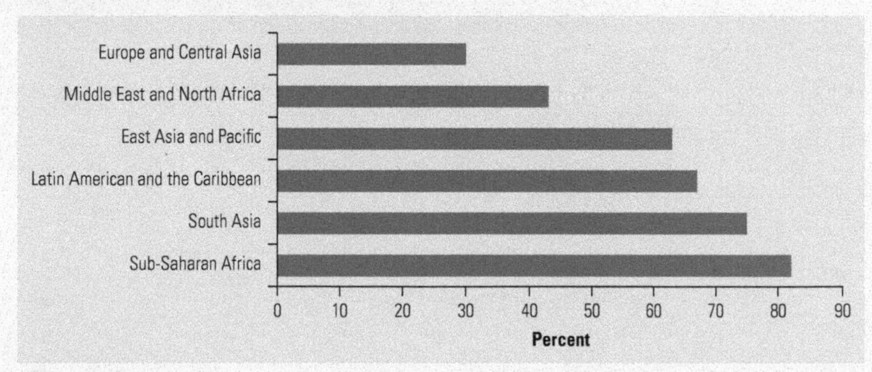

Source: GSMA 2015 data.

In 2014, when M-Pesa was launched in Romania, more than 30 percent of Romanians had no access to traditional banking (most people relied on cash) and mobile penetration was high. In 2016, an agreement between Vodafone and MoneyGram was extended to Romania, enabling customers to send and receive money from abroad, within minutes and in a secure way.

The Baltic States are the most advanced countries in Central and Eastern Europe in m-payments. In 2010, the three major Estonian banks—serving 80 percent of bank customers—introduced m-banking. Estonia has since developed m-money in different sectors such as transport (for tickets and monthly passes for buses and trams) and finance (for loans).

Participation in online work represents an opportunity for countries with lower incomes. As chapter 6 shows, online freelancers' earnings are equivalent to at least seven times the minimum wage in developing ECA. The scarcity of English language as well as technical skills, however, can be important barriers to online freelancing. The top three countries hiring, in order, are the United States, the United Kingdom, and France,[12] and none of the top 10 countries hiring is Russian speaking (table 1.3)—the major language in Central Asia and Eastern Europe.

FIGURE 1.15
Affordability of Internet on a mobile device remains a constraint to more intense adoption of services

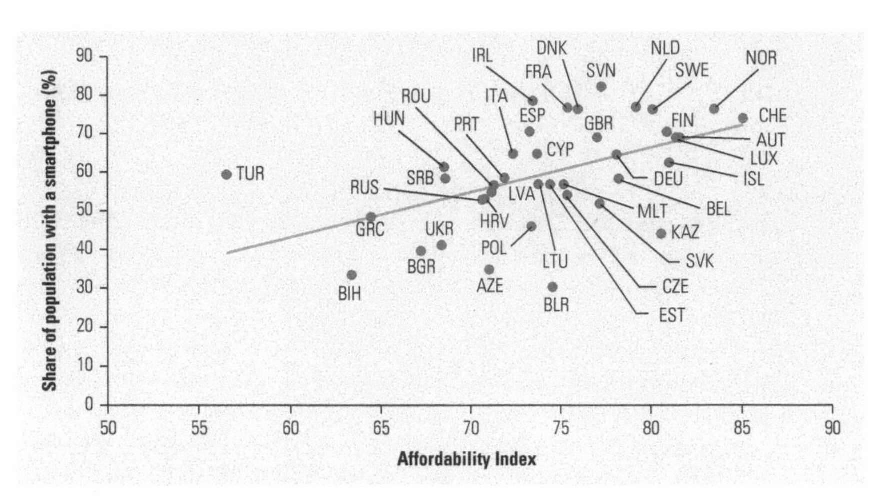

Source: GSMA 2016 data. Mobile Connectivity Index (http://www.mobileconnectivityindex.com), for affordability; GSMA Intelligence (https://www.gsmaintelligence.com), for smartphone connections (adoption).
Note: Smartphone connections are expressed as a share of total connections (excluding machine to machine, M2M), not the total population. The affordability index includes mobile tariff, handset price, income, inequality, and taxation. For the methodology, see http://www.mobileconnectivityindex.com/widgets/connectivityIndex/pdf/ConnectivityIndex_V01.pdf.

TABLE 1.3 Top 10 Countries Hiring Online Freelancers, Based on Spending on Online Work

Rate of growth of spending on online work	Countries
More than 25%	France, Germany, Israel, the Netherlands, Singapore, Switzerland, United Kingdom, United States
10–25%	Australia, Canada

While some ECA countries have already joined the top 10 earning countries (Romania, Russia, and Ukraine), others (countries in Central Asia and Turkey) have yet to catch up. Online work may also offer increased opportunities for groups facing barriers to employment, for example, older workers, women, or adults with young children where daycare services are costly or scarce. Kosovo, for instance, is active in promoting online work model among women in rural areas; the Women in Online Work program (WoW) is an example of such an initiative. However, teleworking is more prevalent among older and female workers in countries, such as those in Northern Europe, with lower barriers to labor force participation (see chapter 6), suggesting that the ability of online opportunities to generate jobs for disadvantaged groups may be limited in the rest of ECA. Online freelancing might also be increasing the rate of informality, as most online freelancers in Eastern Europe are in the informal economy.

Finally, a citizen may receive individual benefits from digital services. Some countries have developed an integrated e-government ecosystem, improving the user experience and transparency of transactions. Examples of e-services include tax payment (Belarus) and e-health record systems (Estonia). As discussed later in this chapter, in some ECA countries access to e-services is facilitated through digital or electronic identity (e-ID) solutions. In countries such as Belgium, Estonia,

Italy, Kazakhstan, Latvia, Luxembourg, the Netherlands, and Portugal, e-ID solutions are used the most for e-services. Box O.3 in the overview and chapter 7 give more examples.

Intensity of Internet Use by Businesses

As discussed in chapter 3, most firms in ECA use the Internet, but only for simple tasks. They do not take full advantage of the technology to replace or to perform more complex tasks. For example, many firms use e-mail to communicate with clients and suppliers, but few have a website, which is more challenging to set up. Chapter 4 finds that few firms in ECA engage in e-commerce, and many firms still use intermediaries instead of online platforms (such as eBay and Amazon) or their own websites to export and to market and sell to customers. Most sell locally in their own country instead of accessing new, foreign markets. The larger firms tend to receive a larger percentage of turnover from e-commerce than smaller firms.

As with consumer services, the intensity (and corresponding benefits) of Internet use by firms has many dimensions (for example, type of applications used) and can depend on a variety of factors (for example, size of the firm, industry sector, and overall needs). With more and more devices connected to the Internet and the growing variety of applications that can boost productivity by performing (and even replacing) more complex tasks, firms have significant opportunities to take advantage of Internet use.

Looking generally at the level of digital intensity in the EU (represented by 12 uses of technology by an enterprise),[13] country scores vary considerably even within the high- and middle-income country groups. The best-performing countries are in Northern Europe, such as Denmark, the Netherlands, and Norway (figure 1.16). Some middle-income countries also score highly in enterprise digital intensity, such as Lithuania and Malta. However, in some countries—for example, Cyprus, Latvia, Poland, and, surprisingly, France—the difference in digital intensity between small and medium enterprises and large enterprises is threefold or even higher.

As of now, high-bandwidth applications[14] have low penetration, but are growing rapidly—for example, the adoption rate for desktop and personal video conferencing is less than 20 percent, but this category is expected to be the fastest-growing Internet service both in ECA and worldwide, according to Cisco forecasts. Regarding enterprises using internal enterprise resource planning (ERP) software, some high-income countries score above the EU average of 35.6 percent (Cyprus, Greece, Italy, Lithuania, and Portugal), whereas some (Hungary, Latvia, and the United Kingdom) score below 20 percent (Eurostat 2015b). Cyprus, Lithuania, and Malta also perform well on the use of customer relationship management (CRM) software, with a score above the EU average of 20.9 percent. The Czech Republic, Estonia, Malta, and the Slovak Republic are above the EU average of 75.3 percent of enterprises having a website or a homepage and above the EU average of 30 percent of enterprises paying to advertise on the Internet, whereas Bulgaria lags behind on these two indicators (Eurostat 2015b). According to one of the DAI business cluster indicators, the proportion of firms with their own

FIGURE 1.16
Northern Europe leads in terms of enterprises with high levels of digital intensity for both small and medium enterprises and large enterprises, 2015

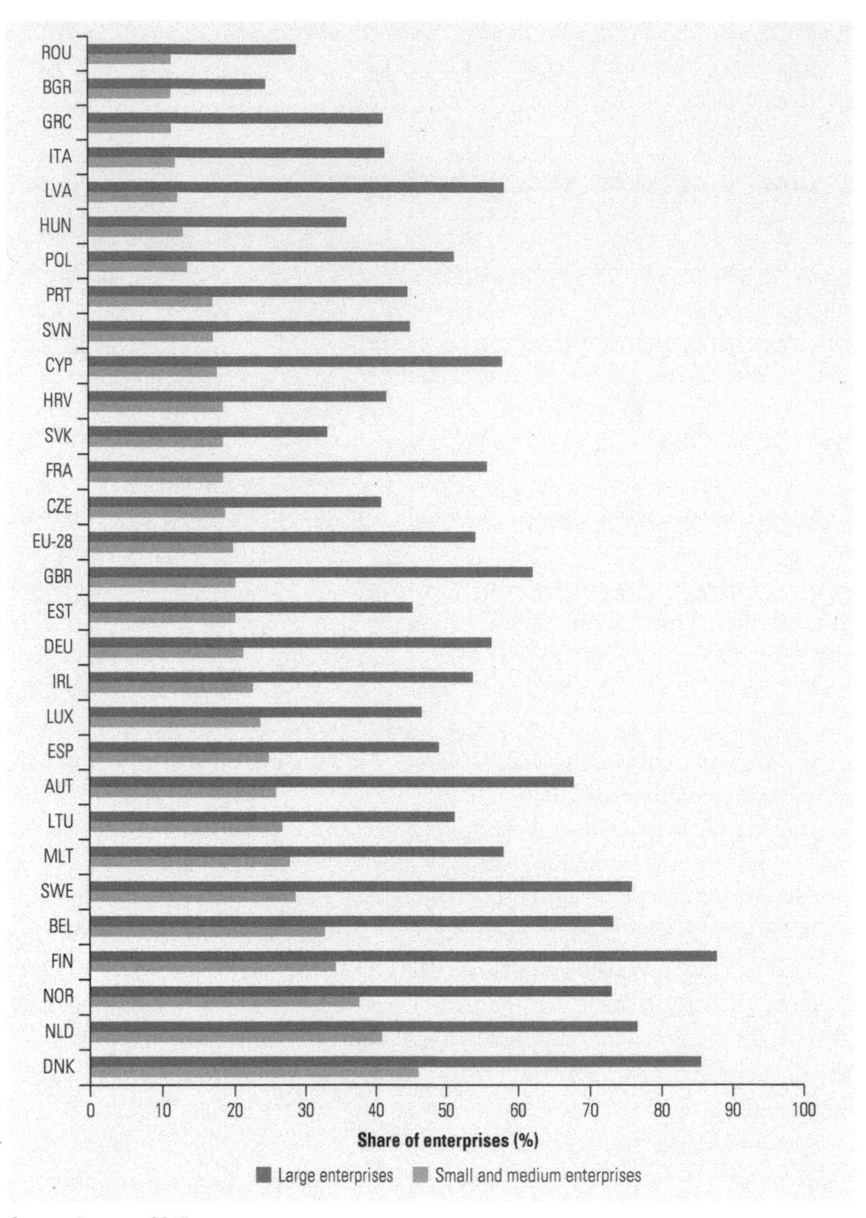

Source: Eurostat 2015a.
Note: High levels are attributed to those enterprises using at least 7 of the 12 digital technologies.

website varies from less than 23 percent in Uzbekistan to more than 90 percent in the Czech Republic, the Slovak Republic, and Sweden. Less than half of businesses in Bulgaria and Romania have a web presence.

Another important trend, especially for medium-size businesses, is the use of cloud services to save costs (for example, by minimizing capital and staffing expenses). According to Cisco, "Cloud services are the number one application class driving an increase in Internet traffic for business use" (Harris 2016). Cisco forecasts that 83 percent of all data center traffic will be through the cloud by 2019, based on the fact that the overall Internet traffic is driven by video. As discussed in chapter 3, few firms in ECA use cloud computing, and those that do use cloud computing use it largely to perform simple tasks such as accessing e-mails rather than complex tasks such as replacing computing power. As of now, Nordic countries score the highest (53 percent in Finland) in firms' use of cloud computing bought via the Internet, and Croatia, Malta, and the Slovak Republic have more than 20 percent of enterprises using this technology (Eurostat 2015a). Central and Eastern Europe and Middle East and North Africa, however, are projected to experience the highest growth in cloud computing. Cisco's Cloud Readiness Tool[15] can estimate the extent to which the concurrent use of basic, intermediate, and advanced applications is possible given the state of a country's network. In ECA, only a handful of countries are "cloud prepared." It will be essential for countries to have in place both infrastructure that meets their requirements as well as policy and regulatory frameworks—the focus of chapter 2.

The role of digital ecosystem players is changing as well. In a digital ecosystem, market players move beyond their traditional lines: for example, telecom operators provide content (AT&T and BT), and digital service providers invest in networks (Google's Loon project in Sri Lanka and Facebook's drones project). The convergence of networks and services is blurring boundaries and creating new issues for authorities with regard to interoperability and common standards, market competition (in countries where broadcasting and telecommunications have traditionally been regulated separately), intellectual property rights, consumer protection, and taxation as well as security and privacy. It is a constant challenge for governments and regulators around the world to keep pace with technological developments and sector evolutions. In May 2016, the European Commission published updated rules for the audiovisual sector and guidance on online platforms (European Commission 2016). Andrus Ansip, vice president for the digital single market, said, "I want online platforms and the audiovisual and creative sectors to be powerhouses in the digital economy, not weigh them down with unnecessary rules. They need the certainty of a modern and fair legal environment: that is what we are providing today. This means not changing existing rules that work, such as those related to the liability of online service providers. It also means deregulating, where necessary, traditional sectors like broadcasting, or extending certain obligations to platforms and other digital players to improve user protection and to reach a level playing field." With the objective of setting "the right environment to attract, retain, and grow new online platform innovators," the framework for online platforms includes applying comparable rules for comparable digital services, facilitating portability of data among different online platforms, as well as launching fact-finding exercises to create a fair and innovation-friendly business environment (box 1.2).

Finally, businesses benefit from digital services through e-government services such as online tax payment. As detailed in chapter 3, some e-government platforms enable business registration and land registry, lowering the cost of market entry for a business.

BOX 1.2 Where Are the European Market Leaders on the Internet?

According to IDATE, the EU-28 countries accounted for just over 20 percent of the global market for Internet services in 2015. But this market is still dominated by firms from the United States and, to a lesser extent, China (table B1.2.1). Among the market-leading Internet firms worldwide, Europe is largely absent. European Internet firms may still be privately held (for example, the Swedish music-streaming service, Spotify), have been acquired by non-European firms (for example, Skype is now owned by Microsoft), have languished (for example, Lastminute.com was briefly worth £768 million when floated in 2000), or may still be "bubbling under" (for example, the German Internet service provider, Rocket Internet, has a market capitalization of US$2.8 billion, but is still well below the threshold for inclusion in table B1.2.1).

Of course, coming up with a list of market-leading companies is fraught with definitional difficulties. If one were to include mobile operators (who generally also provide Internet services), then European companies such as Vodafone (market capitalization of US$82 billion) or Orange (market capitalization of US$36 billion) would certainly be present. But mobile operators still make the majority of their money from voice calls and so are excluded from the list.

So why is Europe still awaiting its Google or Facebook? Language is one reason, with the EU-28 countries alone sharing 24 official languages. Having so many official languages makes it particularly difficult for providers of Internet content and applications, which make most of their money from advertising. By contrast, mobile operators serve a

TABLE B1.2.1 Market Value and Revenue of Top 20 Internet Companies Worldwide, 2015–16
US$, billion

Rank	Company	Country	Mid-2016 market value	2015 revenue
1	Apple	United States	547	235
2	Google/Alphabet	United States	510	75
3	Amazon	United States	341	107
4	Facebook	United States	340	18
5	Tencent	China	206	16
6	Alibaba	China	205	15
7	Priceline	United States	63	9
8	Uber	United States	63	—
9	Baidu	China	62	10
10	Ant Financial	China	60	—
11	Salesforce.com	United States	57	7
12	Xiaomi	China	46	—
13	Paypal	United States	46	9
14	Netflix	United States	44	7
15	Yahoo!	United States	36	5
16	JD.com	China	34	28
17	eBay	United States	28	9
18	Airbnb	United States	26	—
19	Yahoo! Japan	Japan	26	5
20	Didi Kuaidi	China	25	—
	Total		2,752	554[a]

Source: Meeker 2016, based on data from CapIQ, CB Insights, *Wall Street Journal*, and media reports.
Note: — = not available.
a. Based on 15 out of 20 firms only.

(Continued)

BOX 1.2 Where Are the European Market Leaders on the Internet? *(continued)*

series of distinct national markets, rather than regional or global ones, and make money directly from subscribers, not advertisers. Regulation also favors companies from China and the United States. Despite the best efforts of the EU to create a single digital market, in practice rules on advertising, privacy, copyright, and so on vary so much that the cost of serving multiple national markets is prohibitively expensive compared with serving China or the United States. The lack of availability of sufficient venture capital funding in Europe is another often-cited reason. These issues are compounded by the first-mover advantage that makes it hard to compete against a company that already has significant market share in a specific field, such as ride hailing (Didi Kuaidi and Uber) or e-commerce (Alibaba and Amazon).

What could change this picture? Encouraging consolidation of national companies could help, as has happened in the advertising or fashion sectors. Policy measures to harmonize regulatory rules, especially with regard to privacy, could also help, as would scrapping national "windows" for protecting content or easing bankruptcy laws. But ultimately the emergence of entrepreneurial role models such as Bill Gates or Jack Ma is what could make the real difference. And that emergence, in turn, would require a change in cultural as much as economic models.

Intensity of Internet Use by Governments

ECA countries with very different income levels have similar levels of Internet adoption by government. For example, the DAI government cluster[16] shows that governments in Estonia and Kazakhstan use the Internet to a similar extent as governments in the Netherlands and Spain (figure 1.17). Belarus, Tajikistan, and Ukraine have the lowest scores in the ECA region, lower than Turkmenistan. Paradoxically, some "transitioning" countries (according to the World Bank categories), such as Kazakhstan and Russia, score more highly than some "transforming" countries, such as Denmark, Ireland, and the United Kingdom. According to the United Nations' (2014a) e-government development index (EGDI),[17] all Central Asia countries (except Kazakhstan), as well as Bosnia and Herzegovina and the former Yugoslav Republic of Macedonia (upper-middle-income countries) rank in the middle range (between 0.25 and 0.50), while Armenia and Georgia (lower-middle income countries) score more highly (between 0.50 and 0.75).

Exploring some of the differences within these groups, France has the highest score in the world based on the online service delivery subindex, with the Netherlands and Spain being the only other countries in the high-income group in ECA with the highest possible rating. France's high score results from efforts at "better integration of e-services, expanded rollout of mobile applications, and provision of opportunities for e-participation" (United Nations 2014b). Kazakhstan is the only ECA country in the top 10 middle-income countries globally based on service delivery. Looking at the subindexes in more detail, Belarus, Bulgaria, Croatia, the Czech Republic, FYR Macedonia, and Ukraine rank highly in telecommunication infrastructure and human capital, but perform poorly in services, suggesting significant potential for improvement in e-government services.

FIGURE 1.17 Digital Adoption Index for governments

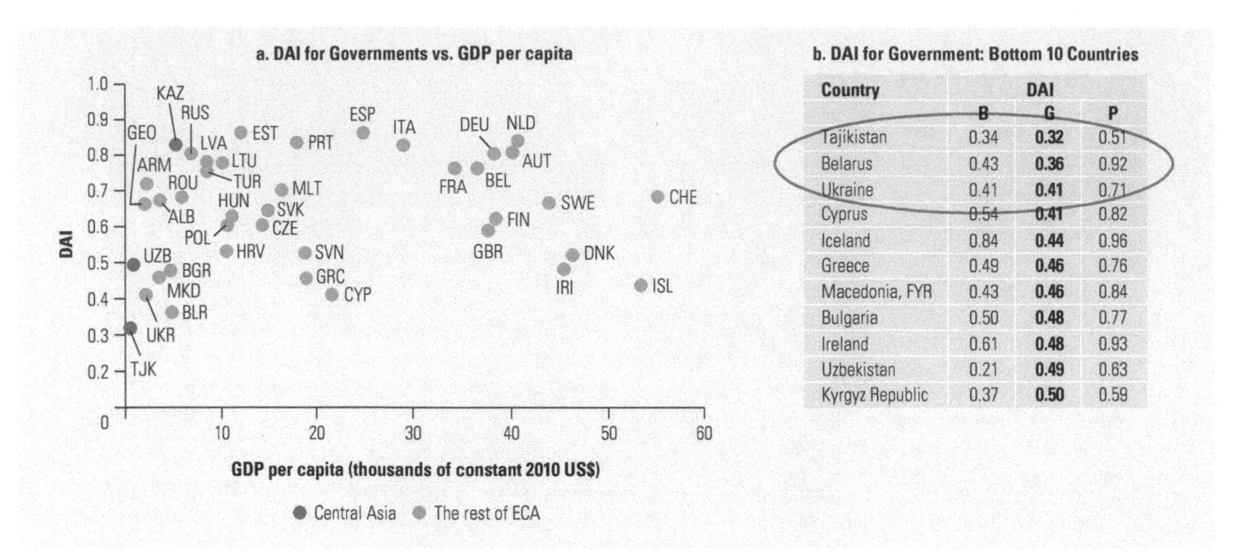

b. DAI for Government: Bottom 10 Countries

Country	DAI		
	B	G	P
Tajikistan	0.34	**0.32**	0.51
Belarus	0.43	**0.36**	0.92
Ukraine	0.41	**0.41**	0.71
Cyprus	0.54	**0.41**	0.82
Iceland	0.84	**0.44**	0.96
Greece	0.49	**0.46**	0.76
Macedonia, FYR	0.43	**0.46**	0.84
Bulgaria	0.50	**0.48**	0.77
Ireland	0.61	**0.48**	0.93
Uzbekistan	0.21	**0.49**	0.63
Kyrgyz Republic	0.37	**0.50**	0.59

Source: GDP per capita—2014 World Bank data, thousands of constant 2010 US$.
Note: B = business; DAI = Digital Adoption Index; G = government; P = people.

In South Caucasus, the situation is reversed, with, especially, Armenia and Georgia performing better in online services than in telecommunication infrastructure.

Most lower-middle-income countries provide "emerging" types of information services (stage 1), rather than being proactive in involving citizens in government activities (figure 1.18). By contrast, Georgia scores more highly than other countries on the provision of "transactional" services, and Armenia has the highest sophistication in the income group.

Within the upper-middle-income group of countries, Albania, Kazakhstan, and Turkey are the most sophisticated within stage 4, similar to some countries in Western Europe. Surprisingly, Germany, Hungary, and Poland, three high-income "transforming" Organisation for Economic Co-operation and Development (OECD) countries, have relatively low scores on stage 4 compared with the rest of the region as well as with countries in other income groups.

Only a few countries in ECA have developed "transactional" services (stage 3): France and Kazakhstan and, to some extent, Estonia, Italy, the Netherlands, Russia, Spain, and the United Kingdom. This category of services requires some form of electronic authentication of the citizen's identity as well as robust online security, payment systems, and channel coordination (United Nations 2014b).

According to the Identification for Development (ID4D) database,[18] biometric e-ID has only been introduced in 15 ECA countries so far (figure 1.19). Currently, e-ID (including digital signature and remote online services) is used for most e-services in Belgium, Estonia, Italy, Kazakhstan, Latvia, Luxembourg, the Netherlands, and Portugal. The picture is mixed in the rest of ECA: Tajikistan and Ukraine for example, use static barcode cards; Kosovo and Bosnia and Herzegovina use smart cards with biometrics but for unique identification only; Belarus and Turkmenistan do not yet have an e-ID solution. Belarus, however, is planning to implement an e-ID initiative in 2017 under the State Program for the Digital Economy and Information Society 2016–2020.

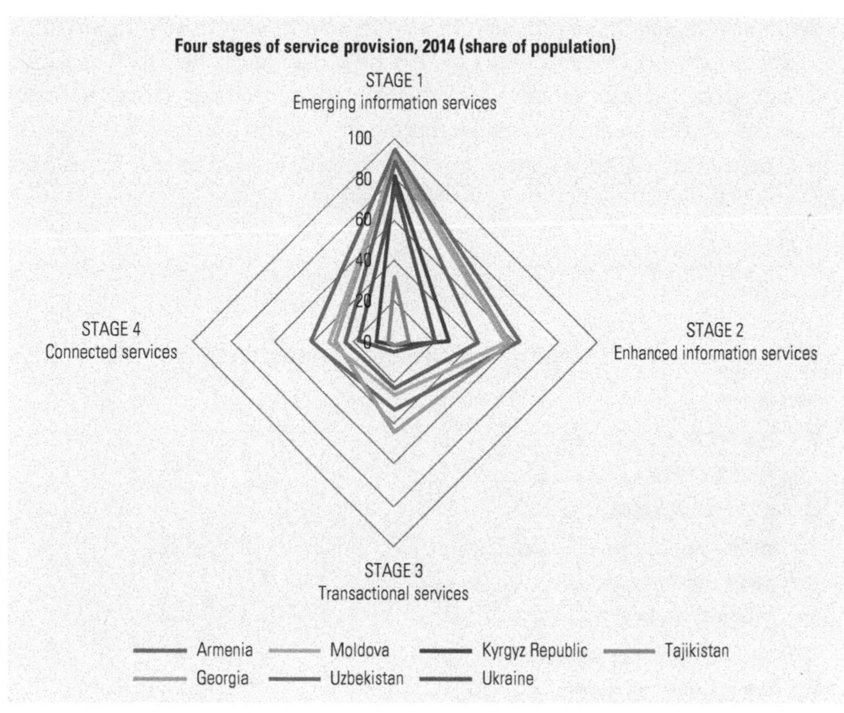

Source: United Nations 2014b.

FIGURE 1.18
Emerging lower-middle-income countries in ECA have much lower sophistication of online government services provision

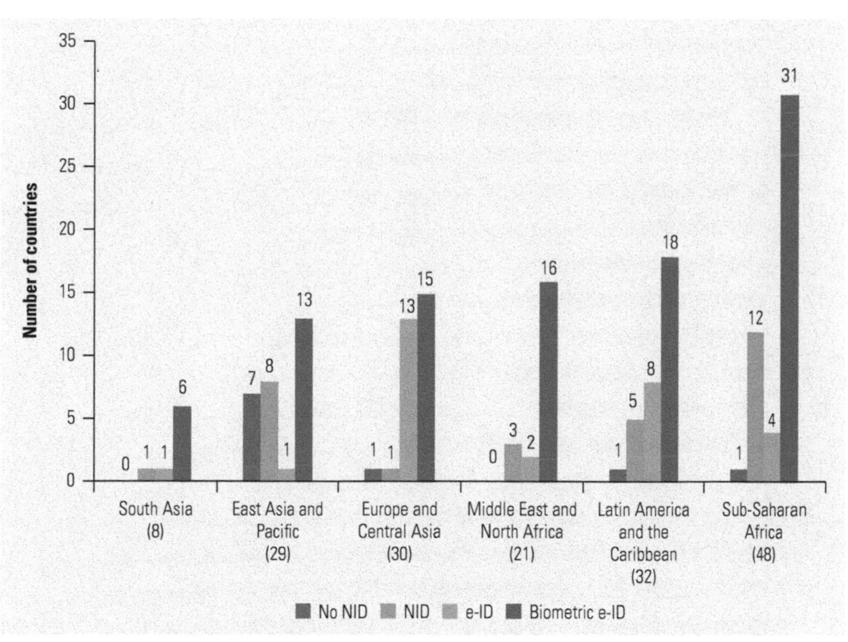

Source: Identification for Development (ID4D) Global Dataset, January 2016.
Note: e-ID = electronic identification card; NID = national identification card. Numbers in parentheses represent the number of countries in the region.

FIGURE 1.19
Biometric e-ID has only been introduced in 15 ECA countries

Beyond considering the provision of e-government services, it is important to consider the uptake of these services. Within the EU, Bulgaria, the Czech Republic, and Romania have the lowest rates of both overall interaction with the government and online submission of completed forms (figure 1.20). In Romania, for instance, only 5 percent of individuals used the Internet to download official forms from public websites in 2015.

FIGURE 1.20
The take-up of e-government services varies within the EU

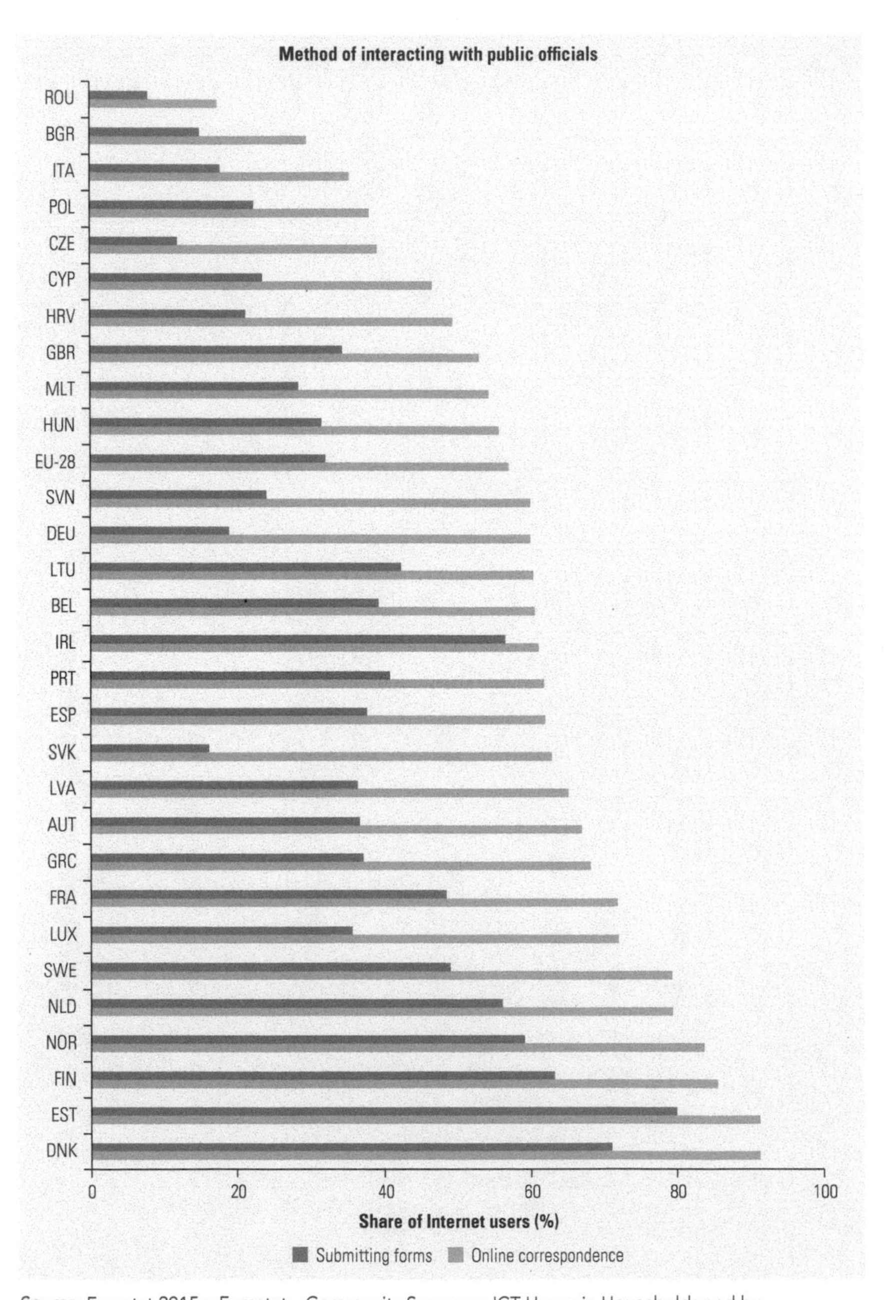

Source: Eurostat 2015a; Eurostat—Community Survey on ICT Usage in Households and by Individuals, 2015.

Conclusion

Substantial differences in the breadth of Internet adoption (uptake and connectivity) and in the depth of Internet adoption (intensity of use) exist within the ECA region. The level of development can, in part, explain these differences, although in some respects Internet use and intensity vary within income groups. Disparities within countries are also tied to level of income, as well as gender, age, and population density. Level of income is positively correlated with broadband penetration, use still differs between genders, younger people tend to have a higher frequency of use than older ones, and urban areas generally have a higher level of connectivity than rural areas. Government use of the Internet is particularly developed in a few of the higher-income countries in ECA, but the level and types of services vary considerably. Following this analysis of the demand for Internet services in ECA, chapter 2 examines the supply side and discusses the possible policy and regulatory frameworks that could improve Internet adoption in the region.

Annex 1A. ECA Country Classification

World Bank country classification, shown in table 1A.1, includes 58 countries in the Europe and Central Asia region (World Bank list of economies, as of September 2016). In this book, 10 of the 57 countries have not been considered because of their particular geographic situation: Channel Islands, Faroe Islands, Gibraltar, Greenland, Iceland, Isle of Man, Liechtenstein, Monaco, San Marino, and Switzerland. Andorra is handled only in chapter 2.

TABLE 1A.1 ECA Country Classifications

Code	ECA country	WB ECA subregion	WB income level	WDR16 country category	EU member
KGZ	Kyrgyz Republic	Central Asia	Lower middle income	Emerging	No
TJK	Tajikistan	Central Asia	Lower middle income	Emerging	No
TKM	Turkmenistan	Central Asia	Upper middle income	Emerging	No
UKR	Ukraine	Other Eastern Europe	Lower middle income	Emerging	No
UZB	Uzbekistan	Central Asia	Lower middle income	Emerging	No
ALB	Albania	Western Balkans	Upper middle income	Transitioning	No
ARM	Armenia	South Caucasus	Lower middle income	Transitioning	No
AZE	Azerbaijan	South Caucasus	Upper middle income	Transitioning	No
BLR	Belarus	Other Eastern Europe	Upper middle income	Transitioning	No
BIH	Bosnia and Herzegovina	Western Balkans	Upper middle income	Transitioning	No
BGR	Bulgaria	Central Europe	Upper middle income	Transitioning	Yes
HRV	Croatia	Central Europe	High income	Transitioning	Yes
CYP	Cyprus	Southern Europe	High income	Transitioning	Yes
CZE	Czech Republic	Central Europe	High income	Transitioning	Yes
GEO	Georgia	South Caucasus	Upper middle income	Transitioning	No
GRC	Greece	Southern Europe	High income	Transitioning	Yes
ITA	Italy	Southern Europe	High income	Transitioning	Yes

(Continued)

TABLE 1A.1 ECA Country Classifications *(continued)*

Code	ECA country	WB ECA subregion	WB income level	WDR16 country category	EU member
KAZ	Kazakhstan	Central Asia	Upper middle income	Transitioning	No
XKX	Kosovo	Western Balkans	Lower middle income	Transitioning	No
MKD	Macedonia, FYR	Western Balkans	Upper middle income	Transitioning	No
MDA	Moldova	Other Eastern Europe	Lower middle income	Transitioning	No
MNE	Montenegro	Western Balkans	Upper middle income	Transitioning	No
PRT	Portugal	Southern Europe	High income	Transitioning	Yes
ROU	Romania	Central Europe	Upper middle income	Transitioning	Yes
RUS	Russian Federation	Russian Federation	Upper middle income	Transitioning	No
SRB	Serbia	Western Balkans	Upper middle income	Transitioning	No
SVK	Slovak Republic	Central Europe	High income	Transitioning	Yes
SVN	Slovenia	Central Europe	High income	Transitioning	Yes
TUR	Turkey	Turkey	Upper middle income	Transitioning	No
AUT	Austria	Western Europe	High income	Transforming	Yes
BEL	Belgium	Western Europe	High income	Transforming	Yes
DNK	Denmark	Northern Europe	High income	Transforming	Yes
EST	Estonia	Northern Europe	High income	Transforming	Yes
FIN	Finland	Northern Europe	High income	Transforming	Yes
FRA	France	Western Europe	High income	Transforming	Yes
DEU	Germany	Western Europe	High income	Transforming	Yes
HUN	Hungary	Central Europe	High income	Transforming	Yes
IRL	Ireland	Western Europe	High income	Transforming	Yes
LVA	Latvia	Northern Europe	High income	Transforming	Yes
LTU	Lithuania	Northern Europe	High income	Transforming	Yes
LUX	Luxembourg	Western Europe	High income	Transforming	Yes
NLD	Netherlands	Western Europe	High income	Transforming	Yes
NOR	Norway	Northern Europe	High income	Transforming	No
POL	Poland	Central Europe	High income	Transforming	Yes
ESP	Spain	Southern Europe	High income	Transforming	Yes
SWE	Sweden	Northern Europe	High income	Transforming	Yes
GBR	United Kingdom	Western Europe	High income	Transforming	No

Notes

1. There is no single definition of Internet adoption. For the purposes of this book, adoption is defined as the use of services delivered over an Internet Protocol (IP)–based network.
2. Offline population represents the difference between the total population and fixed Internet and 3G/4G users. ECA has the fourth largest offline population of any region, after Asia (2.4 billion), Africa (845 million), and Latin America (more than 284 million).
3. Information society statistics at the regional level from Eurostat (2016).
4. 2014 data are from the World ICT/Telecommunication Indicators database (www.itu.int/ti).
5. When European Union is not specified, Europe means "geographic Europe" in this chapter.
6. MVNO operators resell data and voice services at lower prices and higher flexibility. They do not own the wireless network infrastructure, but can contribute significantly to the customer base of mobile operators.
7. Figure 1.7 presents the gap between households' broadband penetration in urban and in rural areas. Urban areas are defined as densely populated areas (with at least 500 inhabitants per square kilometer). Rural areas are defined as sparsely populated areas (with fewer than 100 inhabitants per square kilometer). Data are not available for non-EU countries.

8. For the methodology used by the DAI, see http://www.digitaladoptionindex.org/methodology.html.
9. The business cluster of the DAI is a simple average of four normalized indicators: the percentage of businesses with websites, the number of secure servers, the download speed, and the 3G coverage in the country.
10. According to Cisco (http://www.cisco.com/c/dam/assets/sol/sp/vni/sa_tools/vnisa-highlights-tool/vnisa-highlights-tool.html).
11. According to Nielsen (http://marketingland.com/nielsen-time-accessing-Internet-smartphones-pcs-73683).
12. According to Elance (https://www.elance.com/q/online-employment-report).
13. The 12 uses of technology include use of the Internet by a majority of workers, access to ICT specialists, access to fixed broadband speed at more than 30 Mbit/s, use of mobile devices by more than 20 percent of employed persons, use of a website, use of some sophisticated functions on the website, presence on social media, use of enterprise resource planning software, use of customer resource management software, electronic sharing of supply chain management information, e-sales that account for at least 1 percent of turnover, and exploitation of business-to-customer opportunities of web sales.
14. Cisco, for instance, disaggregates applications into basic (web browsing, web conferencing, Voice over Internet Protocol [VoIP]), intermediate (enterprise resource planning or customer resource management, voice over Long-term Evolution [VoLTE], high-definition video streaming, personal content locker, standard-definition video conferencing), and advanced (telemedicine, connected vehicles safety applications, virtual office, including download and upload speeds and latency requirements).
15. For the Cloud Readiness Tool, see http://www.cisco.com/c/en/us/solutions/service-provider/cloud-readiness-tool/index.html. The tool measures network performance indicators such as download and upload speeds and latency.
16. The DAI government cluster is the simple average of three subindexes: core administrative systems (World Bank), online public services (United Nations online service index), and digital identification (World Bank).
17. The EGDI consists of online services (online service index), the development status of telecommunication infrastructure (telecommunication infrastructure index), and the inherent human capital (human capital index).
18. For the ID4D database, see http://www.worldbank.org/en/programs/id4d.

References

Cisco. 2016. *Visual Networking Index: Global Mobile Data Traffic Forecast Update, 2015–2020.* White Paper. http://www.cisco.com/c/en/us/solutions/collateral/service-provider/visual-networking-index-vni/mobile-white-paper-c11-520862.html.

Ericsson, Arthur D. Little, and Chalmers University of Technology. 2013. "Socioeconomic Effects of Broadband Speed." https://www.ericsson.com/res/thecompany/docs/corporate-responsibility/2013/ericsson-broadband-final-071013.pdf.

European Commission. 2016. *Europe's Digital Progress Report.* Brussels: European Commission. https://ec.europa.eu/digital-single-market/en/news/europes-digital-progress-report-2016. European Commission. 2016b.

Eurostat. 2014. "Digital Economy and Society Statistics, Households and Individuals." Eurostat, Luxembourg.

———. 2015a. "Community Survey on ICT Usage and eCommerce in Enterprises, 2015." Eurostat, Luxembourg.

———. 2015b. "Eurostat Enterprises Survey." Eurostat, Luxembourg. http://ec.europa.eu/eurostat/statistics-explained/index.php/Internet_access_and_use_statistics_-_households_and_individuals.

———. 2016. "Information Society Statistics at the Regional Level." Eurostat Statistics Explained. http://ec.europa.eu/eurostat/statistics-explained/index.php/Information_society_statistics_at_regional_level.

Eurostat. http://ec.europa.eu/eurostat/statistics-explained/index.php/Internet_access_and_use_statistics_-_households_and_individuals.

FCC (Federal Communications Commission). 2014. "Household Broadband Guide." https://www.fcc.gov/research-reports/guides/household-broadband-guide.

GSMA (Global System for Mobile Association). 2015. "Bridging the Gender Gap: Mobile Access and Usage in Low- and Middle-Income Countries." GSMA, London.

Harris, M. 2016. "Staying Ahead of the Business Bandwidth Curve." White Paper, Time Warner Cable Business Class. https://business.timewarnercable.com/content/dam/business/pdfs/resource-center/white-papers/staying-ahead-of-bandwidth-curve-wp-2016.pdf?elq=d7935034306948cab213116c736575e7&elqCampaignId=469&elqaid=686&elqat=1&elqTrackId=a8f95aa48a5c4a3c86240d72ec75e80a.

ITU (International Telecommunication Union). 2013. "WSIS Forum 2013: Measuring Gender and ICT." ITU, Geneva, May 14. https://www.itu.int/en/ITU-D/Statistics/Documents/events/wtis2013/014_E_doc.pdf.

———. 2014. *Measuring the Information Society 2014.* Geneva: ITU. https://www.itu.int/en/ITU-D/Statistics/Documents/publications/mis2014/MIS2014_without_Annex_4.pdf.

———. 2015. *Measuring the Information Society 2015.* Geneva: ITU.

Kongut, C., I. K. Rohman, and E. Bohlin. 2014. "The Economic Impact of Broadband Speed: Comparing between Higher and Lower Income Countries." European Investment Bank Institute and Institute for Management of Innovation and Technology, Gothenburg. http://institute.eib.org/wp-content/uploads/2014/04/EIB_broadband-speed_120914.pdf.

Meeker, M. 2015. "Internet Trends 2015." http://www.kpcb.com/internet-trends.

———. 2016. "Internet Trends 2016." http://www.kpcb.com/internet-trends.

TeleGeography. 2015. "Global Internet Geography." https://www.telegeography.com/research-services/global-internet-geography/index.html.

United Kingdom Cabinet Office. 2014. "Government Digital Inclusion Strategy." https://www.gov.uk/government/publications/government-digital-inclusion-strategy/government-digital-inclusion-strategy.

United Nations. 2014a. "E-Government Development Index (EGDI)." United Nations, New York. http://unpan3.un.org/egovkb/Portals/egovkb/Documents/un/2014-Survey/E-Gov_Complete_Survey-2014.pdf.

———. 2014b. "E-Government Survey 2014." United Nations, New York. https://publicadministration.un.org/egovkb/portals/egovkb/documents/un/2014-survey/e-gov_complete_survey-2014.pdf.

World Bank. 2016. *World Development Report 2016: Digital Dividends.* Washington, DC: World Bank.

World Development Indicators. http://data.worldbank.org/data-catalog/world-development-indicators/.

2

Supply of Internet Networks and Services

Europe and Central Asia is a very heterogeneous region with regard to levels of Internet penetration. For instance, people living in Austria, Luxembourg, and Switzerland have almost universal Internet access. At the other extreme, the majority of residents of the Kyrgyz Republic, Tajikistan, and Uzbekistan have not used the Internet. These two groups of countries are completely different in their historical pathway and level of development today. However, they do share a similarity. All of them are landlocked, which is one of the most important barriers to good and affordable Internet access.

The contrasting experiences of these groups of countries show that achieving universal Internet access is a complex task, as it depends heavily on many factors such as having a skilled labor force and a friendly business environment. It also shows that geographic barriers could eventually be overcome. For landlocked countries, having good diplomatic relationships with their neighbors is crucial to achieving a good Internet connection. This chapter argues that the European Union has been a major force of change in this direction for the region and also a catalyst of telecommunications reform. Could regional coordination have a positive impact in the rest of the region?

Introduction and Main Messages

This chapter examines the supply of Internet networks and services in Europe and Central Asia (ECA). The *World Development Report: Digital Dividends* (WDR16; World Bank 2016) sets the goal of ensuring that the Internet is universal, affordable, open, and safe, and this chapter explores the first two of those objectives—coverage and affordability.[1] An important goal is to explain some of the differences in Internet adoption and use described in chapter 1.

A key message is that competition matters. Drawing on lessons from WDR16, the chapter confirms the important role played by competitive markets, private sector participation, and effective, independent regulatory agencies in enhancing access to broadband—as expressed by better connectivity and more affordable prices. A competitive market generates traffic, and the volume of traffic and its rate of growth attract investment to connect a national market with the global Internet backbone market.

Geography and history also matter. The countries of Western Europe are generally well served with undersea cables, which are less expensive than the cross-border terrestrial linkages providing the bulk of services to countries in Central and Eastern Europe. And after World War II, the countries in Western Europe developed on the basis of decentralized markets, while those in Eastern Europe pursued a model of centrally planned growth, which was disrupted during the transition from socialism in the 1990s. The different trajectories and speeds of development and integration make it difficult to compare the experience of different countries, including their telecommunications markets.

ECA lags behind other developed regions in the adoption of fiber technology, which can provide faster transmission and better quality than cable or copper line networks. Having a mix of cable, copper, and fiber networks also can improve competition by providing alternatives to dominant service providers. An efficient regulatory framework can facilitate competition by permitting competitors to access copper line telephone networks via regulated wholesale broadband offers or local loop unbundling (LLU), which enables more than one telecommunications firm to provide services using the connection from the telephone exchange to the customer's premises.

Global prices for fixed broadband fell by two-thirds between 2008 and 2013. Prices had declined earlier in Europe, but still fell more than 20 percent between 2009 and 2014. ECA (excluding Central Asia) has the lowest prices for entry-level fixed broadband as a percentage of gross national income (GNI) per capita, although prices are higher in comparison with per capita income in Central Asia.

This chapter describes the diversity of the ECA regulatory landscape according to the three categories of countries (emerging, transitioning, and transforming) defined in WDR16. Unlike the WDR16, this chapter defines four Central Asian countries and Ukraine as emerging, as they are ranked significantly below the rest of the region on both the level of technology and complementary issues such as regulation, institutional accountability, and skills. These lower-middle-income countries (except Turkmenistan) are all landlocked, have relatively weak international and domestic connectivity, and have yet to liberalize their international gateway, which remains with the incumbent operators, some of which are state owned.

The transitioning countries are rated higher on the strength of the regulatory framework, according to the International Telecommunication Union's Regulatory Tracker. Most of them have privatized their former national telecommunications operator, and infrastructure sharing is authorized and even encouraged. The 22 transforming countries are all high income, and in 2007 all scored highly in the four regulatory categories surveyed in the Regulatory Tracker.

This chapter also highlights the impact of regional institutions (or lack thereof). The European Union has been an important driver of regulatory reform and competition. However, the lack of a regional institution following the collapse of the Soviet Union has slowed progress in Central Asia.

The chapter begins with a discussion of the drivers of connectivity, at both international and national levels, with particular focus on the pricing of Internet services as well as connectivity challenges of Central Asia. It then covers policy and regulation governing the supply of Internet networks, based on policy frameworks adopted in the WDR16.

Connectivity in ECA

International Connectivity: Why Geography Matters

International Internet connectivity refers to a country's international linkages with the rest of the world or the "first mile" of access.[2] The ECA region has some of the highest levels of international connectivity in the world, notably in the Netherlands, where each inhabitant can enjoy speeds of at least 1.5 megabits per second (Mbit/s) (figure 2.1). High capacity is generally correlated with faster speeds (map 2.1). The Netherlands' world-leading position is thanks, in part, to Amsterdam's role as an international hub, or Internet exchange point (IXP), in competition with Frankfurt, London, Paris, Stockholm, and elsewhere. But ECA is also home to some of the most poorly served countries, such as in Central Asia, where international capacity per person can be less than 1 kilobit per second (kbit/s), barely sufficient to receive text messages, let alone high-definition video. International connectivity is perhaps less important in large, populous countries that generate their own content in their own language (such as the Russian Federation), where domestic connectivity is more important. But everywhere else, international connectivity is a critical bottleneck and a key factor in determining how ordinary users experience the Internet.

The first-mile connection is typically provided for coastal countries via undersea cables, which are less expensive than cross-border territorial linkages or satellite. This is an important reason why international connectivity is high in Western Europe. The countries of Western Europe are generally well served with undersea cables, mainly heading across the Atlantic (map 2.2). Most of these cables were constructed in the late 1990s in the lead-up to the bursting of the Internet bubble around 2001 and still provide abundant capacity. Only now, with new projects such as Hibernia Atlantic (linking landing stations in Halifax in Canada with Southport in the United Kingdom and Dublin in Ireland) and AE Connect (linking Shirley, Long Island, in the United States with Killala in Ireland), is new transatlantic capacity

FIGURE 2.1
ECA countries have a wide variation in international Internet capacity

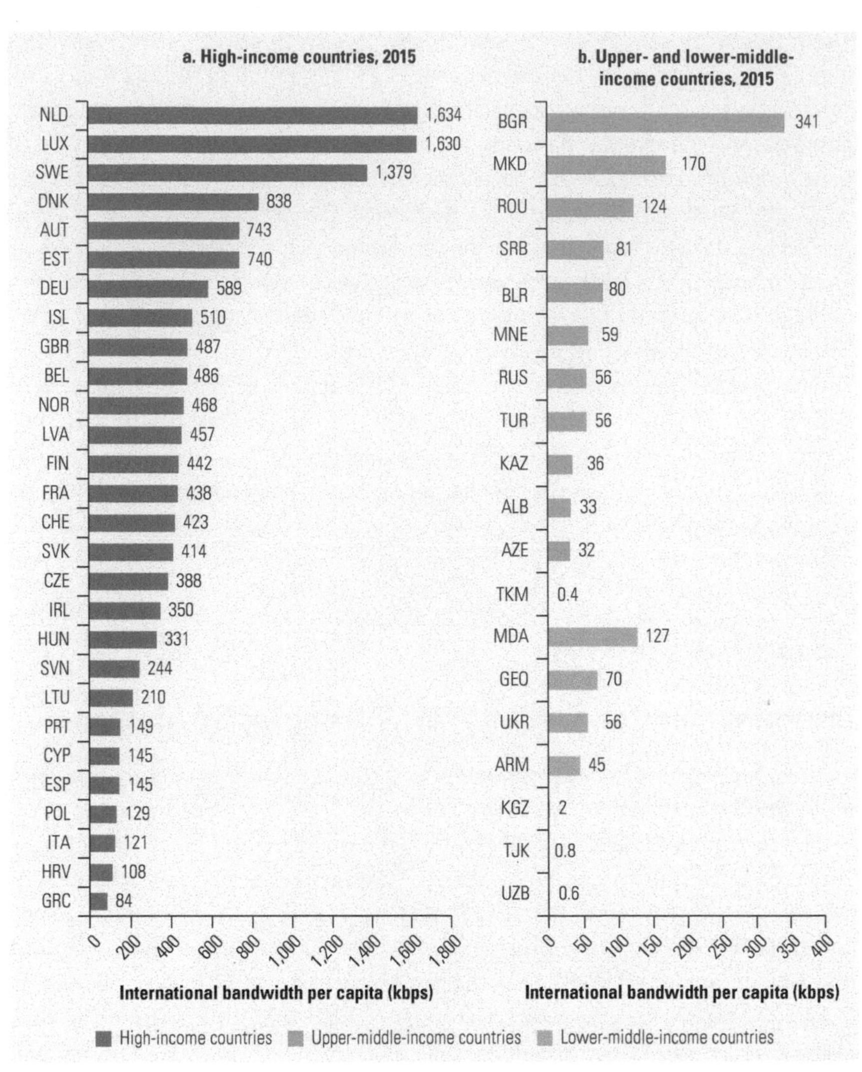

Sources: International bandwidth: TeleGeography data; population: World Bank data.

being planned and built. South Europe will soon be connected to America. The 6,600-kilometer Marea cable, owned by Microsoft and Facebook, will connect Virginia Beach in the United States with Bilbao in Spain. Recent years have also seen a burst of new capacity heading south from Europe toward the Middle East and Africa via the Suez Canal, with Marseille in France emerging as a regional hub. In broad geographic terms, the transatlantic focus of connectivity in the 1990s, which persisted into the 2000s, is now evolving into a more diverse, multidirectional model of connectivity heading south and east as well as west.

Countries in Central and Eastern Europe rely mostly on terrestrial cables. The Caucasus Cable System, which links Bulgaria with Georgia, is one of the few submarine cables serving the east of the region, and it has already been upgraded three times since it was constructed in 2008. The Europe-Persia Express Gateway is a terrestrial cable linking Europe with the Middle East. Put in place in 2012, it goes from Frankfurt, across Eastern Europe, Russia, Azerbaijan,

MAP 2.1 International Internet traffic flows, ECA region (Gbit/s)

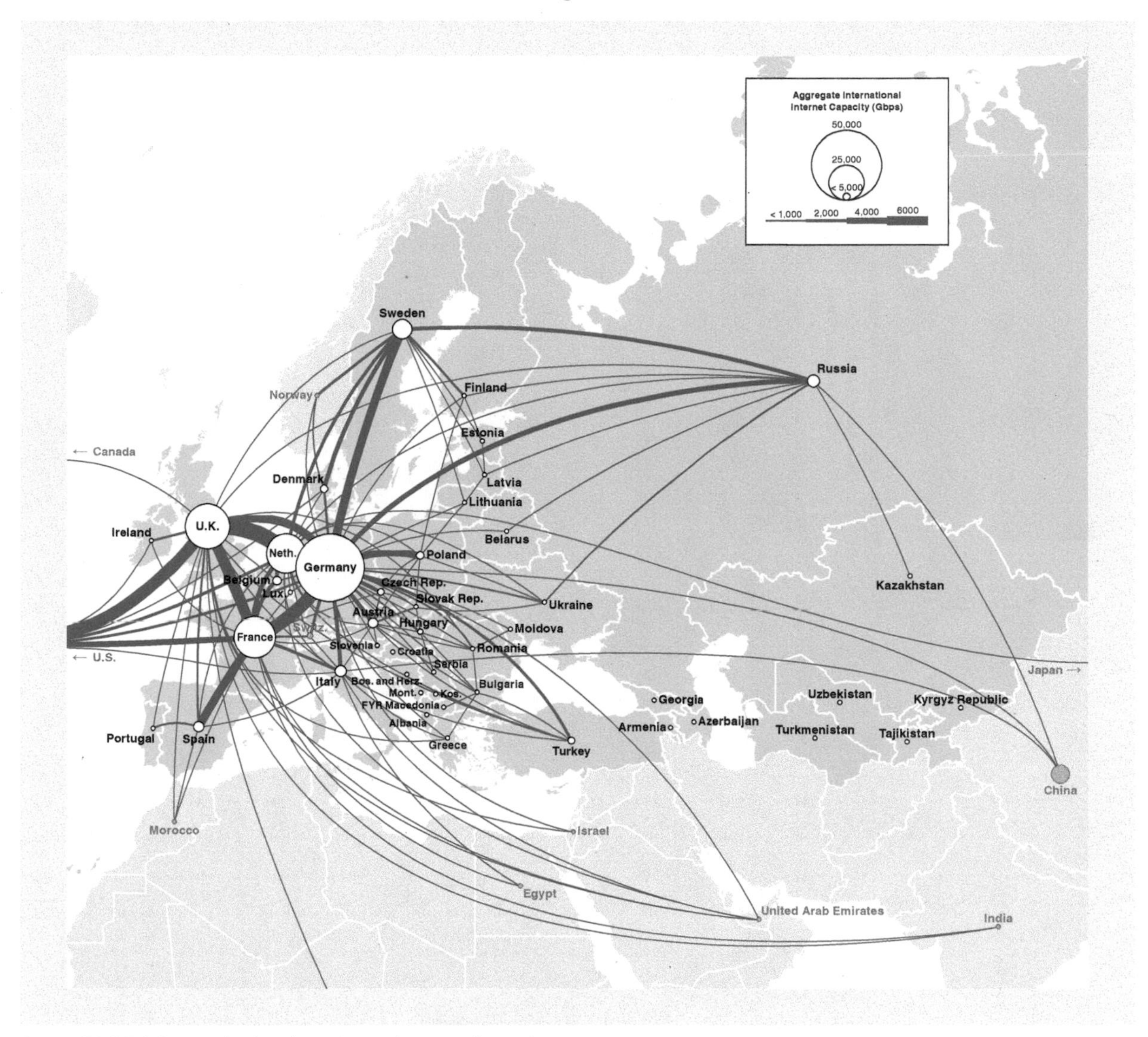

Source: 2015 TeleGeography data (http://www.telegeography.com).

the Islamic Republic of Iran, and the Persian Gulf to Barka, in Oman.[3] It offers an alternative to the Red Sea route, via the Suez Canal, for linking Europe to Asia.

In general, countries with the lowest international bandwidth and lowest average connection speeds in the region are landlocked, although this relationship does not always hold, as the cases of Luxembourg and, more modestly, Moldova, illustrate. Landlocked countries that are relatively poorly interconnected across borders are doubly disadvantaged because they can be held hostage by Internet transit providers in neighboring countries. There are, for instance, only three fiber connections into Turkmenistan, while Armenia's borders with several of its neighbors are effectively closed for transit because of strained political relations.

MAP 2.2 Existing submarine networks, ECA region

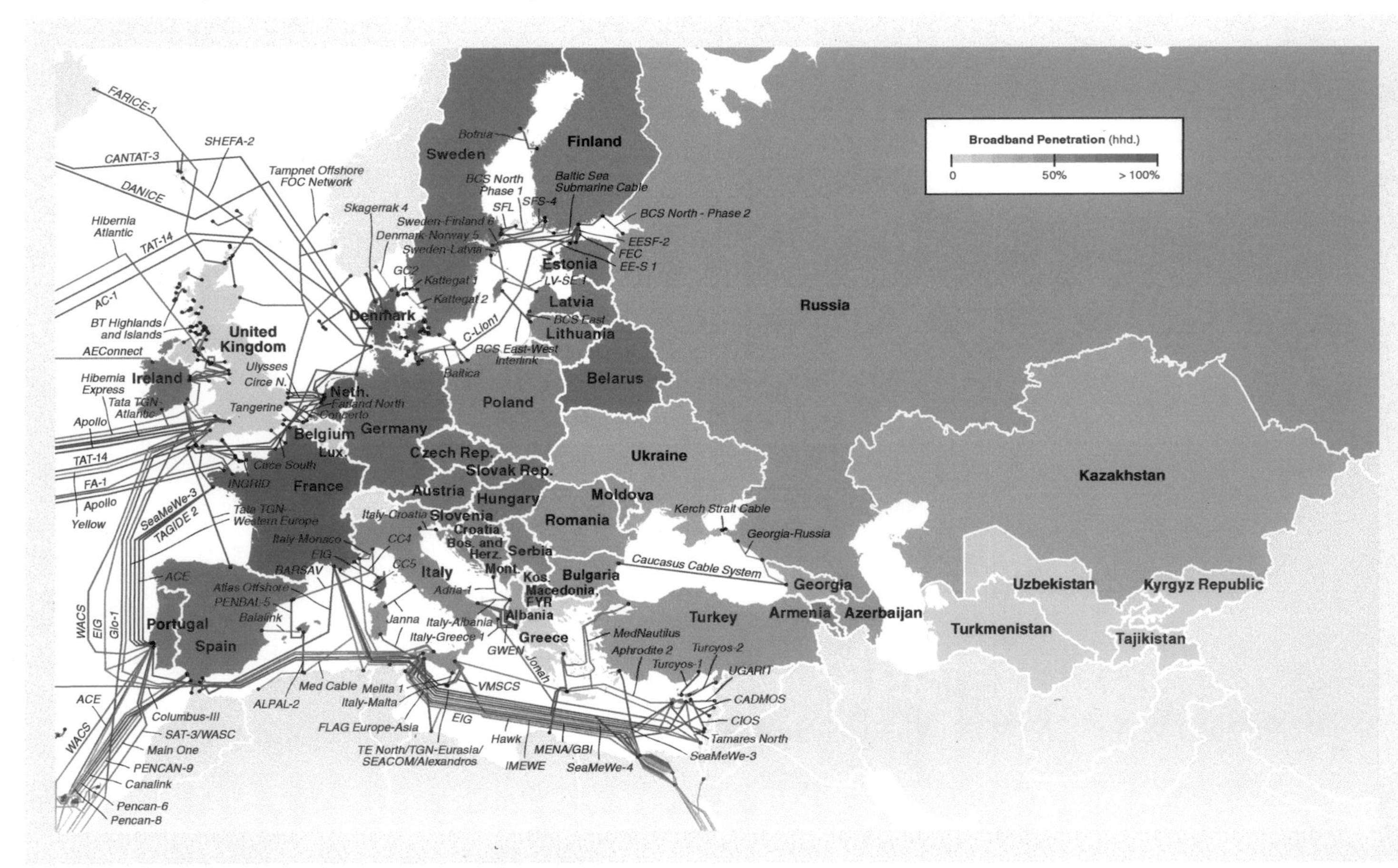

Source: 2015 TeleGeography data (http://www.telegeography.com).

A key factor in understanding the evolution of fixed-line connectivity in the region is the development of Internet exchange points.[4] An IXP is a network facility that enables its interconnected users—Internet service providers (ISPs)—to exchange Internet traffic through mutual "peering" agreements, contributing to competition between ISPs. IXPs promote development of the Internet at the national level, as exchanges between users do not have to pass through distant infrastructure any-more. In general, IXPs are well developed in the west of the region, but poorly developed in the east, both in their number and in the volume of traffic they exchange.[5] Europe has 103 IXPs in 31 countries: 14 in Germany and 13 in France, but only 2 in Kazakhstan, 1 in Armenia, and none in the other Central Asia and South Caucasus countries. In countries lacking an IXP, the incumbent operator is in a strong position. For countries seeking to improve the Internet market, setting up an IXP is an easy fix, which can be relatively expensive but bring quick returns. Setting up local caches (temporary storage spaces or memory that allows fast access to data) and encouraging the in-country hosting of content developed overseas can help to improve Internet services in countries with limited international connections.

National Fixed Connectivity

Broadband Internet can be delivered over three kinds of fixed lines (wireless is discussed in the next section): repurposed copper lines of the public telecommu-nications operators (which now support digital subscriber line [DSL] technologies), coaxial cable laid mainly by cable television operators, or the new generation of fiber infrastructure. Fiber provides high-bandwidth rates over long distances (unlike DSL and coaxial, where speeds tend to decay with distance), offers capacity that can be expanded (using electronic light pulse technology), is more secure, and requires less energy. With more and more devices connected to the Internet, fiber should be able to keep up with growing demand and constantly evolving applications, at least for consumers, for the foreseeable future. Several countries are already investing in fiber networks capable of providing speeds above 10 gigabits per second (Gbit/s), including Denmark and Turkey. Other countries, such as the United Kingdom, are revisiting their definition of the minimum speed that should be provided to households.

ECA is lagging behind in the adoption of fiber technology and, consequently, in the speed and quality of broadband services. Even in some high-income coun-tries, for example, Belgium, Germany, and Greece, less than 2 percent of broad-band connections are fiber subscriptions, which compares unfavorably with world leaders such as Japan and the Republic of Korea (above 70 percent; see figure 2.2). In terms of countries with the highest levels of fiber penetration in ECA are Lithuania (36.8 percent of broadband subscriptions), Latvia (36.2 percent), and Sweden (35.2 percent), while countries with the highest number of fiber connec-tions are Russia (15 million subscribers), followed by Spain (2.6 million), France (2.4 million), and Romania (2.3 million) (Fibre to the Home Council Europe, 2016). Several high-income countries (Croatia, France, Greece, and Italy) have lower levels of fiber penetration than Eastern European countries (Bulgaria, Romania, and Slovenia), which is reflected in lower Internet speeds. Domestic fiber optic networks in Central Asia remain limited.

FIGURE 2.2
Fiber in the connectivity diet

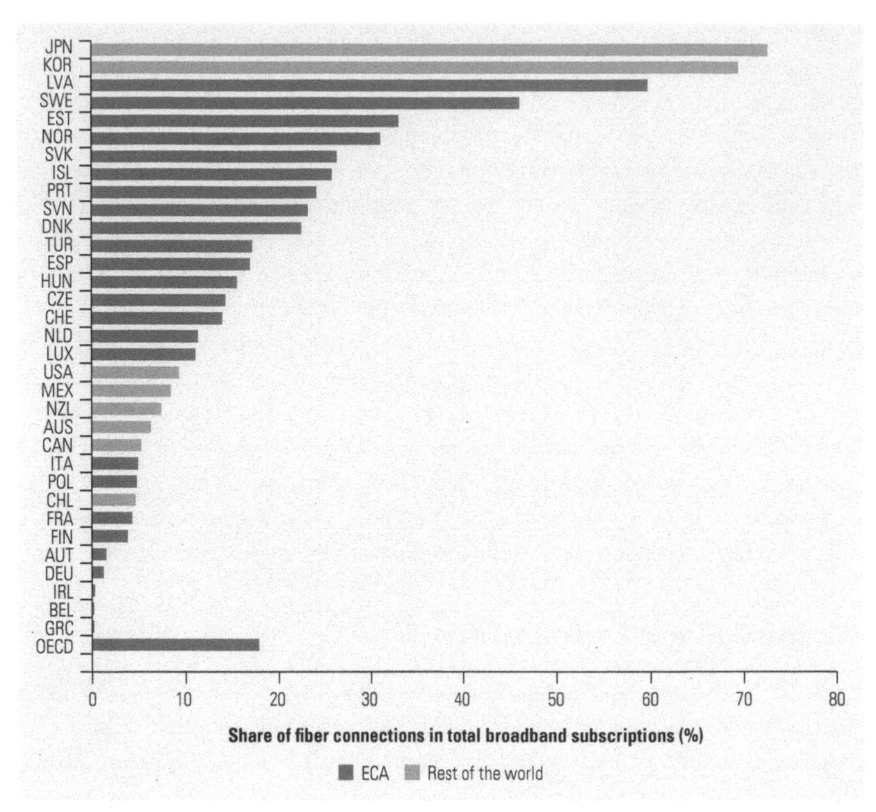

Source: OECD 2015.

Existing legacy investments in copper networks that have been done in these countries can be capitalized through G.fast development. This DSL protocol standard for local loops can reach performance speeds between 150 Mbit/s and 1 Gbit/s and can bring copper access into the gigabit era.

The degree of competition in the provision of broadband Internet over fixed lines, and thus often the level of service quality and pricing, is greatly affected by the availability of different types of infrastructure technology for delivering services. In some ECA countries, the share of DSL infrastructure is high, and the incumbent operator has a dominant market share. While firms providing broadband services can compete by offering different end products, it is difficult to compete by using more efficient infrastructure. By contrast, in countries where broadband Internet is provided through a mix of DSL, fiber, and cable infrastructure, competitors can access copper line telephone networks via regulated wholesale broadband or local loop unbundling, meaning that more than one telecommunications firm can provide services through the connection from the telephone exchange to the customer's premises, which allows for a much higher degree of flexibility in the design of end products for customers (Gelvanovska, Rogy, and Rossotto 2014). At the same time, it requires greater levels of investment and maintenance since it involves operating network components (European Commission 2016c).

In ECA, the share of DSL infrastructure exceeds two-thirds in Croatia, Cyprus, France, Germany, Greece, Ireland, Italy, Luxembourg, and the United Kingdom (figure 2.3). In countries that recently joined the European Union (EU), competitors

FIGURE 2.3
Fixed broadband subscriptions, by technology market shares, July 2015

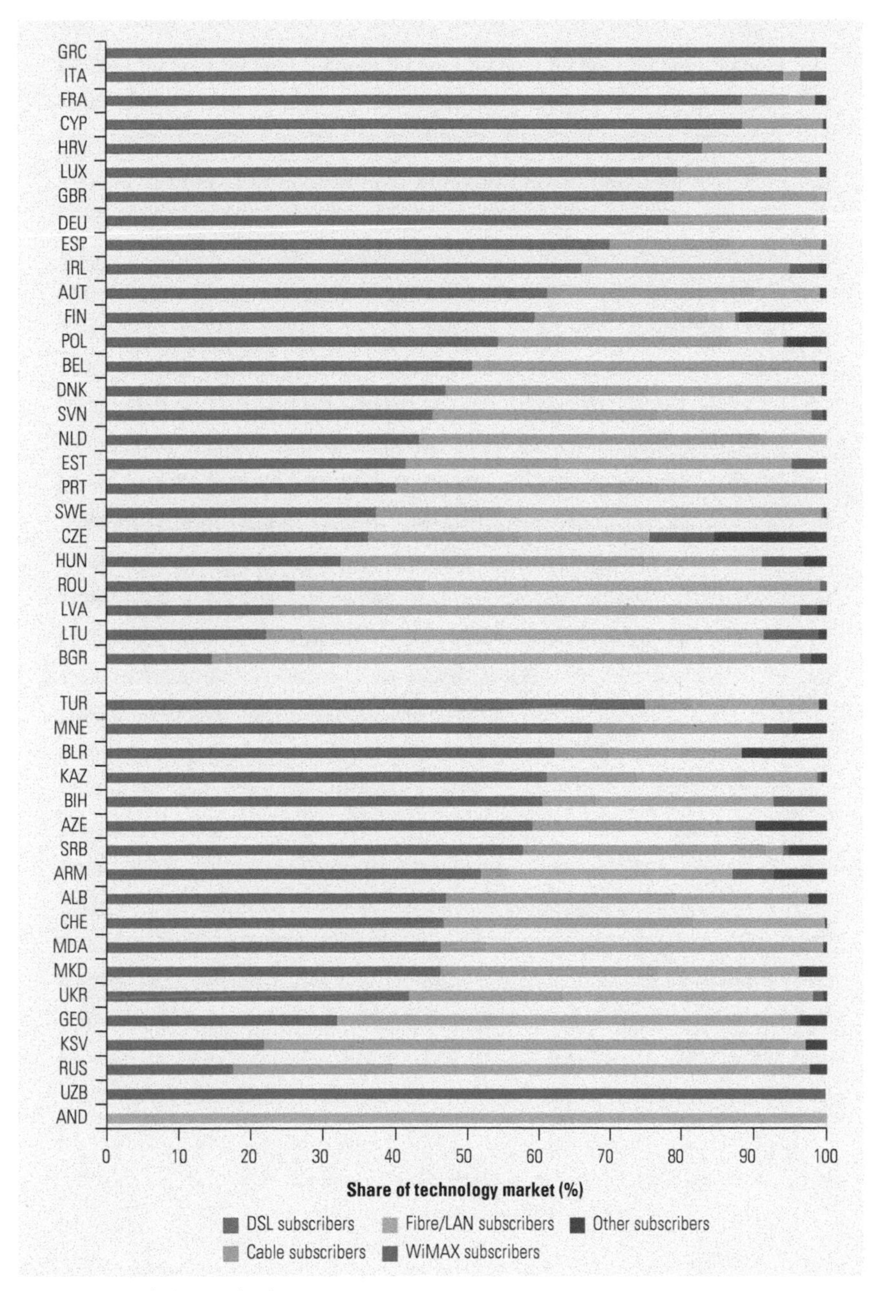

Source: 2015 TeleGeography data.
Note: DSL = digital subscriber line; LAN = local area network; WiMAX = Worldwide Interoperability for Microwave Access.

tend to have a higher degree of flexibility because the main period of network build-out was generally more recent, and infrastructure-based competition is more likely. For example, in Bulgaria and Romania new entrants are able to use their own infrastructure rather than that of the incumbent (figure 2.4). This ability often translates into higher speeds, suggesting that some of the EU's newest members are "leapfrogging" the older members (box 2.1). This development augurs well for the ability of the eastern countries to catch up with the western ones in the region.

FIGURE 2.4
Self-reliance: Proportion of new entrant networks that is self-built, by number of subscriptions

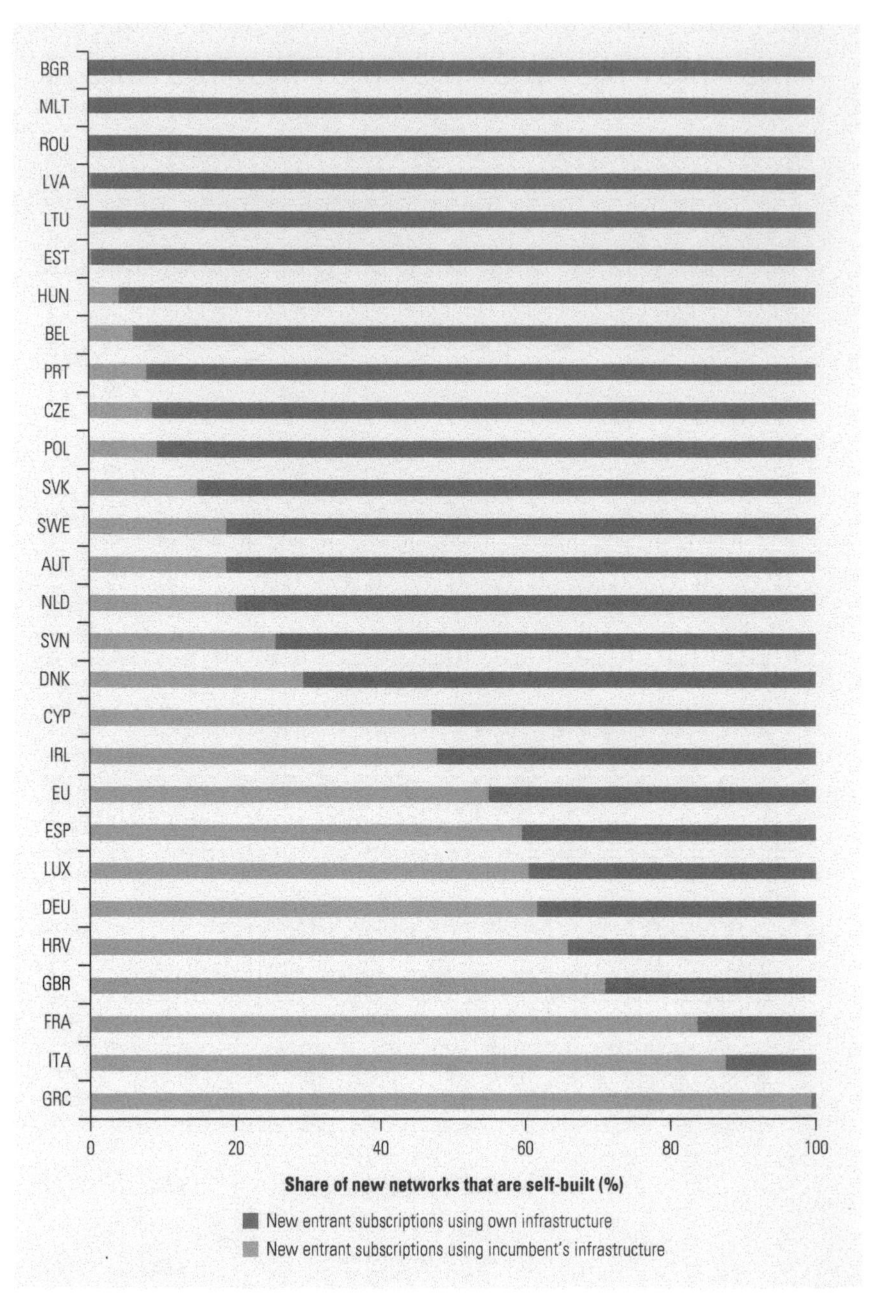

Source: European Commission 2016b (2015 data).

Outside the EU, Azerbaijan, Belarus, Kazakhstan, Montenegro, and Turkey have relatively high levels of DSL in the infrastructure mix (figure 2.3). Incumbent operators in these countries tend to have high market shares, notably in Kazakhstan (74.6 percent) and Turkey (67.1 percent), likely as a result of inadequate policy and regulatory frameworks. Turkmenistan, for instance, delivers broadband Internet only over copper (DSL) infrastructure, and there is only one operator.

BOX 2.1 Digital Leapfrogging: Countries That Recently Joined the EU Often Have Faster Broadband Speeds

The Digital Single Market Strategy for Europe sets the following target for 2025: "All European households, rural or urban, should have access to networks offering a download speed of at least 100 Mbps, which can be upgraded to 1 gigabit." The European broadband landscape is diversified in terms of speed, and countries with the highest speeds are not always those one would expect. Indeed, some EU newcomers have leapfrogged Western European countries with regard to broadband speed.

In a sample of 19 countries, 6 countries (Czech Republic, Hungary, Latvia, Poland, Romania, and the Slovak Republic) that have joined the EU since 2004 each have higher average broadband speed than long-standing members, such as France, Italy, and Spain, despite having a lower gross domestic product (GDP) per capita (figure B2.1.1). Indeed, the Czech Republic, Latvia, and Romania each perform better than the EU average (11.6 Mbit/s), with broadband speeds comparable with those of Scandinavia. Moreover, their next-generation access coverage is usually higher than the EU average: in Latvia, for instance, 91 percent of households have access to next-generation networks (NGNs) as opposed to 71 percent in the EU as a whole.[a]

These EU newcomers built their national broadband networks more recently than most Western European countries, enabling them to jump straight

FIGURE B2.1.1 GDP per capita versus average broadband connection speed

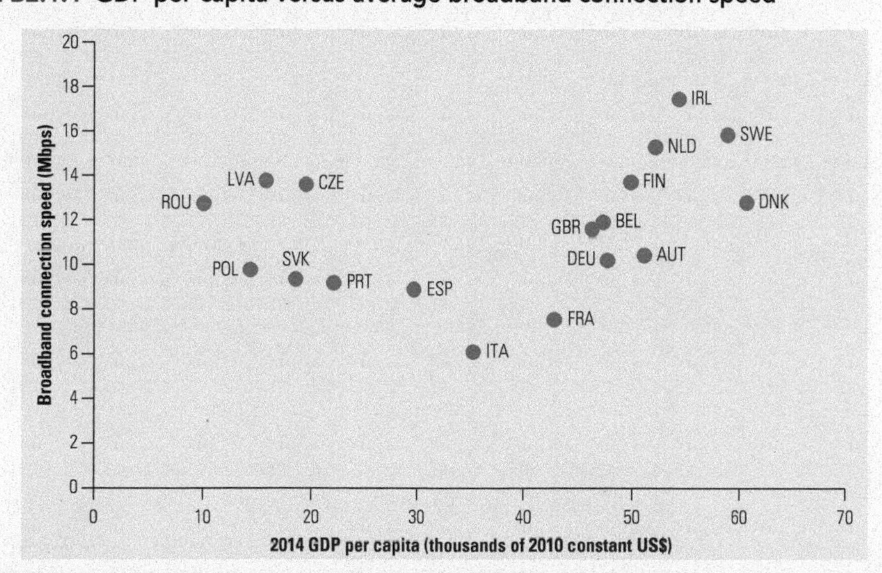

Sources: GDP: World Bank data; broadband speed: Akamai 2015. GDP = gross domestic product.

(Continued)

BOX 2.1 Digital Leapfrogging: Countries That Recently Joined the EU Often Have Faster Broadband Speeds *(continued)*

FIGURE B2.1.2 Year of entry to the European Union versus percentage of fiber connections in total broadband subscriptions, June 2015

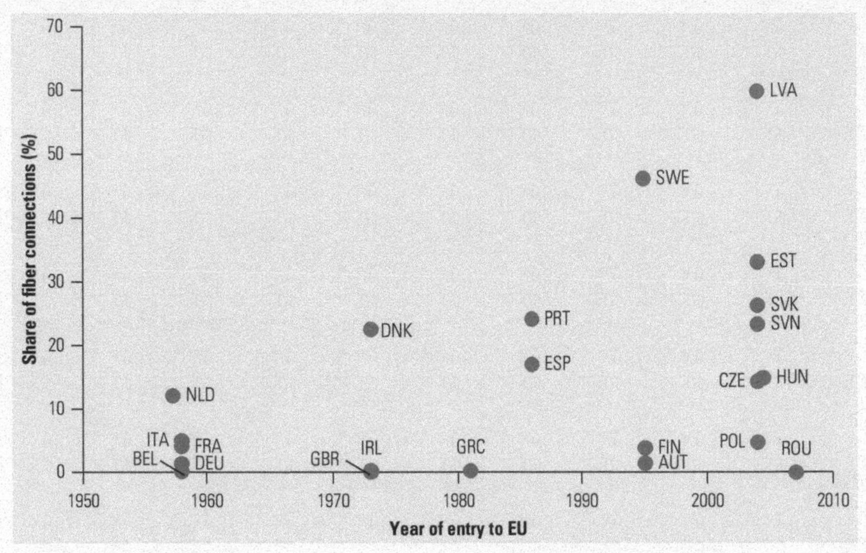

Source: 2015 OECD data. EU = European Union.

to a fiber network and to benefit directly from the advantages of speed and economies of scale. Furthermore, in several countries, such as Latvia and Romania, new market entrants have built their own fiber-based networks rather than relying on the incumbent's copper local loops (figure B2.1.2).

Digital leapfrogging seems to be a viable strategy. Can it be repeated in the far eastern part of the continent?

a. NGNs can support services with speeds exceeding 30 Mbit/s (in the definition of the European Commission 2014) or 50 Mbit/s (in the definition of OECD 2014a).

Competition is feasible even in countries where DSL accounts for a high share of broadband subscriptions (figure 2.4). One approach is to foster service-based competition, where new entrants use the incumbent's access network to provide Internet services. In Greece, competition is fully based on regulated access to the incumbent's network. This approach leads to lower service prices. In France, Free has emerged as a formidable competitor to the incumbent, Orange, mainly through favorable arrangements for leasing the incumbent's local loop capacity. The opposite approach, also taken in some EU countries, is for new entrants to develop their own networks and thus compete both on infrastructure and on services. This is the case in Bulgaria, Malta, and Romania, where challengers have made major inroads into the incumbent's market share by relying on their own-built infrastructure. This approach usually has a positive effect on innovation.

Prices for Internet Services

According to the International Telecommunication Union (ITU), the average global price of a fixed broadband plan fell two-thirds between 2008 and 2013. In Europe, broadband prices had declined earlier, but still fell more than 20 percent between 2009 and 2014. The highest prices for fixed Internet are found in Azerbaijan, Tajikistan, and Turkey (more than US$110 per Mbit per month, in purchasing power parity [PPP]) and in Albania, Kazakhstan, Montenegro, Serbia, and Slovenia (more than US$25, PPP). While lower prices could have impaired service quality due to higher levels of network congestion, in reality speeds continue to increase. Nevertheless, despite recent declines, Internet prices remain a barrier to adoption in many countries. The large disparities in Internet adoption, even within the same income group (in 2014, lower-middle-income Uzbekistan and Armenia had broadband penetration of 7 and 62 percent, respectively, and upper-middle-income Turkmenistan and Belarus had broadband penetration of 0.1 and 69 percent, respectively), may be driven partially by high prices for broadband services, notably at the retail level, where customers are most affected, but also at the wholesale level, where competition is lacking.

As shown in figure 2.5, ECA countries mostly fall into two distinct quadrants: those with low prices and high penetration (for example, Belarus) and those with high prices and low penetration (for example, Albania and Turkmenistan). Some ECA countries, in particular, Ukraine and Uzbekistan, deviate from this pattern, implying that other factors are also in play.

Significant differences also remain for high-speed Internet. While the overall median price of stand-alone Internet service of between 30 and 100 Mbit/s decreased from €43 (US$59) in 2009 to €34 (US$41) in 2014, it varies between €22 and €23 (about US$25) in Latvia, Lithuania, and Romania and €71 and €102 (US$86 to US$124) in Cyprus, Malta, and Slovenia. In Cyprus, these speeds represent fewer than 5 percent of subscriptions—similar to Croatia, Greece, and Italy. In most of the European countries, the prices for service packages offered at

FIGURE 2.5
Price as a driver of Internet adoption (2014 data)

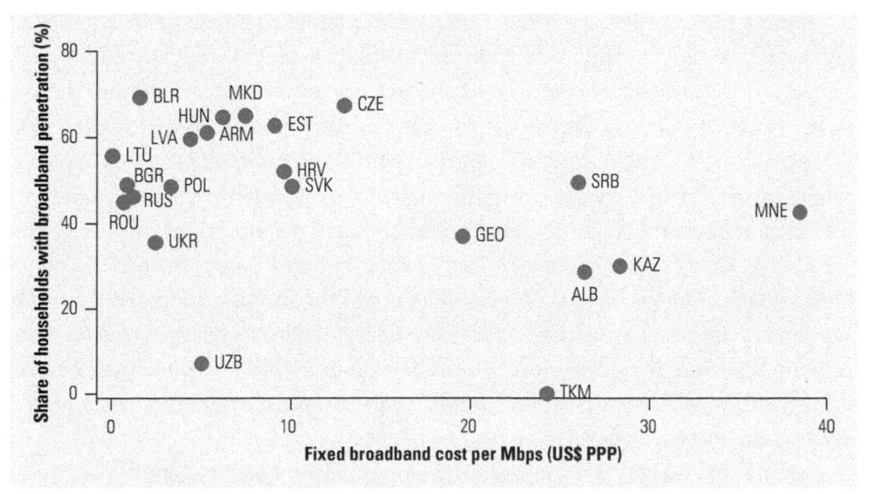

Sources: Data on prices collected in 2014 (from operators' websites) and 2014 TeleGeography data.
Note: Azerbaijan (US$262), Tajikistan (US$269) and Turkey (US$115) do not appear in the figure because their respective broadband prices would be to the extreme right of the figure.
PPP = purchasing power parity.

12–30 and 30–100 Mbit/s do not differ much, the exceptions being Austria, Belgium, Cyprus, Malta, the Netherlands, and Slovenia, where the difference is more than twofold. In Central Asia, high-speed packages are rare and expensive.

Prices alone, however, tell only half the story—the distribution of income and affordability thresholds provide a different perspective (A4AI 2015). ECA (excluding Central Asia) has the lowest prices for entry-level, fixed broadband prices (that is, 256 kbit/s or the lowest speed available) as a percentage of GNI per capita compared with other regions. In 2011, the United Nations Broadband Commission set a target price for fixed broadband services of less than 5 percent of a country's GNI per capita (Broadband Commission for Digital Development 2015), an indicator of affordable broadband. Turkey (7.4 percent), Azerbaijan (19.2 percent), and Tajikistan (92.7 percent) do not meet this target, but EU member states and most others in the region do. In Tajikistan and Turkmenistan, the cost of a basic subscription package accounts, respectively, for 16 percent (US$27.64 for 512 kbit/s) and 24 percent (US$60.74 for 256 kbit/s) of the average monthly ratio of per capita income to consumption expenditure (2011 US$, PPP). The European Union has the most affordable mobile broadband, with all countries meeting the threshold of below 2 percent of GNI per capita; in addition, in 80 percent of these countries, mobile broadband is also cheaper than basic fixed broadband Internet. When considering the distribution of income, more than 80 percent of the population in Armenia, Georgia, and the Kyrgyz Republic would need to spend more than 10 percent of their household expenditure on a basic mobile plan. These countries, however, have different levels of GNI per capita, and their entry speed offers differ as well.

Regulatory initiatives could reduce the price of broadband services, as well as address the pricing of access devices and taxation on broadband. According to a recent ITU report, "Different regulation may account for almost 10% of the differences in prices observed across countries" and "different competition levels may explain around 6 per cent" (ITU 2014). Increasing market competition for broadband could help to reduce retail prices for subscribers, while maintaining the level of service quality.

In some ECA countries, the incumbent ISP continues to enjoy significant market power, which can contribute to high prices for consumers. The entry of a challenger with low-priced Internet plans can force competitors to reduce their prices, as with the 2012 entry of Free in the mobile market in France. Free proposed low-price plans to its subscribers and offered an interesting bundle of Internet services, television, and phone calls in competition with the incumbent operator, Orange.

Policy makers might also want to enforce open access on international and backbone networks and make local loop unbundling mandatory. Some ECA countries still lack an LLU framework or reference prices for wholesale offers. Policy makers and regulators also could set up clear interconnection rules in order to stimulate competition. Requiring operators to interconnect at an IXP or to share their passive infrastructure, whenever possible, would help to reduce operators' costs, in the aggregate, and to lower prices.

Experience suggests that, in countries where different services are bundled into packages (as in the United States), making it difficult to disentangle the prices of different services, per unit prices at the bottom of the market tend to be much higher. Without needing to regulate prices directly, regulators can do more to encourage price transparency, promote honesty in advertising, and ensure that consumers have the option to buy unbundled services (for example, Internet alone) at reasonable prices. Some countries, like Portugal, are setting deadlines for terminating service through copper lines, in order to encourage operators to invest in fiber networks and enable the move to higher broadband speeds. Policy makers can also use the development of a national broadband plan to carry out benchmarking and to set targets for future broadband prices. Some ECA countries, such as Finland and Ireland, have started to make broadband a legal right for their citizens.

National Mobile Cellular Connectivity

Mobile broadband offers an increasingly attractive complement to and, in some cases, a substitute for fixed broadband networks.[6] Mobile broadband technology is typically described in terms of "generations," although in practice each generation is also progressively upgraded:

- Second-generation (2G) networks refer to narrowband digital mobile networks that mainly follow the Global System for Mobile (GSM) standards and were implemented from 1991 onward. They are optimized for voice and text. Globally, 2G coverage is about 96 percent.
- Third-generation (3G) networks introduced higher speeds that are suitable for Internet use and were launched in 2002. In Europe, they mainly use Wideband-Code Division Multiple Access (W-CDMA) and are often upgraded to higher capacity using High Speed Downlink Packet Access (HSPDA) and related technologies. Globally, 3G coverage is about 78 percent.
- Fourth-generation (4G) networks are optimized for high-speed data and follow the Long-Term Evolution (LTE) set of standards, first introduced in 2009. 4G is capable of providing typical speeds of between 12 and 30 Mbit/s for download and between 2 and 5 Mbit/s for upload (2016 GSMA data). More recent variants of LTE such as LTE-advanced using carrier aggregation of multiple-frequency bands can deliver much higher speeds, with peak throughputs well above 100 Mbit/s. Globally, 4G coverage is about 35 percent.
- Fifth-generation (5G) networks provide speeds as high as 10 Gbit/s and with very low latency, thus creating an opportunity for growth in applications and devices. 5G mobile technology has two key components: (1) "improved" orthogonal frequency-division multiplexing (OFDM) technology to provide an air interface that is better able to combine the diverging requirements of mobile broadband services, Internet of Things (IoT)/machine-to-machine (M2M) services and applications, and vehicle-to-infrastructure (VtI) services and applications, such as traffic light information systems; and (2) spectrum in the centimeter and millimeter bands for very high-capacity but short-range mobile communication. The ITU World Radio Conference 2019 will seek to harmonize future bands for mobile communication above 6 GHz. This development toward 5G will be driving the need for more spectrum for mobile services, not only in the usual bands between 400 and 6000 MHz but also in much higher bands.

Mobile coverage varies widely in the ECA region. Belarus has a rare chance to lead Europe in 3G, but is lagging in 4G. At the other end of the scale, Kosovo is moving straight to 4G. In between, Belgium and Denmark have close to universal coverage in 3G and 4G. In general, rural areas lag behind urban ones in coverage. In France, for instance, rural 4G coverage is stuck at around 5 percent, while urban coverage is approaching 80 percent. Only Denmark achieves parity in urban and rural coverage. Actual adoption of 4G services lags well behind coverage. The GSMA estimates, for instance, that within Europe only a third of users who could theoretically access 4G services actually do so. This gap will narrow as individual users upgrade their phones and as prices for 4G subscriptions fall.

In the EU, average download speeds vary, but overall exceed 12 Mbit/s. This is higher than in the United States (10 Mbit/s), but lower than in Korea, which averages 29 Mbit/s, exceeded in Europe only by Romania (30 Mbit/s) (2016 GSMA data). Making available suitable spectrum is a key driver of 4G rollout, together with the timing of spectrum awards, typically by auction (box 2.2). In particular, the so-called "digital dividend" spectrum below 1 gigahertz (GHz), previously used by analog television and radio broadcasters, is the most valuable, as this generally enables networks to be built at the lowest cost and offers the best coverage.

Some of the ECA economies have started to roll out 4G only recently:

- Although 4G rollout in Kosovo only began in 2013, the mobile operator IPKO claims to cover 99.7 percent of the population, compared with just 35 percent coverage for 3G.
- Belarus reports 15 percent 4G coverage, although it claims almost universal 3G coverage. The first LTE network went live at the end of 2015 and uses shared services in the 1800 MHz band. A slow start to LTE deployment and the use of a frequency above 1 GHz both affect population coverage.
- Serbia reports 4G coverage at 21 percent. Initially, LTE was offered in the 1800 MHz band, with licenses in the 800 MHz band awarded in November 2015.

BOX 2.2 EU Spectrum Regulation

In the European Union, spectrum bands have been defined for harmonized use across the member states (European Commission 2015). For instance,

- The 800 MHz band, designated for LTE
- The 900 MHz band, originally used for GSM, but now technology-neutral and used for 2G, 3G, and 4G services
- The 1800 MHz band, again originally used for GSM, but now technology-neutral and mainly used for 2G and 4G services
- The 1900 MHz and 2100 MHz bands, originally used for 3G, but now becoming technology-neutral and enabling 4G services
- The 2.6 GHz band, originally intended for 3G expansion, but typically now used for 4G and, in some countries, for Worldwide Interoperability for Microwave Access (WiMAX).

The "second digital dividend," the 700 MHz band, is also important for providing additional spectrum for mobile broadband (LTE). In most countries, the 700 MHz band is typically used for digital television services. A refarming of the digital television service is required to release the 700 MHz band for mobile broadband services.

In addition to a slow start, some ECA economies face additional challenges:

- Turkmenistan reports 57 percent 3G coverage. The country is large and has a low population density. The 3G market was effectively supplied by a single provider network from 2010 to 2014, as the second mobile operator was forced to shut down its 3G network because of licensing issues. As for 4G (30 percent coverage), the current LTE network uses the higher frequency bands (2600 MHz), and, as of the end of 2015, only TMCell was providing LTE service.
- In Uzbekistan, mobile penetration was about 65 percent in mid-2015, compared with more than 80 percent in 2011, the consequence of major turmoil in the wireless market. Russia's Mobile TeleSystems (MTS) returned to the market in December 2014, after it had been controversially ejected in mid-2012. Also in 2012, Vimpelcom and TeliaSonera, who own market leaders Beeline and Ucell, respectively, were accused of paying bribes to acquire their licenses, a cloud that continues to hang over them.

Mobile Data Offloading and Wi-Fi

With mobile data traffic growing at an estimated 71 percent in 2015 in Central and Eastern Europe and 52 percent in Western Europe (Cisco 2016), mobile networks are coming under increasing pressure. Globally, mobile offload exceeded cellular traffic for the first time in 2015. In 2015, 51 percent of total mobile data traffic was offloaded onto the fixed network through Wi-Fi or femtocell. At least 80 percent of Internet data traffic passes over unlicensed spectrum via Wi-Fi (Kinney 2016; Mobile Experts 2016), principally for Over-the-Top (OTT) services such as YouTube, WhatsApp, Skype, and social networking. Globally, the use of unlicensed bands may constitute 90 percent of data traffic by 2020. Thus, a key issue for the mobile industry is the relationship between mobile data networks (using licensed spectrum) and Wi-Fi hubs and hotspots using license-exempt bands. The latter are especially important in dense urban environments.

Mobile operators and their suppliers, such as Deutsche Telekom, are now testing LTE for unlicensed bands as an alternative technology (as LTE-U with several variations). The economic impact of having more unlicensed spectrum could be significant. A study for the European Commission in 2012 (Forge, Horvitz, and Blackman 2012) highlights the advantages of greater use of license-exempt bands in the region below 1 GHz.

Connectivity Challenges in Central Asia

The countries of Central Asia are experiencing significant challenges in the supply of Internet services, which explains why they have some of the lowest rates of Internet take-up in the ECA region (as discussed in chapter 1). The region is held back by constraints regarding both capacity and prices. The Kyrgyz Republic, Tajikistan, Turkmenistan, and Uzbekistan all charge high Internet Protocol (IP) transit prices of more than US$100 per Mbit/s per month (Terabit Consulting 2013). As a result, end users pay some of the highest prices in the world (UNESCAP 2014) for low-quality services. Moreover, capacity issues lead to limited bandwidth per capita, below 1 kbit/s per inhabitant in four of the five countries, and thus to poorly performing Internet service markets. A vicious circle results, whereby slow speeds mean that demand (mainly for data services) is low and fails to generate incentives for infrastructure investment.

Geographic issues partially explain why these Central Asian countries experience connectivity challenges. Being landlocked, they cannot access international bandwidth directly through a submarine cable, but only through cross-border terrestrial connections or satellite. Besides mountain and desert terrain, the low population densities and large rural areas of this region raise per capita investment requirements. Thus, Central Asian countries are highly dependent for connectivity on two of their large neighbors—China and Russia. Central Asian countries also charge high transit fees to one another (Terabit Consulting 2013). Moreover, even when connectivity exists, there is a shortfall in the ability to pay. The cost of a gigabit of mobile data, for instance, ranges between US$9.31 (Uzbekistan) and US$37.26 (Turkmenistan) measured in PPP exchange rates (table 2.1).

Political and governance hurdles also explain the lack of regional connectivity. Most incumbents are still state owned, and most regulatory bodies are state agencies and thus not independent. The apparent reluctance of the Turkmen government to foster the development of the broadband market as well as the Tajik government's concerns over security issues on the southern route through Afghanistan are examples of political caution regarding development of the Internet. Geopolitical issues, such as the difficult relationship between Tajikistan and Uzbekistan, also hamper schemes to improve regional connectivity, as is the case elsewhere in ECA. This has resulted in dependence on a limited number of international linkages and higher transit prices. A shortage of investment financing and a lack of technical skills compound the issue of Internet access.

Nevertheless, several competing regional programs have been proposed in recent years to address these issues:

- The "One Belt One Road" strategic initiative, begun by China in 2013, is a modern adaptation of the Silk Road. Its objective is to improve trade relations by creating infrastructure networks through China to the West, including railroads, highways, pipelines, broadband, and sea routes. The "Information Silk Road" aims to strengthen information infrastructure by ensuring interconnection between trade partners in 50 countries.
- The Trans-Eurasia Information Network (TEIN) project connects research and education communities across Europe and Asia. Cofunded by the European Union, its first generation was launched in 2000, with success

TABLE 2.1 Central Asia Benchmarks for Internet Access

Country	Fixed broadband: price per Mbit/s (US$ PPP)	Fixed broadband: annual subscription as % of GNI per capita	Mobile broadband data: price per Gbit/s (US$ PPP)	International bandwidth (Gbit/s)	International bandwidth per capita (kbit/s)
Kazakhstan	28.17	1.60	10.33	436.2	26.8
Kyrgyz Republic	12.82	1.60	33.29	7.9	1.4
Tajikistan	189.02	92.70	31.91	4.8	0.6
Turkmenistan	24.26	1.20	37.26	1.2	0.2
Uzbekistan	5.00	1.10	9.31	13.1	0.4

Sources: 2014 data from International Telecommunication Union, TeleGeography, and World Bank 2016.
Note: Gbit/s = gigabit per second; GNI = gross national income; kbit/s = kilobit per second; Mbit/s = megabit per second; PPP = purchasing power parity.

involving e-learning, telemedicine, and weather forecasting. A significant upgrade is planned.

- The Trans-Eurasian Information Superhighway (TASIM) was proposed by Azerbaijan in 2008 in order to connect 20 countries, passing through China, Kazakhstan, Azerbaijan, Georgia, and Turkey en route to Germany. This project, however, remains not implemented till now.
- Digital CASA (Central Asia/South Asia Fiber Optic Network) is a World Bank regional broadband project aimed at improving connectivity between Central Asia, South Asia, and China by encouraging private investment and modernization of state regulations. Leveraging existing infrastructure, such as the World Bank CASA-1000 energy project or TASIM, is a possible scenario. Digital CASA would improve international connectivity in the region by linking Afghanistan, the Kyrgyz Republic, and Tajikistan to Pakistan. It offers an opportunity to create a digital hub between Europe and East Asia as well as a reliable alternative to submarine cables.
- A second phase of the Far East Submarine Cable System is under construction by Russian state-backed Rostelecom and Huawei Marine connecting the far eastern peninsula of Kamchatka with the island of Sakhalin. The 900-kilometer section will form the second phase of Rostelecom's new cable route connecting Kamchatka-Sakhalin-Magadan. The first phase (Sakhalin-Magadan) was completed in 2015, alongside terrestrial network extensions in Kamchatka.

Promoting infrastructure deployment through regional coordination, as well as stimulating market competition and demand via effective regulation, will help the Central Asia region to benefit from the opportunities offered by the Internet. Developing cost-effective international networks within the region will benefit end users (by reducing consumer prices and improving broadband reliability) and will contribute to economic growth (by stimulating demand and supply, including in industries other than information and communication technology [ICT], and creating opportunity for new products and innovations). It will also increase government revenues (increasing tax revenues from ICT investment and license awards) and promote regional stability (encouraging transborder social initiatives like health care or education) (Terabit Consulting 2013).

Policy and Regulatory Tools for the Supply of Internet Networks and Services

Achieving universal coverage will require a combination of financial and nonfinancial measures. The literature distinguishes between gaps in sustainable coverage (areas where service providers will enter if the policy and regulatory barriers are removed and where the service would be affordable if provided by an efficient market) and gaps in universal coverage (areas where financial intervention from the government will be needed to complement this effort) (Navas-Sabater, Dymond, and Juntunen 2002). From the regulatory perspective, measures will be determined largely by a country's stage of development as well as the state of its current regulatory framework.

This section looks at regulatory frameworks across the region in more detail and compares the findings with analysis in chapter 1. It is based on the threefold categorization of countries proposed in WDR16—emerging, transitioning, and transforming—and is presented in table 1A.1 in annex 1A. This follows the policy framework set out in the WDR16, which identifies four stages of the network value chain (World Bank 2016, 206). This value chain stretches from the point where the Internet enters a country (the first mile), passes through that country (middle mile), and reaches the end user (last mile). In addition, certain hidden elements that are vital to ensuring the integrity of the value chain, such as spectrum, cybersecurity, numbering plan, and so on, may be referred to as the "invisible mile." As stated in WDR16, "The supply side of the Internet is conditioned by rules on market competition, shaped by the respective roles of the public and private sectors, and mediated by the degree to which regulation of the sector is independent of government and the operators." In general terms, countries that are geographically disadvantaged (for example, landlocked, with low population density) need the most help in the first and middle mile, while countries that have achieved a good measure of Internet penetration may need help with the last mile and the invisible mile.

The European Union's regulatory framework for electronic communications applies to its 28 member states, which include some transitioning countries but mainly comprise transforming countries (European Commission 2016e). Applied since 2003 and updated in 2009, it includes, among other things, directives on access, authorizations, universal service, net neutrality, and privacy. As part of the Digital Single Market Strategy, adopted in May 2015, proposals for a major overhaul of the regulatory framework were issued in September 2016 (European Commission 2016a). The European Commission (2016d) put forward the following strategic connectivity objectives for 2025:

- "All main socio-economic drivers, such as schools, universities, research centres, transport hubs, all providers of public services such as hospitals and administrations, and enterprises relying on digital technologies, should have access to extremely high—gigabit—connectivity."
- "All European households, rural or urban, should have access to connectivity offering a download speed of at least 100 Mbps, which can be upgraded to Gbps."
- "All urban areas as well as major roads and railways should have uninterrupted 5G coverage. As an interim target, 5G should be commercially available in at least one major city in each EU Member State by 2020."

The EU's regulatory framework is a major driver of regulatory reform, not only in its member states, but also in countries seeking to enter the EU, such as Albania, the former Yugoslav Republic of Macedonia, Montenegro, Serbia, and Turkey. Moreover, in 2014 the EU signed association agreements with Georgia, Moldova, and Ukraine that create a framework for close economic and political cooperation and alignment of their regulatory frameworks with those of the EU.

Emerging Countries in ECA

According to the definition of the WDR16, no ECA countries are in the emerging category. Nevertheless, five countries—the Kyrgyz Republic, Tajikistan, Turkmenistan, Ukraine, and Uzbekistan—score significantly lower than the rest of the region on both level of technology and complementary issues (regulation, institutional accountability, skills), even compared with countries in the same income group. They are therefore considered as emerging countries in this book. All of them (except Ukraine) are in Central Asia; they are all lower-middle-income countries (except Turkmenistan) and landlocked (although some have access to an inland sea). But these are not the only factors that unite them. These countries have relatively weak international and domestic connectivity and similar features in their policy and regulatory frameworks. Also, the lack of independent regulators in some of these countries (despite their World Trade Organization commitment to create such regulators) might become a constraint for the sector.

In the first mile, all five countries have yet to liberalize their international gateway, which remains with the incumbent operators, some of which are state owned. This means that prices for international calls are likely to remain high. In the middle mile, too, countries in Central Asia are characterized by the continuing dominant role of incumbent operators in the market. In Tajikistan, Turkmenistan, and Uzbekistan, incumbents still enjoy a monopoly on fixed-line services provision. In Turkmenistan, the incumbent operator is the only provider of Internet services (as of January 2016), and the highest speed it offers is just 1 Mbit/s at a monthly price of US$303.72 (table 2.1). A new operator was formed in 2015 by the Ministry of Communications (which owns 30 percent of shares) and the incumbent (which owns 60 percent), but it will only provide services in the capital. In the Kyrgyz Republic, the government confirmed that the sale of Kyrgyz Telecom shares was ruled out in September 2013. In Uzbekistan, the government is planning to retain 45 percent of shares, with the rest being sold to foreign investors. The timeline, however, remains uncertain.

In the middle mile, in 2013 the Kyrgyz Republic moved to improve the use of infrastructure, in particular, infrastructure sharing, and to support a more competitive market, by introducing the concept of significant market power (SMP)[7] and proposing a means of enforcement (ITU 2015). At the same time, operators are not required to publish a reference interconnection offer,[8] and interconnection prices have not been made public, which reduces transparency. In addition, the Kyrgyz Republic went through major sector governance reform during the past year. The reform has weakened the independence of the sector regulator and reduced its competences in the areas of price regulation and the adoption of secondary legislation and enforcement.

In the last mile, as of January 2016, there were no known plans for local loop unbundling in any of the countries in this category. The absence of LLU reduces the scope for competitive market entry because it obliges potential market entrants to build their own networks from scratch rather than initially renting lines from the incumbent. Infrastructure-based competition is generally quicker to develop where market entrants can compete by offering services through existing connections first.

With regard to the last mile, even the mobile sector remained under monopoly service provision in some countries. In Turkmenistan, the 3G market was effectively a single-provider network between 2010 and 2014, when the second mobile operator was forced to shut down its new 3G network over licensing issues. MTS Turkmenistan returned to the market in 2012 and relaunched its 3G network in September 2014. 4G services were only provided by TMCell as of the end of 2015. In the Voice over Internet Protocol (VOIP) segment, with telecom licenses not offered on a technology-neutral basis, companies seeking to offer VOIP services must apply for a specific Internet telephony concession. No such licenses have been awarded so far.

Also, some emerging countries have not authorized satellite dishes for consumer use. In Turkmenistan, for instance, residents of multistory apartment buildings have been ordered to take down satellite dishes on the basis that they ruined the view of the city. Authorities told residents they could instead get cable television packages from the government or via state satellite antennas (Human Rights Watch 2015). In Tajikistan, satellite dishes are allowed, but the State Committee on Television and Radio is selective in giving licenses to new agencies (Freedom House 2015).

Transitioning Countries in ECA

The quality of the regulatory framework differs somewhat among the 24 countries of the "transitioning" category, which cuts across all income levels and includes countries from different subregions.[9] The ITU Regulatory Tracker rates all countries according to four "clusters" of characteristics: regulatory authority (cluster 1), regulatory mandate (cluster 2), regulatory regime (cluster 3), and competition framework (cluster 4). Ratings on each of these categories are then aggregated to generate an overall country rating, referred to as "generations" of regulation.[10] In 2013 (the latest date for which comprehensive data are available), most transitioning countries were categorized as G3 (total score between 70 and 84) or G4 (total score between 85 and 100). G3 includes "markets that enable investment, innovation, and access and usually have a dual focus on stimulating competition in service and content delivery as well as consumer protection." G4 includes "countries with integrated regulation, led by economic and social policy goals." Only Azerbaijan and Kazakhstan still had G2 (total score between 40 and 69) type of regulation (basic reform, partial liberalization, and privatization across the layers). Belarus was the only country in the "transitioning" category that still had a G1 (total score between 0 and 39) type of regulation (generally markets with regulated public monopolies and a command and control regulatory approach). In Russia, efficient sector regulation is not fully implemented.[11]

All of the transitioning countries (except Belarus) made progress on all four criteria between 2007 and 2013. Azerbaijan had low scores on clusters 1, 2, and 3, but scored well for its competition framework (cluster 4). Yet, the country is still lacking an interdependent sector regulator. Over the same period, Kazakhstan strongly improved its competition framework, and Italy made significant progress with regard to regulatory authority (cluster 1).

The WDR16 framework (World Bank 2016, 206) sets out the priorities for policy and regulation in transitioning countries. Regarding the liberalization process, there are big disparities. Most of the transitioning countries have privatized their telecommunication operator. Some of them plan to privatize it soon (for example, Azerbaijan and Cyprus). Some incumbent operators are still majority state owned (such as BH Telecom in Bosnia and Telekom Srbija in Serbia), sometimes after attempted privatizations have failed (such as Telekom Slovenije in Slovenia). Belarus, Kosovo, and Moldova still have a 100 percent state-owned incumbent, but maintain competition on both mobile and fixed markets, with two to five competitors.

Infrastructure sharing is authorized and even encouraged in most transitioning countries. Most of them have active companies with towers providing cellular service (for instance, Armenia, Cyprus, Kazakhstan, and Slovenia). But countries differ in their type of infrastructure sharing and in their regulation and practice. Greece includes sharing of passive infrastructure in its broadband strategy. Bulgaria and Croatia allow telecommunications operators to use the infrastructure facilities of other industries (energy, gas, and so on) to roll out their own network. Russia started LTE frequency sharing in 2015, the Czech Republic has used sharing of the Radio Access Network since 2012 (between O2 and T-Mobile), and Romania has done so since 2013 (between Vodafone and Orange).

Most of the transitioning countries (for example, Bulgaria in 2004, Georgia in 2005, and Albania in 2009) have integrated LLU in a legal framework or in their national law on communication, sometimes after public consultation, as in Bosnia (2008) and Kosovo (2012). Azerbaijan and Belarus have no legislation on LLU, and LLU is made on the basis of private agreements. In Kazakhstan and Russia, LLU does not exist, and no steps have been taken to implement it.

Most of the transitioning countries have a regulatory framework for mobile virtual network operators (MVNOs), which are companies that provide wireless communication services over another firm's wireless infrastructure (MVNOs help to promote competition in the market for wireless content). The MVNO market ranges from 1 virtual operator (as in Montenegro and Serbia) to 12 (in Russia) and 16 (in the Czech Republic). Albania, Azerbaijan, and Moldova have permitted the entry of MVNOs in the market, but no MVNO has been launched so far. And the policy and regulatory framework in some transitioning countries (such as Armenia, Belarus, Georgia, and Kazakhstan) have no specific provision for a regulatory framework on MVNOs.

A majority of transitioning countries have not yet launched a national broadband plan. Governments that have developed such a strategy often support the introduction of ICT into public services, such as e-government, e-health, and e-education projects. In Romania, the 2015 National Plan for Next-Generation Network Infrastructure Development and issues like e-government and e-health are part of the National Strategy on the Digital Agenda for Romania 2020. In Azerbaijan, the government has promoted e-government and e-education since 2010 and has launched a national action plan for promotion of open government (2012–15). In 2015, the first phase of a national broadband network project, launched by the incumbent AzTelekom, forms part of Azerbaijan's 2020 concept launched by the president. In Greece, national broadband projects include the 2014 Digital School project and the Development of Broadband Infrastructure in Rural White Areas of Greek Territory. The Bulgarian National Strategy for Development of Broadband Access was adopted in 2009, and the 2013 unified state network has the objective of providing the necessary infrastructure for e-governance (Digital Bulgaria 2015).

Finally, the national representation of transitioning countries in international organizations is part of the framework for effective regulation. In particular, countries have been benchmarked against representation in four international organizations—namely, the ITU, the Internet Corporation for Assigned Names and Numbers, the World Trade Organization, and the United Nations Economic and Social Commission for Asia and the Pacific. Being a member gives visibility on the international stage as well as access to discussions on best practices. Some 75 percent of transitioning countries are represented in three of these four organizations, and four (Armenia, Kazakhstan, Russia, and Turkey) are represented in all of them. Kosovo and FYR Macedonia are not represented in any of them, but FYR Macedonia is a member of the Body of European Regulators for Electronic Communications.

Transforming Countries in ECA

The 22 transforming countries are all high income,[12] and in 2007 all scored highly in the four clusters of the Regulatory Tracker (between 30 and 40 on clusters 1 and 2 with a maximum score of 42; above 20 on cluster 4, with a maximum score of 28), so there was little need for further progress over the period 2007–13. Only Denmark was lagging behind on clusters 1, 2, and 3 in 2007 (Denmark already scored highly in the competition framework in 2007), but had caught up with the other transforming countries by 2013. This is particularly noteworthy for regulatory regime (cluster 3), specifically on infrastructure sharing, quality of service monitoring, number portability, and national broadband plan. Andorra is the only transforming country that still has a first-generation type of regulation.

In the last mile, transforming countries are generally more advanced than transitioning and emerging countries: LLU is part of their national regulatory framework and part of EU policy. Nevertheless, it took some adjustments for some EU members to comply with the EU requirements. For instance, in Estonia (in 2007) and in Lithuania (in 2011), the European Commission required the national regulator to undertake a market analysis to determine the SMP of an operator.

Open access rules—whether for the national backbone (middle mile) or the local loop (last mile)—remain a challenge and a subject of discussion. The Body of European Regulators on Electronic Communications emphasized the importance of open access for facilitating broadband rollout, particularly with regard to the rollout of next-generation access (NGA) networks and the provision of additional current-generation broadband services in underserved areas (BEREC 2011). A regulatory consensus has emerged on the necessity of open access to national Internet infrastructure, as duplicating infrastructure is commercially or economically not viable (except in densely populated areas).

Also in the last mile, most transforming countries ensure that all citizens have universal access. EU member states are required to comply with the Universal Service Directive as part of their national communication law.

In the middle mile, transforming countries have a definition and criteria for determining SMP, which is confirmed by the ITU Regulatory Tracker, with high scores for competition framework (cluster 4). For example, under an article of its Law on Electronic Communications, Lithuania defines SMP as having "a position of economic strength affording it a power to behave to an appreciable extent independently of competitors, customers, and ultimately consumers." When SMP is designated in a market analysis, the Lithuanian regulator is empowered to require transparency, nondiscrimination, access, accounting separation, and price control.

In the invisible mile, data protection and privacy guidelines have been included in most transforming countries. In the EU, member states must comply with the

Data Protection Directive. In April 2016, the EU adopted the General Data Protection Regulation, which will come into force in April 2018. The new regime is binding on all member states and extends the scope of the data protection law to all foreign companies processing data of EU residents.

Conclusion

In summary, being landlocked may hinder the international bandwidth and the average connection speed of a country. This explains part of the connectivity challenges of Central Asian countries. But other drivers obviously affect Internet supply in ECA countries, contributing to the diversity of connectivity and regulatory landscape of the region.

Besides regional coordination through, for instance, regional infrastructure programs, policy and regulatory initiatives need to be promoted, depending on the level of technology and digital complements (regulation, institutional accountability, skills).

As described in this chapter, most *emerging countries* have weak international and national connectivity. Most still have state-owned incumbents that play a dominant role in some markets. Policy frameworks in emerging countries are needed to strengthen market competition in fixed and mobile markets, coordinate regionally to promote infrastructure deployment, and bring enhanced connectivity to landlocked countries.

Transitioning countries have developed regulatory frameworks for infrastructure sharing and MVNO markets, and most of them have privatized their incumbent operator. Some of them have no legislation on LLU and no national broadband plan yet. An independent, light touch regulation of the sector and development of national broadband plans via public-private partnerships would contribute to improving connectivity in transitioning countries.

For *transforming countries*, universal access for all citizens, the notion of significant market power, and regulation on data protection and privacy have all been included in laws but remain challenges. A policy framework is needed to ensure provision of open access networks (including LLU and MVNO licensing) and other measures to limit the degree of market dominance, particularly for digital platforms.

Notes

1. The United Nations 2030 Agenda for Sustainable Development also sets the goal of a universal and affordable Internet.
2. For more information on the global Internet infrastructure, see OECD (2014b).
3. For information on the Europe-Persia Express Gateway, see http://www.epegcable.com/#page-about-the-project.
4. For more information on IXPs and the Internet backbone market, see Weller and Woodcock (2013). For a directory of IXPs, see https://prefix.pch.net/applications/ixpdir/index.php.
5. See TeleGeography's Internet Exchange Map for geographic distribution of IXPs in ECA (http://www.internetexchangemap.com/#/).

6. Mobile broadband cannot be considered as a substitute for fixed broadband in dense urban areas, as it is limited by the number of simultaneous users who can be accommodated within a cell. However, to some extent, this limitation may be mitigated by offloading data to Wi-Fi. A further limitation is that mobile broadband is rarely made available for unlimited (that is, unmetered) use. But with newer generations of mobile technology, such as fourth-generation technology, mobile broadband can provide an effective substitute.
7. As defined by the OECD (2008), "Under the current EC Directives, an operator designated as having SMP is subject to specific obligations such as the requirement to produce an RIO [reference interconnection offer] and the obligation to have cost-oriented tariffs (except for mobile operators). An operator is presumed to have SMP if it has more than 25 percent of a telecommunications market in the geographic area in which it is allowed to operate."
8. As defined by OECD (2008), "Under the existing EC directives, a firm which is designated as having SMP must regularly produce a document [stating] the terms and conditions at which it will provide access to specified services. This document must be approved by the regulator."
9. Transitioning countries include Western Europe (Cyprus, Greece, Italy, Portugal), Central Europe (Bulgaria, the Czech Republic, Romania, the Slovak Republic, Slovenia), South Caucasus (Armenia, Azerbaijan, Georgia), Albania, Bosnia and Herzegovina, Belarus, Croatia, Kazakhstan, Kosovo, FYR Macedonia, Montenegro, Moldova, Russia, Turkey, and Serbia.
10. Kosovo and Russia are not included in the ITU Regulatory Tracker.
11. See "A Sector Assessment: Broadband in Russia": http://www.worldbank.org/en/country/russia/publication/broadband-in-russia.
12. Transforming countries include countries in Western Europe (Andorra, Austria, Belgium, Denmark, Finland, France, Germany, Iceland, Ireland, Liechtenstein, Luxembourg, the Netherlands, Norway, Spain, Sweden, Switzerland, and United Kingdom), Estonia, Hungary, Latvia, Lithuania, and Poland.

References

Akamai. 2015. "The State of the Internet." https://www.akamai.com/uk/en/our-thinking/state-of-the-internet-report/

A4AI (Alliance for Affordable Internet). 2015. *Affordability Report 2015/16.* Washington, DC: A4AI. http://a4ai.org/affordability-report/report/#the_affordability_index.

BEREC (Body of European Regulators for Electronic Communications). 2011. "Open Access." BEREC, Riga. http://berec.europa.eu/eng/document_register/subject_matter/berec/reports/212-berec-report-on-open-access.

Broadband Commission for Digital Development. 2015. *The State of Broadband 2015: Broadband as a Foundation for Sustainable Development.* Geneva: Broadband Commission for Digital Development. http://www.broadbandcommission.org/Documents/reports/bb-annualreport2015.pdf.

Cisco. 2016. "Visual Networking Index: Global Mobile Data Traffic Forecast Update, 2015–2020." White Paper, Cisco, San Jose. http://www.cisco.com/c/en/us/solutions/collateral/service-provider/visual-networking-index-vni/mobile-white-paper-c11-520862.html.

European Commission. 2014. *Digital Agenda for Europe 2010–2020.* Brussels: European Commission. http://europa.eu/pol/pdf/flipbook/en/digital_agenda_en.pdf.

———. 2015. "Implementation of the EU Regulatory Framework for Electronic Communication—2015." Staff Working Document SWD (2015) 126 final, European

Commission, Brussels. http://ec.europa.eu/transparency/regdoc/rep/10102/2015/EN/10102-2015-126-EN-F1-1.PDF.

———. 2016a. "Digital Single Market." *Digital Single Market*, May 12. https://ec.europa.eu/digital-single-market/en/digital-single-market.

———. 2016b. *Europe's Digital Progress Report*. Brussels: European Commission. https://ec.europa.eu/digital-single-market/en/news/europes-digital-progress-report-2016.

———. 2016c. "Infrastructure and Service-Based Competition." *Digital Single Market*, July 25. https://ec.europa.eu/digital-single-market/en/infrastructure-and-service-based-competition.

———. 2016d. "State of the Union 2016: Commission Paves the Way for More and Better Internet Connectivity for All Citizens and Businesses." Press release, European Union, September 14. http://europa.eu/rapid/press-release_IP-16-3008_en.htm.

———. 2016e. "Telecom Rules." *Digital Single Market*, September 14. https://ec.europa.eu/digital-single-market/en/telecoms-rules#Article.

Forge, S., R. Horvitz, and C. Blackman. 2012. "Perspectives on the Value of Shared Spectrum Access." Final report for the European Commission, Brussels, February. http://ec.europa.eu/digital-agenda/sites/digital-agenda/files/scf_study_shared_spectrum_access_20120210.pdf.

Freedom House. 2015. "Tajikistan." Freedom House, Washington, DC. https://freedomhouse.org/report/freedom-world/2015/tajikistan.

Gelvanovska, N., M. Rogy, and C. M. Rossotto. 2014. *Broadband Networks in the Middle East and North Africa: Accelerating High-Speed Internet Access*. Directions in Development—Communication and Information Technologies. Washington, DC: World Bank. https://openknowledge.worldbank.org/handle/10986/16680.

GSMA (Global System for Mobile Association). 2016. GSMA Intelligence Database. London: GSMA. http://www.gsmaintelligence.com/.

Human Rights Watch. 2015. "World Report 2015: Turkmenistan." Human Rights Watch, New York. https://www.hrw.org/world-report/2015/country-chapters/turkmenistan.

ITU (International Telecommunication Union). 2014. *Measuring the Information Society 2014*. Geneva: ITU.

———. 2015. "ITU Regulatory Tracker." ITU, Geneva. http://www.itu.int/en/ITU-D/Regulatory-Market/tracker/Pages/default.aspx.

———. n.d. World ICT/Telecommunication Indicators Database. Geneva: ITU. www.itu.int/ti.

Kinney, S. 2016. "Wi-Fi Carries 80% of Mobile Data Traffic." *RCR Wireless News*, July 7. http://www.rcrwireless.com/20160707/network-infrastructure/wi-fi/analyst-wi-fi-carriers-80-mobile-data-tag17.

Mobile Experts. 2016. "Carrier Wi-Fi and LTE-U 2016." Mobile Experts, Campbell, CA. http://mobile-experts.net/Home/Report/61.

Navas-Sabater, J., A. Dymond, and N. Juntunen. 2002. "Telecommunications and Information Services for the Poor: Towards a Strategy for Universal Access." Discussion Paper 432, World Bank, Washington, DC, April.

National Program "Digital Bulgaria 2015." https://www.mtitc.government.bg/en/politiki/cifrova-bulgariya/national-programme-digital-bulgaria-2015.

OECD (Organisation for Economic Co-operation and Development). "Glossary of Statistical Terms." https://stats.oecd.org/glossary/detail.asp?ID=6751.

———. 2014a. "A Broadband and Internet Economy Metric Checklist: A New Approach." OECD, Paris. http://www.oecd.org/sti/ieconomy/48255721.pdf.

———. 2014b. "International Cables, Gateways, Backhaul, and International Exchange Points." Digital Economy Paper 232, OECD Publishing, Paris. http://dx.doi.org/10.1787/5jz8m9jf3wkl-en.

———. 2015. "Digital economy facts and figures." http://www.oecd.org/newsroom/world-must-act-faster-to-harness-potential-of-the-digital-economy.htm.

TeleGeography. 2014. 2015. 2016. "Global Comms: All Regions." Online resource (subscription), www.telegeography.com.

Terabit Consulting. 2013. *Broadband Infrastructure in North and Central Asia: Markets, Infrastructure, and Policy Options for Enhancing Cross-Border Connectivity*. Cambridge, MA: Terabit Consulting.

UNESCAP (United Nations Economic and Social Commission for Asia and the Pacific). 2014. "An In-Depth Study of Broadband Infrastructure in North and Central Asia." Working Paper, UNESCAP, Bangkok. http://www.unescap.org/sites/default/files/Broadband%20Infrastructure%20in%20North%20and%20Central%20Asia%20FINAL%20_English_0.pdf.

Weller, D., and B. Woodcock. 2013. "Internet Traffic Exchange: Market Developments and Policy Challenges." Digital Economy Paper 207, OECD Publishing, Paris.

———. 2016. *World Development Report 2016: Digital Dividends*. Washington, DC: World Bank.

3

Firm Productivity and Dynamics

Mr. Terzić owns a small clothing manufacturing firm in Sarajevo, Bosnia and Herzegovina. In the early 2000s, he installed computers and information technology software programs after noticing that other clothing companies were receiving new orders from Europe through e-mail. The new technology made a large improvement in how he ran his business: the word processing and spreadsheet programs allowed him and his sales team to complete and track their orders electronically, and e-mails and websites made it easier for him to connect and communicate with his suppliers and customers.

More than 10 years later, new technologies have emerged that could improve his business operations even further. But the options are overwhelming for Mr. Terzić. Should he use enterprise resource planning software to manage his product planning, inventory, sales, shipping, and finance activities? Should he introduce programs to send and receive electronic invoices to his suppliers and customers? Should he move his e-mail and business software to the cloud? Since he knows that most of his competitors are not using these technologies, he has very little incentive to upgrade his systems.

This story highlights the main focus in this chapter: why firms should adopt the new Internet technologies and how governments can encourage more adoption.

Introduction and Main Messages

The Internet can make an enormous contribution to productivity by improving the efficiency of individual firms, encouraging the entry and growth of more productive firms, and encouraging the exit of less productive firms. Growth accounting exercises find that information and communication technology (ICT), a broad term that includes the Internet, has contributed 17–45 percent to the increase in labor productivity in European Union (EU) countries and 33–73 percent in the United States (for a discussion of the growth accounting method and these studies, see Cardona, Kretschmer, and Strobel 2013).

The Internet can raise firm productivity in a wide variety of ways. E-mail can facilitate communication, and, except for in the least developed countries, many firms in Europe and Central Asia (ECA) use e-mail to communicate with their clients. Internet banking and getting customers to perform their own tasks (through automatic checkout kiosks in retail stores) also can raise productivity. Even greater efficiency is generated by more sophisticated Internet processes, such as communicating with clients through a website, sharing information within the firm using enterprise resource planning (ERP) software, automating billing and purchasing processes using electronic invoices, and tracking goods for production, sales, and customer service purposes using radio frequency identification (RFID). Particularly the last three can reduce the resources devoted to manually intensive processes, decrease the likelihood of human error, and increase the speed of processing. However, the costs of setting up these more sophisticated Internet uses can be high, so that adoption tends to be greater in larger and more established firms. As the benefits to the firm of e-invoicing and RFID depend, to a large extent, on other firms adopting them, their use in ECA is more prevalent in the more developed countries.

Cloud computing, where firms rely on a network of remote servers to store and process their data on the Internet rather than making their own information technology (IT) investments, can have transformational effects when it is used to deploy customer relationship management (CRM) software and finance and accounting software or to purchase computing power for running in-house software. This can greatly increase firms' flexibility in scaling their operations up or down in response to changes in demand and enable firms to access information anywhere. However, most firms that use cloud computing use it only for e-mail, particularly in countries with unreliable or expensive broadband access.

The Internet also improves aggregate productivity by increasing the entry of new firms. Government websites can reduce the costs of information and registration for businesses. And replacing a storefront with sales through online platforms, investments in IT with cloud computing, or office workers with remote workers can lower the costs of starting a business and thus encourage greater entry. The Internet makes it easier for new firms to enter traditional sectors such as transportation, finance, and retail (for example, Uber and Lyft in taxi services). And the Internet of Things (IoT), a network of physical objects embedded with sensors and electronics that are connected through the Internet to each other and to a centralized system, allows firms to enter new markets by transforming the sale of goods into a service (for example, by installing sensors in machinery to warn about service needs). Nevertheless, the aggregate effects of the Internet on firm entry may be limited by other factors in ECA, as there is no clear cross-country correlation between changes in Internet penetration and changes in firm entry.

The increase in productivity and firm entry due to the Internet increases competition, which can reduce the ability of weaker firms to survive. This churning of firms can increase overall productivity by increasing the capital and labor available to more productive firms. However, barriers to exit—for example, high costs of fixed assets, redundancy, or government protection in the form of subsidies or entry barriers—can result in the continued devotion of capital and labor to relatively unproductive activities, thus lowering overall productivity. Firm exit also can increase transitory unemployment, which can cause hardship and have undesirable social effects. The implications of the Internet for outsourcing and offshoring of work are considered in chapter 4, and the implications for wage inequality and for automation and job displacement are considered in chapter 5.

The use of the Internet and the intensity of use vary considerably among ECA countries and among firms within countries. A lack of appropriate skills or weak management can limit reliance on the Internet. As the integration of Internet processes into business operations often requires extensive changes in organization and strategy, competent management is important to reaping the productivity gains. Nevertheless, the competitive environment is probably the most important determinant of Internet adoption, and there is a strong positive correlation between use of the Internet among firms and level of competition in ECA. Government can contribute to a more competitive environment by implementing appropriate domestic regulation and lowering the barriers to foreign trade and investment, creating the incentives for incumbent firms to adopt new technology in order to remain profitable and for entrants to use new technology to modernize traditional sectors.

ECA countries can be classified into three groups: emerging, transitioning, and transforming, in order of the level of competition. Emerging digital countries should focus on reducing the barriers to trade and investment and the difficulties in starting a business. Transitioning digital countries should focus on implementing a strong regulatory framework for competition policy and on establishing an agency to enforce these rules. Transforming digital countries should focus on responding to new regulatory issues arising from how the Internet has changed business models and how firms operate in markets. Key issues on the frontier include establishing an appropriate competitive framework for multisided markets (for example, eBay and Amazon), where concentration of firms may be efficient due to network effects; limiting abuses from the use of detailed data to target differentiated products to particular consumers; coping with cybersecurity threats, particularly with the IoT; and setting standards to encourage participation in Internet markets, such as the use of e-invoices, where the benefits increase with the number of users.

This chapter first considers how the Internet can be used to improve the productivity of individual firms and then discusses how the Internet can be used to improve aggregate productivity by encouraging the entry of more productive firms and the exit of less productive firms. It then reviews the policy priorities for governments, which depend on the level of competition, followed by the emerging challenges for policy.

The Internet Can Increase Firm Productivity

The ICT sector is often the most innovative and fast-changing sector in many countries. Firms in the ICT sector garner attention, as they are on the technology and innovation frontier, providing many high-skilled jobs and earning high profits. For example, building on its traditions as the "Silicon Valley" of the Eastern Bloc, Bulgaria has a large ICT sector—10 percent of gross domestic product (GDP)—and among the largest number of certified IT professionals in the world. The sector is attractive and features in the development plans of many countries.

The ICT sector, however, contributes only a small percentage to total value added. Even among the Organisation for Economic Co-operation and Development (OECD) countries, the ICT sector only accounts for 6 percent of total value added (figure 3.1). The share is even smaller in Southern and Central European countries. Among ECA countries, the ICT sector makes the largest contribution to value added in Ireland, due largely to the presence of the global or regional headquarters of large ICT companies such as Airbnb, Amazon, LinkedIn, and Google.

FIGURE 3.1
ICT sector contributes only 6 percent to value added among OECD countries

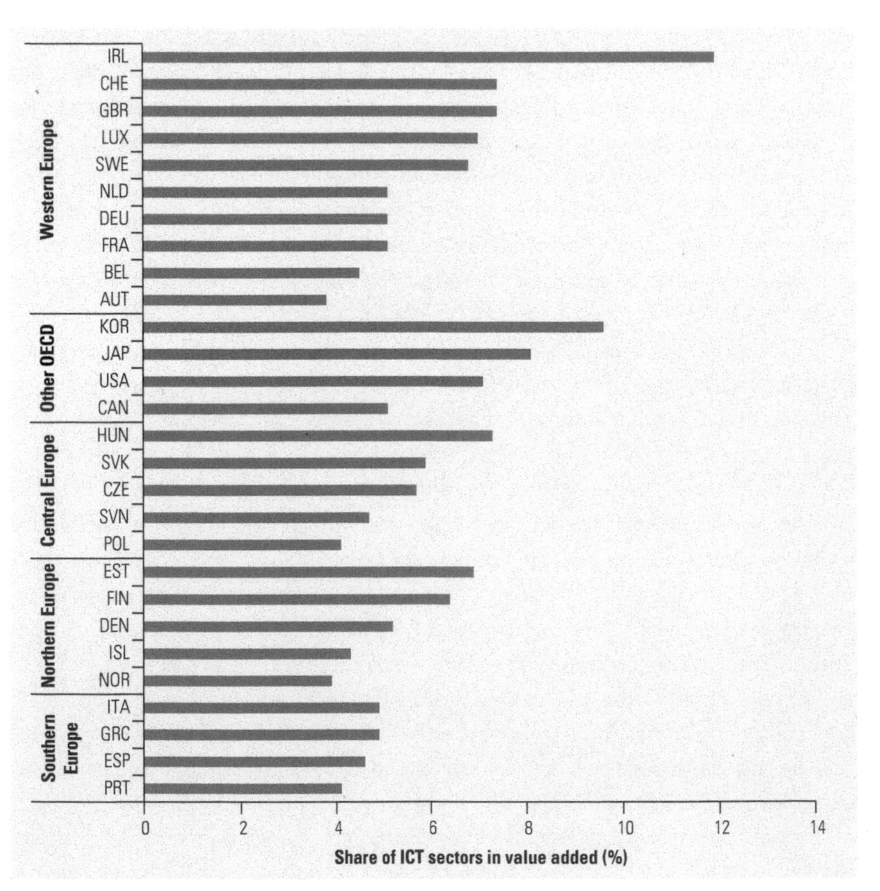

Source: OECD 2011.
Note: ICT value added is the difference between the ICT sector gross output and intermediate consumption. The aggregate of information industries here includes ISIC rev. 4 Division 26 (Manufacture of computer, electronic, and optical products) and Section J (Information and communication), that consists of Divisions 58–60 (Publishing and broadcasting industries), 61 (Telecommunications), and 62–63 (Computer programming, and Information service activities). ICT = information and communication technology; ISIC = International Standard Industrial Classification.

While the ICT sector may contribute a relatively small proportion to the economy, it is an important catalyst for change and an important input in the production function of firms. As a sector on the technology frontier, the ICT sector can bring new technology, ideas, and processes to firms in other sectors. Firms can use technology to be more productive and enter new markets.

Many studies, mostly on firms in Europe and North America, find that more ICT is associated with higher firm productivity (for a summary, see Cardona, Kretschmer, and Strobel 2013; Draca, Sadun, and Van Reenen 2007). An analysis linking different micro-level data sets in 15 EU countries shows that an increase in the share of broadband Internet-enabled employees was related to an increase in the labor productivity of firms between 2001 and 2010 (Eurostat 2012, 2013; figure 3.2).

The relationship between Internet and firm productivity, however, may reflect the fact that productive firms are more likely to adopt the Internet rather than the impact of Internet adoption on productivity (see, for example, Haller and Siedschlag 2011). Nevertheless, many firm-level studies, using a variety of estimation methods and types of data, have shown that ICT has had a positive impact on firm productivity (see table 3A.1 in annex 3A). Perhaps the most convincing approach considers exogenous changes in government policies that increase the availability of the Internet. For example, one study finds that Norwegian municipalities that benefited from infrastructure investments during the 2000s that increased broadband access also experienced higher increases in the wages and productivity of skilled workers than other municipalities (Akerman, Gaarder, and Mogstad 2015). The impact of the Internet on firm productivity depends on the kinds of tasks for which it is used, ranging from simple (using e-mail and

FIGURE 3.2
Increasing the share of employees using broadband Internet increases labor productivity

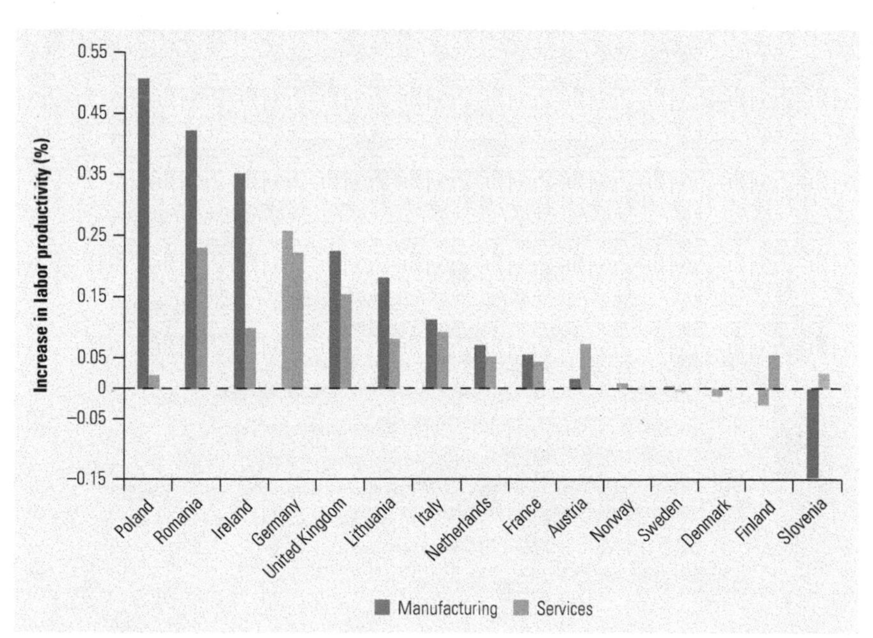

Source: Eurostat 2012, 2013.
Note: Increases in labor productivity are calculated from a 1 percent increase in the share of broadband Internet-enabled employees. The regressions are done on pooled firm-level regressions for the period 2001–10, except for the estimate for Romania, which is obtained from the analysis in Eurostat (2012) and covers the period 2001–09. Manufacturing sectors are sectors 15 to 37 and services sectors are sectors 50 to 74 in ISIC Rev. 3 classification. Most estimates are significant at the 5 percent level, with the ones that are not significant shaded in a lighter color. ISIC = International Standard Industrial Classification.

smartphones) to intermediate (setting up websites and using social media) to complex (using ERP software).

More than 80 percent of ECA firms used e-mail to communicate with their clients and suppliers in 2013 (figure 3.3). Certain ECA countries, however, still have relatively low rates of enterprise use of e-mail: only 44 percent of firms use e-mail in Uzbekistan, 51 percent in Albania, 57 percent in Tajikistan, and 62 percent in Moldova. The firms in these four countries are hampered by poor Internet access and high prices. Use of the Internet to perform higher-order tasks, such as providing customer service, is much lower, at only 11 percent of firms in ECA.

There is a large difference between the share of firms using e-mail and the share using websites to communicate with customers and suppliers. A website is a better communication tool, as it can provide information, customer service, and e-commerce services, but creating a website requires more advanced IT skills and resources. While there is a 23 percentage point difference, on average, in ECA

FIGURE 3.3
Most firms use e-mail to communicate with clients and suppliers, but fewer use websites

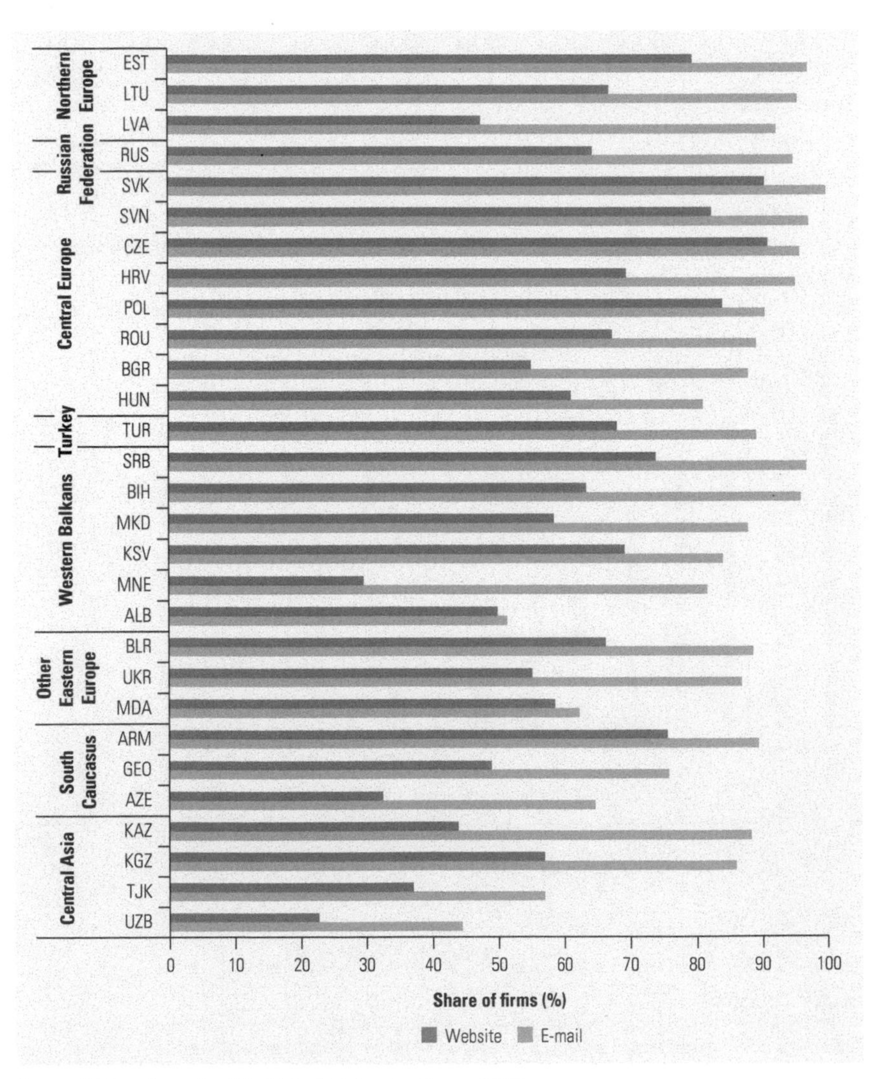

Source: Elaboration using World Bank Enterprise Survey 2013.
Note: All figures are calculated using 2013 data except for the Russian Federation, which uses 2012 data.

between the share of firms using these two technologies, the difference is higher than 30 percentage points in a third of the 29 ECA countries with adequate data: the Russian Federation (30 percentage points), Azerbaijan and Ukraine (32 percentage points), Bosnia and Herzegovina and Bulgaria (33 percentage points), Kazakhstan (44 percentage points), Latvia (45 percentage points), and Montenegro (55 percentage points). Across all ECA countries, firms with more than 50 employees are more likely to use websites to communicate with their clients and suppliers (figure 3.4). Firms that have a male manager are more likely to use websites to communicate than those with a female manager. The difference between male-managed and female-managed firms that use websites ranges from 1 to 25

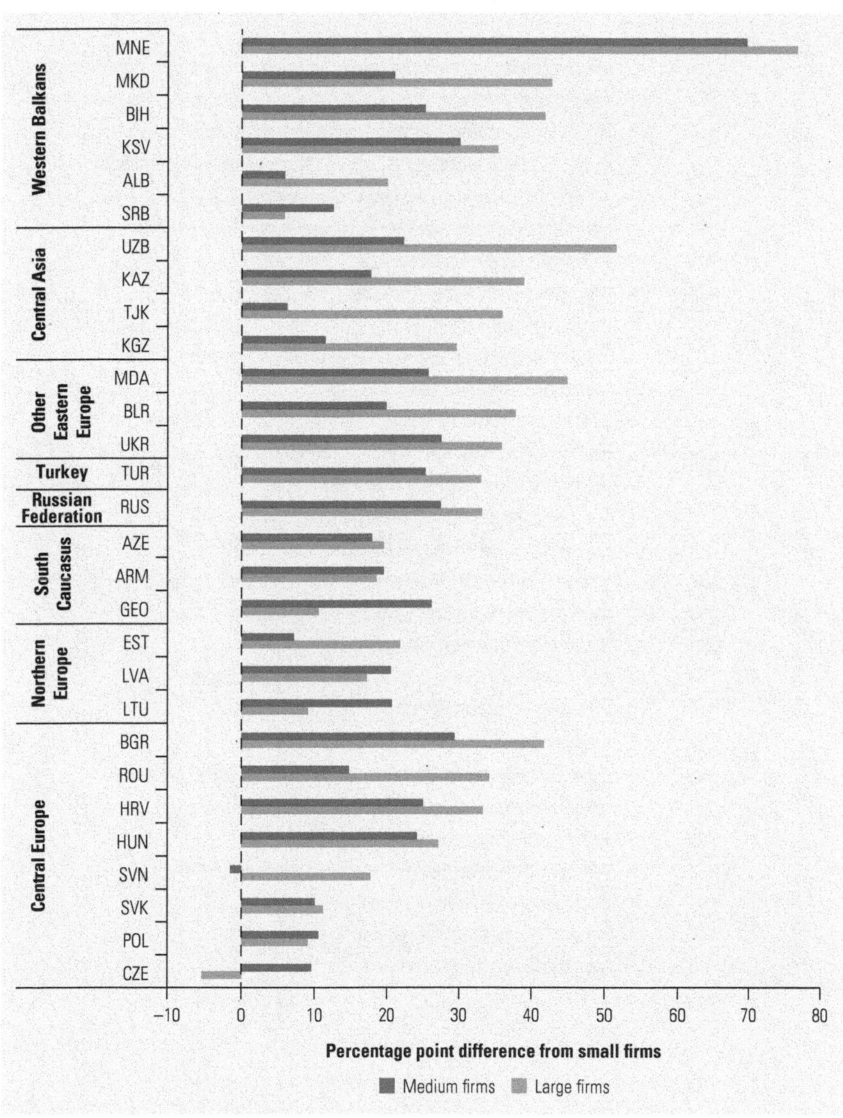

FIGURE 3.4
Medium- and large-size firms are more likely to use websites to communicate

Source: Elaboration using World Bank Enterprise Survey 2013.
Note: Firm size is measured as number of employees: small (<50), medium (51–250), and large (>250). The differences are the percentage of medium- and large-size firms that use websites to communicate with suppliers and clients to the percentage of small firms that use websites, measured in percentage points.

percentage points (figure 3.5). However, in countries in Central Asia, South Caucasus, and Turkey, female-managed firms are more likely to use websites to communicate. As it is difficult for women to be managers in these countries, female managers may be, on average, more educated and receptive to new technology than male managers.

Beyond communication, the Internet can be used to automate tasks and reduce the time and resources spent on them, allowing firms to redirect these resources to more productive uses. Investments in IT-enabled equipment can raise firm efficiency, reduce production and inspection time, and increase the ability of firms to customize their products (Bartel, Ichniowski, and Shaw 2007). At its simplest, firms

FIGURE 3.5
Firms with male managers are more likely to use websites to communicate with clients and suppliers

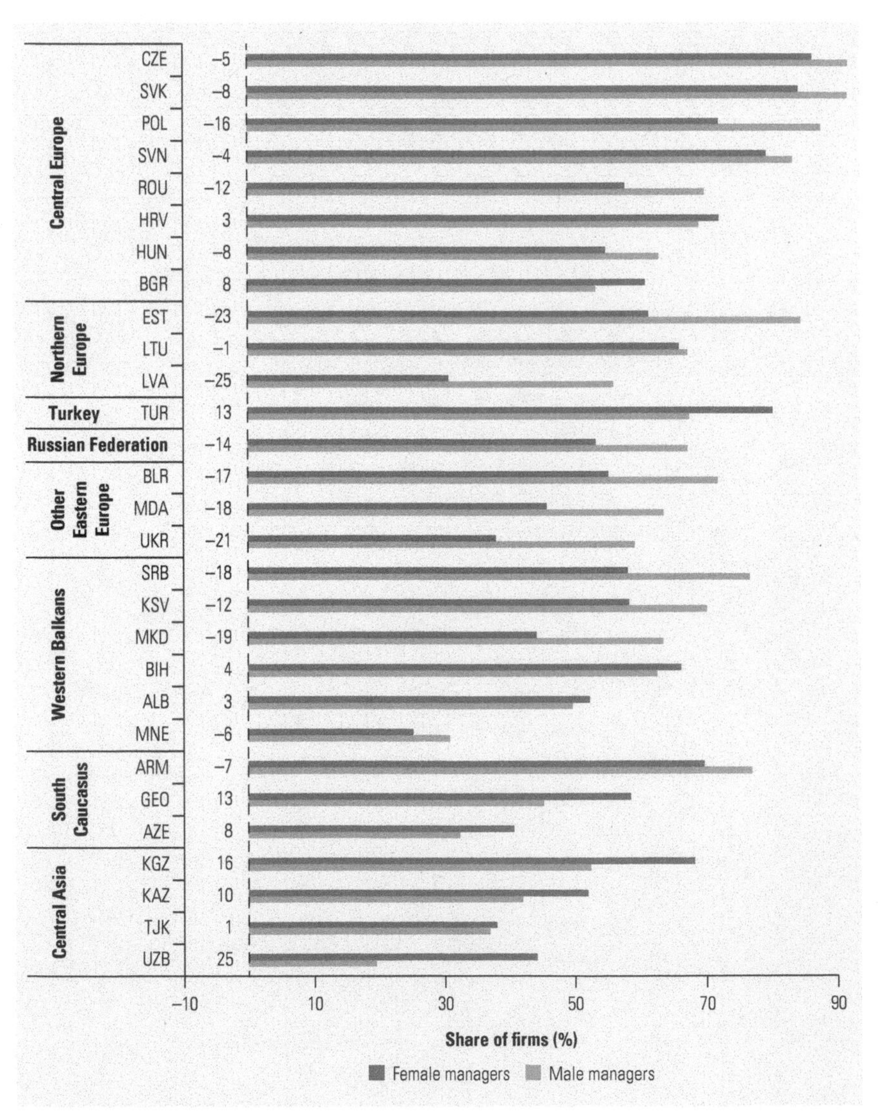

Source: Elaboration using World Bank Enterprise Survey 2013.
Note: The gender premium is measured as the difference between the percentage of male-managed firms that use websites for communication with suppliers and clients to the percentage of female-managed firms that use websites, measured in percentage points. The numbers on the x-axis represent the difference between the percentages of firms managed by male and female managers.

can use the Internet to automate tasks such as banking, which eliminates the time spent at the bank and increases the firm's flexibility in when and how it conducts its finances. Many firms in Central and Eastern Europe have high rates of firms using Internet banking, with 90 percent of firms in the Czech Republic, the Slovak Republic, and Slovenia already using Internet banking in 2008.[1] By contrast, only 29 percent of firms use Internet banking in Azerbaijan, 32 percent in Kazakhstan, and 37 percent in Russia.

Using the Internet to automate more complex tasks can dramatically transform the way a business operates. Firms can use ERP software to share information internally, e-invoices to automate billing and purchasing processes with customers and suppliers, and RFID to track their goods for production, sales, and customer service purposes. These new technologies can reduce costs and increase efficiency for many firms. For example, e-invoice systems eliminate the need for paper-based invoices by converting all exchanges to an electronic format. These technologies can reduce the resources and manpower a firm must devote to this manually intensive process and, which is more important, decrease the likelihood of human error and increase processing speed. Changing from a paper-based to an electronic invoice can reduce the operational costs of billing 70–75 percent (Capgemini Consulting 2007). But using the Internet to automate or replace these complex tasks requires more resources, specialized skills, and organizational capacity.

Firms may lack the resources or the incentive to deploy such technologies in their business operations. Many ECA countries, even among high-income countries in Western Europe, have low rates of firms using these technologies, with more firms using ERP software than RFID and e-invoices (figure 3.6). On average, 14 percent of firms in Western and Northern European countries use ERP software to communicate between business units, compared with about 10 percent in Central Europe.

Firms are more likely to adopt ERP software than e-invoices or RFID because the benefits of the former do not depend on coordination with other firms. That is, ERP software facilitates communication within the firm, while the benefits of e-invoice systems and RFID only increase as more suppliers and customers use the same technology. There is little incentive for firms to adopt an e-invoice system if other firms cannot receive the e-invoice. Government can promote these technologies by establishing a standard to reduce the uncertainty about regulations and use of e-invoices. The EU has introduced a harmonized standard for e-invoices across all countries to facilitate cross-border transactions, but the standard governs only public procurement.[2] Turkey took a different approach by developing and managing the e-invoicing application within its Ministry of Finance and requiring firms operating in certain sectors to use e-invoices in 2014.[3]

Firms can use the Internet in other ways to reduce their operational costs and improve productivity. Firms in the retail and financial services sectors use the Internet to reduce their labor costs by getting their customers to perform their own tasks. For example, self-service checkout kiosks in supermarkets and retail stores allow customers to complete their own purchase, reducing the labor costs for the firm, as one worker can attend to four to six customers at once. Internet and mobile banking has reduced the number of bank branches

FIGURE 3.6
The use of advanced Internet technologies is limited, even among countries in Western Europe

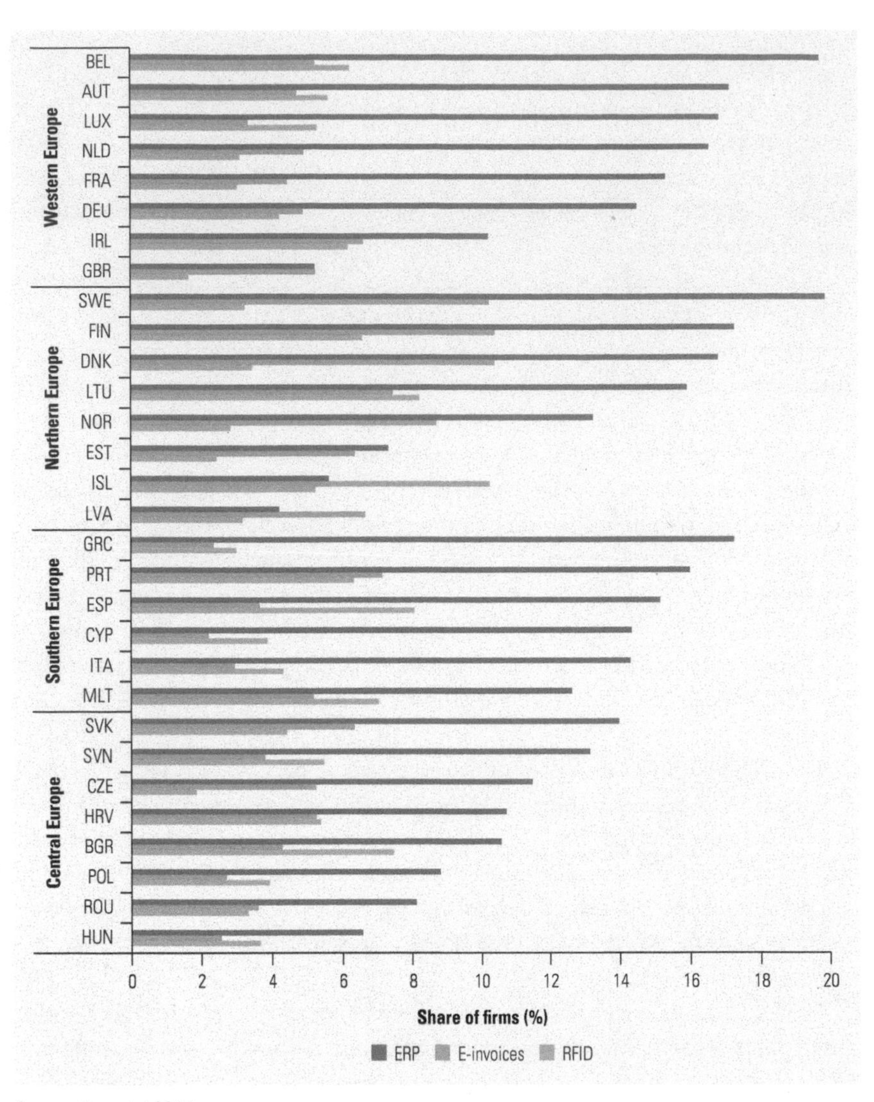

Source: Eurostat 2014.
Note: ERP = enterprise resource planning; RFID = radio frequency identification.

and bank tellers by giving customers the ability to check their accounts and perform simple transactions on their computers and mobile phones. Turkey is leading many banks in other developed countries, such as France, Germany, and the United States, in its offering of mobile banking services, with more than 65 percent of Turkish Internet users using mobile banking to perform banking transactions.[4]

Recent advances in digital technology, such as cloud computing, are providing new ways for firms to increase their productivity. Cloud computing is a technology where firms rely on a network of remote servers hosted on the Internet to store and process their data. Firms do not need to invest in IT capital and labor to manage specialized local servers and computer software. An average of 27 percent of firms in Northern Europe and 22 percent of firms in Western Europe used cloud computing in 2014. Firms in Central Europe, especially in Croatia, the Czech Republic, the Slovak Republic, and Slovenia, approach the levels of use in Western Europe (figure 3.7).

In contrast, less than 10 percent of firms in Bulgaria, Hungary, Latvia, Poland, and Romania are using cloud computing. As cloud computing requires a stable, affordable, and constant broadband access to be a practical technology, this difference in the use of cloud computing might be due to the lower rates of broadband access in these countries (according to the World Bank 2013 enterprise survey data). These countries have the lowest number of firms with broadband access in the EU, but higher levels of broadband access than firms in Central Asia

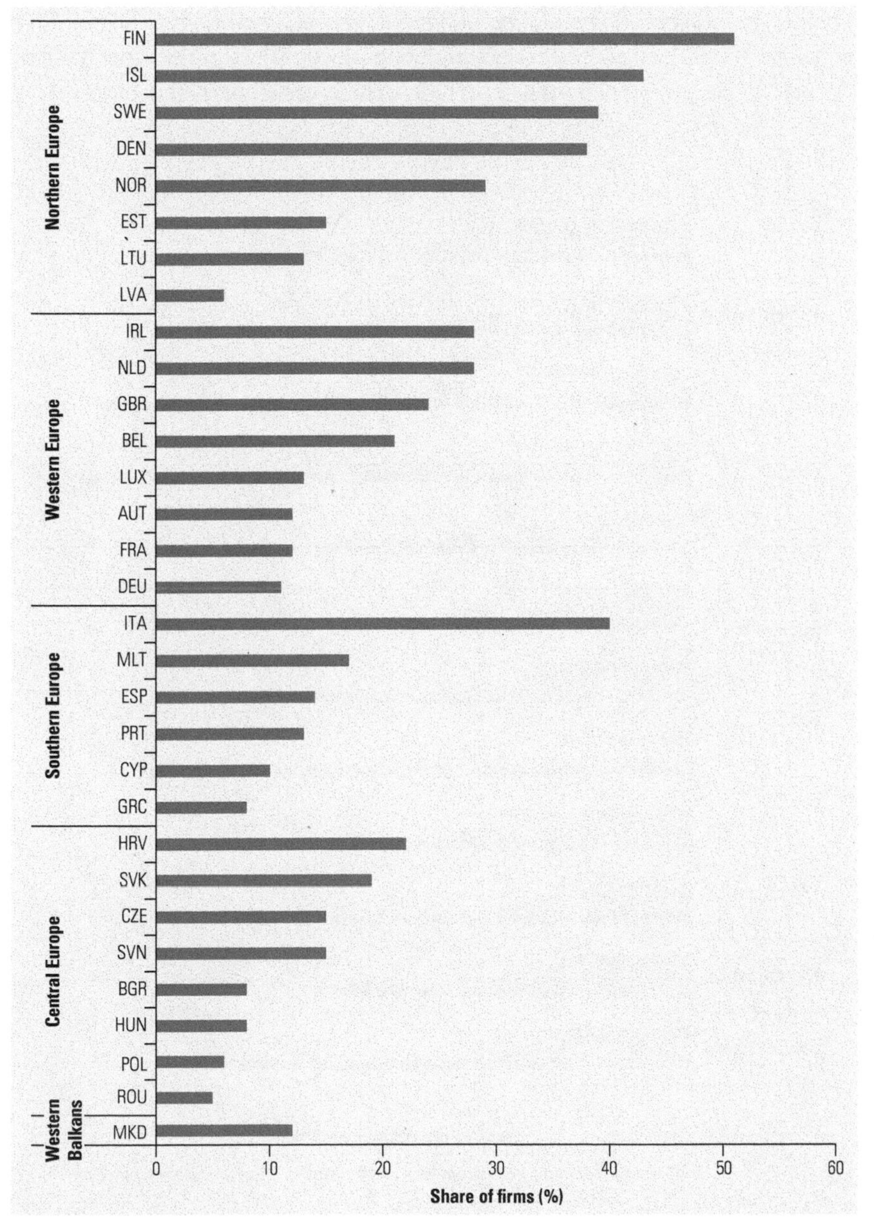

FIGURE 3.7
Many firms in ECA use cloud computing

Source: Eurostat 2014.

and South Caucasus, suggesting that the rates of cloud computing use in these regions may be even lower.

Many firms in ECA that use cloud computing rely on it only for simpler tasks, such as sending and receiving e-mails, accessing office software, and hosting data (figure 3.8). Among firms that use cloud computing, most use it for e-mail. But cloud computing will only have transformational effects on a firm or a sector when it is used for more complex tasks such as deploying CRM software and finance and accounting software or purchasing computing power for in-house software. This would give firms incredible flexibility in scaling their operations up or down and in accessing information anywhere. Despite their higher level of income and development, only 33 percent of firms in Western Europe use cloud computing to access finance or accounting software, 25 percent to access CRM software, and 20 percent to access computing power. This suggests that it will be difficult for firms in other parts of ECA to use cloud computing for complex tasks.

FIGURE 3.8
More firms use cloud computing for simple purposes but not advanced tasks

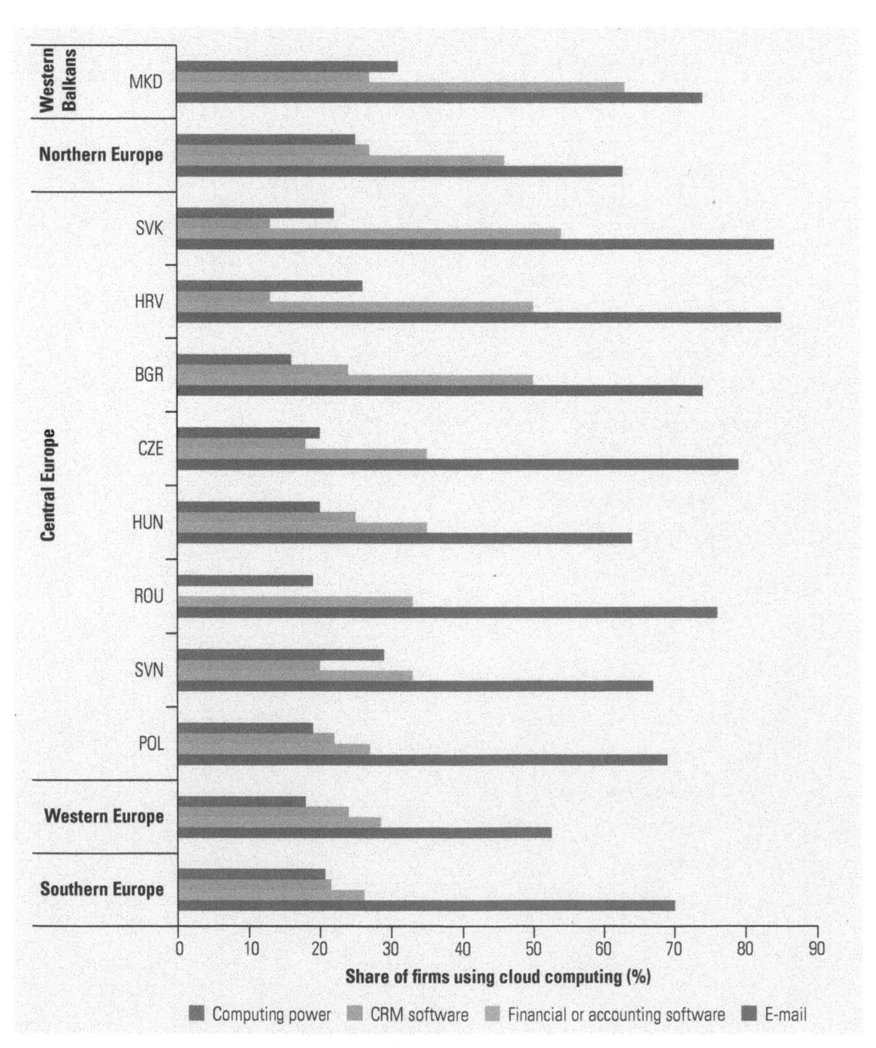

Source: Eurostat 2014. CRM = customer relationship management.

The Internet Can Increase Firm Entry

The Internet can help firms to enter new markets, thus increasing dynamism and entrepreneurship. Restricting entry of new firms can significantly reduce productivity: a cross-country study finds that increasing the costs of entry by 80 percent of income per capita will decrease total factor productivity (TFP) by 22 percent (Barseghyan 2008). In 2015 it cost 5 percent of per capita income to start a business in the ECA region compared with 56 percent in the Africa region. New firms contribute significantly to job creation (Haltiwanger, Jarmin, and Miranda 2013), which is an important agenda in many ECA countries. The Internet can encourage the entry of both formal and informal firms. This section focuses on formal firms. The implications of Internet use for informal activity in the labor market are considered in chapter 5. It can be difficult to measure the independent impact of the Internet on firm entry, as a positive relationship between Internet use and firm entry may simply reflect the greater availability of the Internet in denser, more populated areas that are also attractive to new firms. However, one study shows a causal relationship, where the probability of firm entry in rural parts of Iowa and North Carolina in the United States increased after the introduction of access to broadband Internet (Kim and Orazem 2016).

The Internet can facilitate firm entry through three channels. First, government websites, for example, e-government platforms for business registration and land registry, can reduce the information and transaction costs required for firms to enter new markets. The average cost of starting a business in ECA countries with an online business registration system is 4.6 percent of per capita income, compared with 7.9 percent in ECA countries without this system (figure 3.9).

Online business registration, however, does not compensate for complicated regulations and licensing processes. For example, despite the presence of an online business registration system, the costs of starting a business in Turkey are around 15 percent of per capita income, about the same as in Bosnia and Herzegovina, which lacks an online system. Reducing entry costs by introducing online business registration can encourage new firms to enter the market. Firm entry increased in 9 of the 11 ECA countries that implemented online business registration in the last decade (figure 3.10).

Once established, firms can use the Internet to interact with government departments, which can reduce the administrative costs for firms. Interactions with the government can range from the provision of information and paperwork to more complicated tax filing. Governments can also digitize many of their interactions with firms. More than 70 percent of firms in the Baltic States, Hungary, and Slovenia are already using the Internet to file their value added and corporate taxes, a rate even higher than in Western Europe, where 50 percent of firms file their taxes online (figure 3.11). The share of firms using the Internet to interact with the government in Azerbaijan, Belarus, Russia, and Ukraine is as high as, or even higher than, the average in Western Europe.[5] Even though the data cover a wide range of government interactions (not just tax filings), only 20 percent of Kazakh firms and 2 percent of Kyrgyz firms use the

FIGURE 3.9
The cost of starting a business is lower when there is online business registration

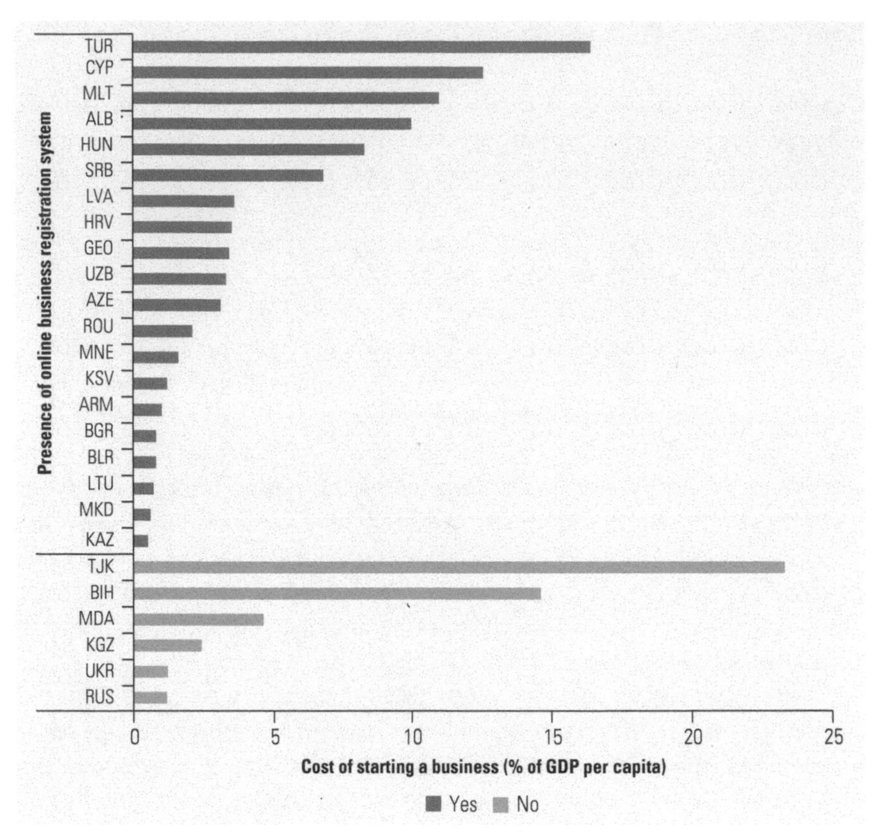

Source: Calculations based on World Bank 2015. GDP = gross domestic product.

FIGURE 3.10 Firm entry rises when countries implement Internet business registration

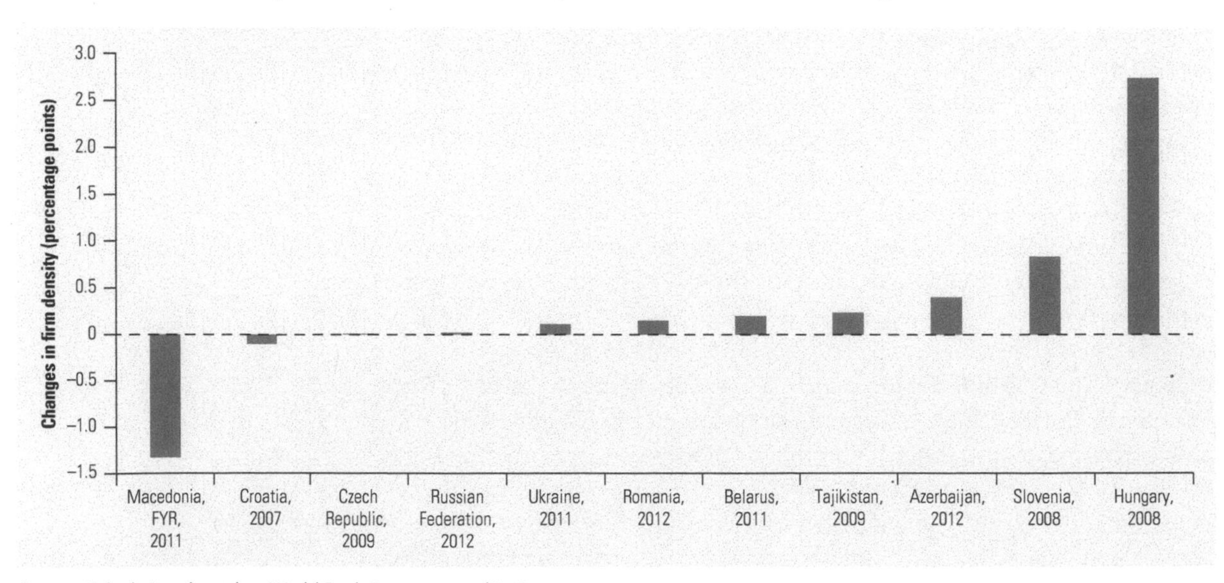

Source: Calculations based on World Bank Entrepreneurship Data.
Note: The years after the country names indicate when Internet business registration began. The average firm densities are compared two years before and two years after the reform year. For Romania, the Russian Federation, Tajikistan, and Ukraine, data were available for only one year after the reform year.

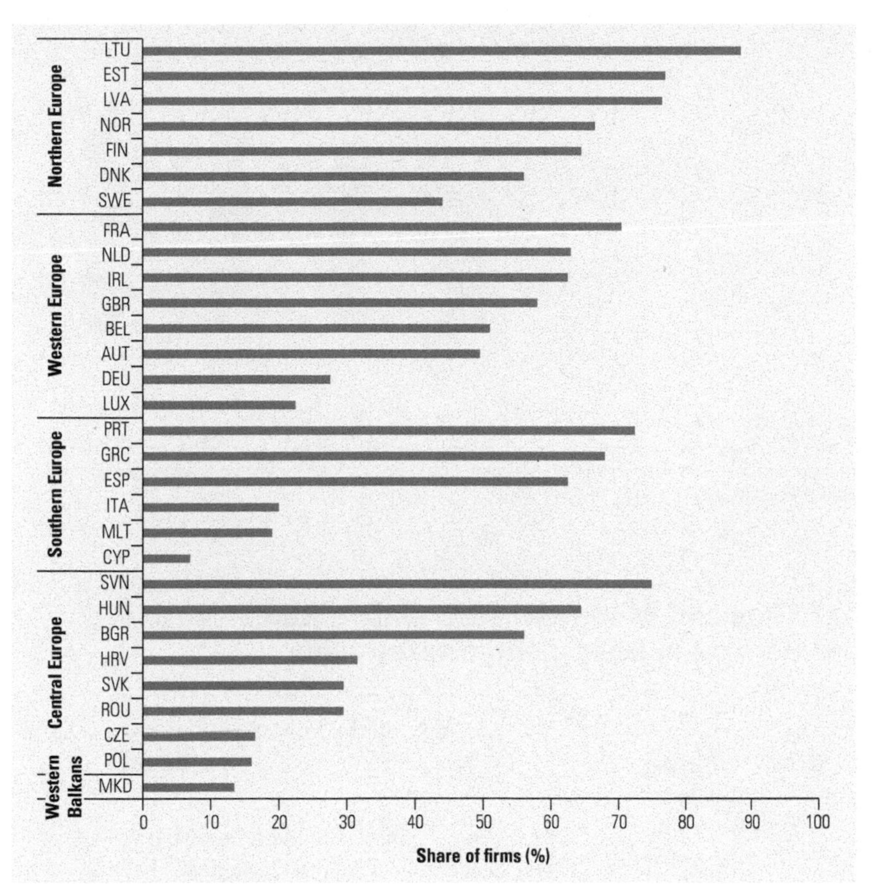

FIGURE 3.11
Many firms in Northern Europe use the Internet to interact with the government

Source: Eurostat, circa 2011.
Note: The data on tax filing is from Eurostat and it is the average of firms that use the Internet to file their value-added tax and corporate tax.

Internet to interact with the government (figure 3.12). The World Bank has worked with ECA governments to implement such electronic systems. For example, Bosnia and Herzegovina implemented an e-register website that exhaustively lists all of the information a firm needs to apply for different licenses and permits, making it easier for firms to obtain information and submit their applications (box 3.1).

Second, the Internet can increase firm entry by altering the cost structure of businesses, changing fixed costs into variable costs. New firms can replace the investment in a storefront with online platforms to sell their products, replace IT equipment and software with cloud computing, and replace offline labor with online workers. By doing so, firms can control the quantity of inputs they need and can scale their operations up or down according to the business cycle. In addition, replacing fixed capital investment with variable costs confers a tax advantage, as variable costs are considered operating expenses, which are tax deductible in most countries.

For a small fee, online platforms such as Amazon and eBay make it easy for firms to list their products and handle the logistics of shipping goods. Thus, the number of new firms, or firm births, is higher in the online market than in the

FIGURE 3.12
Few firms in Kazakhstan and Kyrgyzstan use the Internet to interact with the government

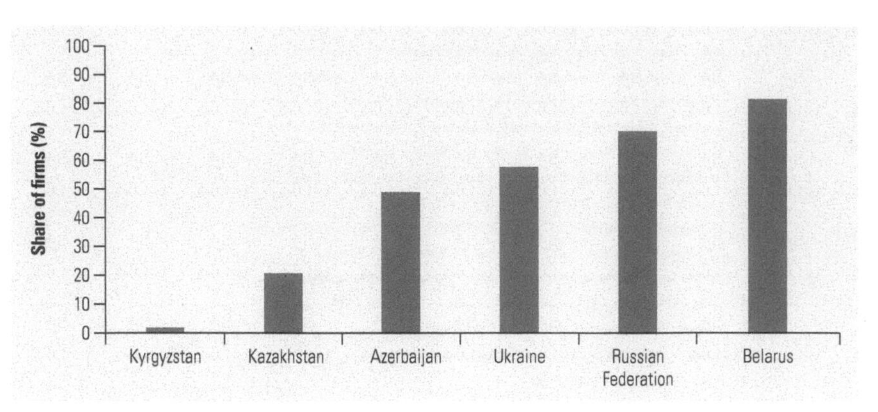

Source: UNCTAD Business Use of ICT.
Note: The data on government interaction does not specify the type of interaction the firm has with the government. The data from UNCTAD come from 2012 or 2013 except for the Kyrgyz Republic, which are from 2009.

BOX 3.1 Saving Firms Time and Money with an Online Registry for Licenses and Permits in Bosnia and Herzegovina

The World Bank Group worked with the government in Bosnia and Herzegovina to simplify their regulations between 2007 and 2014. An e-register website was launched that lists all of the application procedures for permits and licenses, including the required documentation, responsible agencies, deadlines, and fees. The website enhanced the transparency and predictability of administrative procedures and rules for businesses and reduced the scope for administrative discretion of municipal and cantonal governments.

The e-register has helped to reduce the time spent on applications for many firms. The Evaluation Society in Bosnia and Herzegovina (2015) evaluated the project through interviews and surveys of users. The following are two responses on their impressions of the e-registers:

> Always it is better from your house and office to do it than going to municipality ... I would give the grade "A" to such an idea.
>
> *Mr. Darko Drobić, from Eko-Metals firm (Bijeljina)*

> I think the situation should be improved further. Sometimes it is time-consuming to collect all necessary documents and to certify them ... Considering e-register I would give grade "A" to the idea.
>
> *Ms. Indira Huskić, a certified accountant from Tešanj*

The Evaluation Society found that the e-register reduced the costs and time spent for many firms: 29 percent of firms reported that costs were reduced, 66 percent of firms reported that time spent was reduced, and 69 percent of firms reported that the procedures were less complicated after the establishment of the e-register.

offline market. In many countries, especially in Western Europe, there are more firm births on eBay than in the offline market (figure 3.13).

Countries where the cost of starting a business is high tend to have more firms using the Internet to start a business. For example, the cost of starting a business is high in Cyprus (12.6 percent of per capita income), Malta (11 percent), and

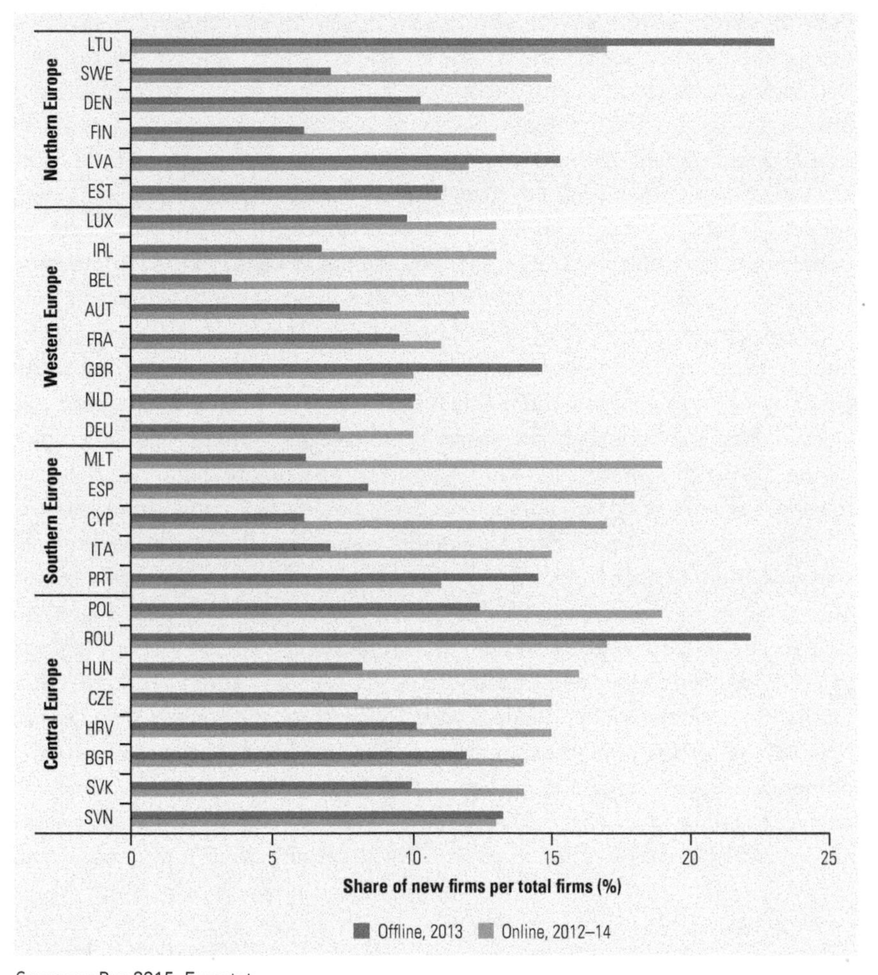

FIGURE 3.13
It is easier for new firms to enter the online market than the offline market

Sources: eBay 2015, Eurostat.
Note: Online firm birth is the share of new firms of all firms on eBay and the new firms are measured as a firm that enters the market and sells more than US$10,000. Offline firm birth is taken from Eurostat and is the share of new firms in the total number of enterprises.

Hungary (8.3 percent) (figure 3.9), and these countries also have among the highest number of firms entering the online market rather than the offline market: Malta with 12.8 percentage points more firm births online than offline, Cyprus with 11 percentage points, and Hungary with 7.8 percentage points. Conversely, the costs of starting a business in Latvia, Lithuania, and Romania are low (between 0.7 and 3.6 percent of per capita income), but, despite fast and affordable Internet access and a high rate of firm Internet use, the rates of firms starting businesses are higher in the offline market. This relationship is more ambiguous in the Western European markets, where the costs of firm entry are low, but the number of firm births is higher in the online market.

While adopting new technologies can be expensive, as new firms have to invest in hardware, servers, software, and specialized staff, the costs of adopting technology are decreasing as hardware and technology services become cheaper. New innovations based on the Internet, such as cloud computing, give firms the flexibility to scale their operations up or down according to business needs. Firms do not need to make large investments in IT capital and equipment that might be wasted capacity if demand shrinks or if the IT capital becomes obsolete. A large share of firms in ECA consider that cloud computing increases operational flexibility and reduces the amount of capital investment. While all firms will benefit equally from this flexibility when substituting cloud computing services for simpler purposes such as e-mail, smaller firms are more likely to benefit most from this flexibility when they use cloud computing for more complex purposes, such as hosting their databases and accessing financial and CRM software. Larger firms already have the resources to develop these services in-house and do not need to rely on cloud computing. Among firms using advanced cloud computing services, more small-size firms than medium and large-size firms realized the benefits of the flexibility to scale their operations up or down (figure 3.14). On average, there is a 10 percentage points difference between the share of small and large-size firms in Central Europe, Estonia, and Lithuania.

Third, the Internet can offer new avenues to enter traditional sectors such as transportation, finance, and retail, thus expanding the boundaries of a sector. Technologies such as IoT allow firms to transform the sale of goods into a service and to manage workers and monitor objects from afar. IoT covers a range of products, from mundane household items such as "smart" light bulbs to futuristic applications like driverless cars. There does not seem to be a limit to how IoT will propagate across sectors, as long as firms discover a customer demand. IoT generates a large amount of data ("big data"), and firms can analyze the data to provide new services, such as predictive maintenance technology, which enables firms to maintain equipment when it is needed and

FIGURE 3.14
More small firms using cloud computing experience the flexibility to scale up or down

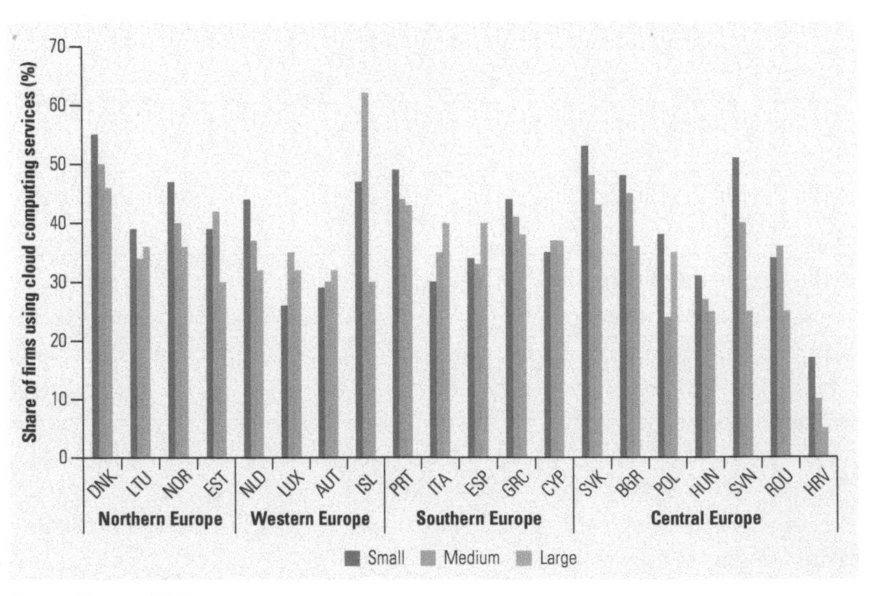

Source: Eurostat 2014.
Note: Small-size firms have 10–49 employees, medium-size firms have 50–249 employees, and large firms have more than 249 employees.

prevent unexpected machine failure. Other examples include home appliance manufacturers that offer grocery services through the IoT sensors in its refrigerators, and Rolls-Royce, which manufactures aircraft engines and leases them to customers on a subscription model, charging customers for the engine's use, repairs, and additional services (Marr 2015).

Some of these technologies will be easier for small firms to implement than others. Accessing government websites and using existing Internet platforms to sell merchandise is relatively inexpensive. By contrast, implementing cloud computing and IoT are costly. For example, cloud computing requires new product hardware, embedded software, electronics to provide connectivity, software for the connected product to communicate with the cloud that is running on remote servers, security tools, external information sources (for information that will affect product capabilities, such as weather and traffic), and integration with internal business systems. A small firm or new start-up may find it difficult to obtain the funding or the skills to make these investments.

Despite these benefits, the aggregate effects of the Internet on firm entry may be limited by other factors in ECA, as there is no clear correlation between changes in Internet penetration and changes in firm entry (figure 3.15). While Internet penetration increased about 20 percent in the following five countries, firm entry decreased 11 percent in Serbia and 18 percent in the former Yugoslav Republic of Macedonia, but increased 7 percent in Turkey, 12 percent in Albania, and 81 percent in Azerbaijan.

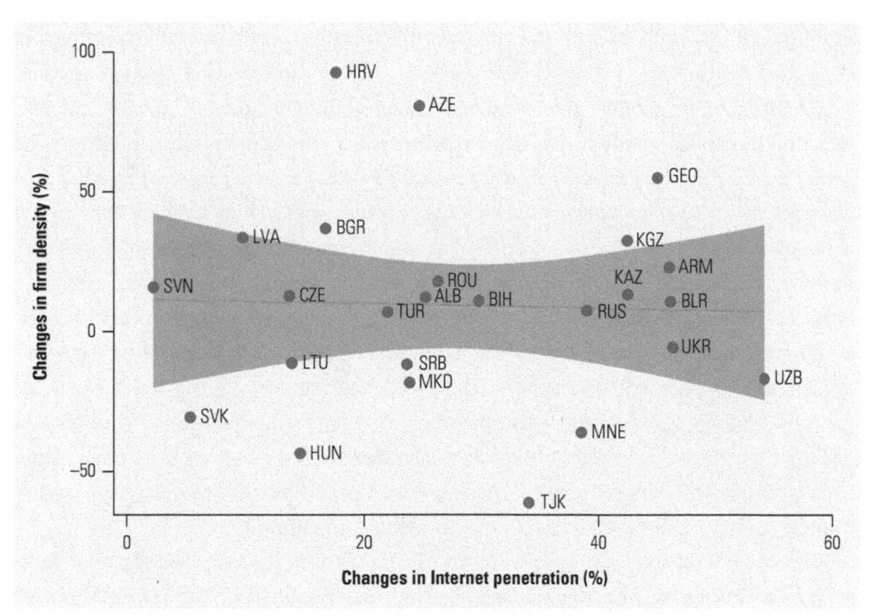

FIGURE 3.15
The correlation between changes in Internet penetration and firm entry is close to zero

Sources: World Bank Entrepreneurship Database and World Development Indicators.
Note: Shaded area represents the confidence interval of the fitted line. Percentage changes are calculated between 2010 and 2014. The most recent data were used for those countries without firm density data in 2010 and 2014. Firm density is measured as number of new firms registered per 1,000 people. Internet penetration is the number of Internet users per 100 people. Countries are represented with ISO alpha-3 codes. ISO = International Organization for Standardization.

Constraints on Firms' Use of the Internet to Increase Productivity and Entry

Two broad sets of issues can reduce the impact of the Internet on the productivity and entry of firms. One relates to factors that are largely within the control of the firm. Strong management is required to ensure that new technologies are incorporated effectively into business processes. Organizational change, as well as a new strategy and vision for the company, is often needed. A study of U.K. firms finds that IT investments have a positive impact on firm productivity, but the effects are larger when this investment is complemented with organizational change (Crespi, Criscuolo, and Haskel 2007). Similarly, the combination of skilled labor *and* firm reorganization explains the returns to ICT investment in Italian manufacturing firms (Bugamelli and Pagano 2004). The extent and type of organizational change will depend on what type of technology the firm is implementing. For example, introducing a website to provide customer service might require modifying the existing customer service functions, while introducing more collaboration through a cloud-based office software requires large changes to processes, procedures, and workflows. A study of French firms shows that adopting an ERP system requires redesigning the organization to focus on core competencies, quality improvements, and a decentralized decision-making structure (Bocquet, Brossard, and Sabatier 2007).

Studies have shown that good management practices are a complement to technology adoption (Bloom and others 2013; Correa, Fernandes, and Uregian 2010). Since the Internet is a skills-biased technology, a firm adopting new Internet technologies needs to replace less skilled workers with more skilled workers and to upgrade the skills of workers who can be retrained. A study of the effects of broadband Internet in Norway finds that a 10 percentage point increase in broadband Internet improved the productivity of skilled workers 0.6 percent and reduced the productivity of unskilled workers 0.2 percent (Akerman, Gaarder, and Mogstad 2015). The reason is that broadband Internet complements nonroutine abstract tasks usually performed by skilled labor and is a substitute for routine tasks performed by unskilled labor.[6] This improvement will require management to set incentives that reward high performers and remove low performers. Similarly, a decentralized decision-making structure is more likely to allow workers to incorporate the Internet into operations in a way that is more suitable to their workflow and context (Acemoglu and others 2007). U.S. multinational firms are found to be more IT intensive and to generate more productivity improvements from their IT capital than non-U.S. multinational and domestic firms in seven European countries because they imported U.S. management practices that reward high performance, remove poor performers, and develop and retain talent.[7]

Governments also can help firms to improve their ability to adopt new technologies. Programs that provide free consulting services to improve management practices have been shown to improve productivity and quality among firms. A study that randomly assigned Indian firms to receive consulting services shows that such programs can increase firm productivity 17 percent and that these firms grow faster than firms in the control group (Bloom, Sadun, and Van Reenen 2013). But management programs need to be tailored for each firm and can be expensive to implement.

Another set of issues relates to factors that are outside the control of the firm. A weak education system or restrictive labor regulation makes skilled professionals scarce or costly. Inadequate infrastructure may impose high costs for using the Internet and related services. Import tariffs or subsidies may reduce the competitive pressures necessary to force firms to adopt new technology. Entry barriers put in place by governments to protect certain interests may also prevent firms from entering the market.

A lack of skilled manpower to introduce and use new technologies is a binding constraint on firms. Firms in the ECA countries often face high costs to hire IT professionals due to a lack of supply of skilled professionals or restrictive labor regulations. Firms require workers with a level of skill and technology proficiency to use the Internet effectively and specialized IT professionals to integrate the technology into business operations. Cumbersome labor regulations can dampen the positive impact of the Internet, as firms are less likely to hire skilled workers and to invest in technology.[8] The negative effects of poor labor market regulations are more pronounced for less productive firms (Paunov and Rollo 2015).

Ultimately, the competitive environment, rather than the availability of skilled labor or the quality of management, is the most important factor in a firm's decision to adopt and use the Internet more intensively. Indeed, the slowdown in European productivity in the late 1990s and early 2000s was due to regulatory barriers rather than a lack of ICT investments (Gordon 2004). Competition creates the incentives for incumbent firms to adapt and innovate, to find opportunities and adopt new technology, and to be efficient and remain competitive. A rich literature examines how competition, especially in the product market, improves the innovative activities of firms and consequently their productivity.[9] And the same competitive pressures that encourage firms to adopt new technology can lead them to adopt better management practices (Bloom and Van Reenen 2007; Bloom, Kretschmer, and Van Reenen 2009). A study examining the importance of organizational change to IT adoption finds that competition affects organizational change (Crespi, Criscuolo, and Haskel 2007).

Greater foreign competition due to a lowering of trade barriers can result in more firms adopting the Internet and using the Internet more intensively.[10] Faced with more import competition from China, firms in 12 Western European countries increased the intensity of IT use and raised productivity levels.[11] An increase in import penetration from China increases the intensity of ICT use of firms by 36.1 percent and TFP by 25.7 percent. The increase in ICT intensity and TFP is still substantial after accounting for industry trends and domestic production. Between 2000 and 2007, the increase in Chinese imports explained 13.5 percent of the growth in ICT intensity and TFP, with most of the growth explained by increases within firms instead of resource allocations between firms and exits from the market. Exporting can also increase productivity, as firms learn through exposure to more demanding customers.

Greater local competition also provides an incentive for firms to adopt the Internet and use it more intensively. When a firm adopts a new technology, it can increase its productivity, sell to its customers at a lower price, and gain market share. Other firms will follow the market leader by adopting the technology or using better technology to reclaim market share. The use of mobile banking in

Turkey is an illustrative example. Turkey has one of the best mobile banking services in the world, ranking above that of many industrial countries with regard to the services offered. Turkish banks face a very competitive market, with 47 deposit banks with more than 11,100 branches.[12] Many banks are constantly innovating and adapting to consumer demands to remain competitive. DenizBank was the first bank in the world to allow its customers to access their credit and debit card information on Facebook in 2012 (Ensor 2012). This innovation was followed up with the iGaranti app, which is the first mobile banking app that combines a digital wallet, savings, and loan application within the social media platform and allows users to send payments to friends through Facebook (Accenture Consulting n.d.).

There is a strong positive correlation between use of the Internet among firms and level of competition in ECA. Countries with higher intensity of domestic competition have higher percentages of firms using the Internet, measured as firms using websites (figure 3.16). There is generally a moderate level of competition in ECA, as most countries score above 3.5 out of 7 for the intensity of local competition. While there is wide variation across countries, a general pattern emerges for the subregions within ECA. Most countries in Western and Northern Europe have high levels of competition and high levels of Internet use by firms. Conversely, countries in Central Asia, South Caucasus, and Western Balkans have low levels of competition and low levels of Internet use by firms.

This relationship between competition and the Internet is not straightforward. First, causation works both ways: competition encourages firms to use the Internet more intensively, and more Internet use by firms can increase the

FIGURE 3.16
Internet use by firms is positively correlated with the intensity of local competition

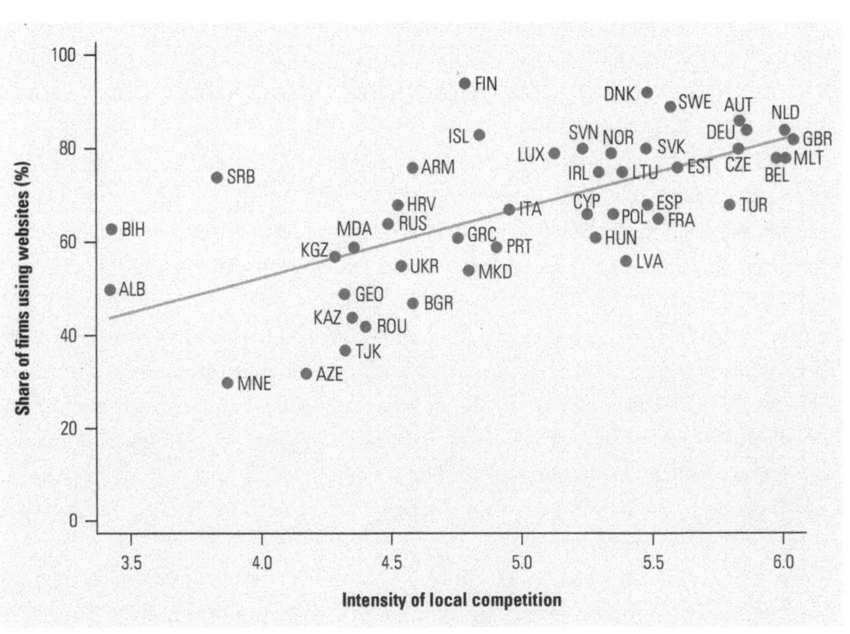

Sources: Calculations based on data from World Bank Enterprise Survey, Eurostat, and WEF Global Competitiveness Report.
Note: The 2013 data are used for most countries, except for Russia, which has Internet use data from 2012, and for Tajikistan, which has competition data from 2014.

level of competition. Second, the relationship between competition and ICT adoption and innovation is influenced by the market share and level of technology of firms. Large incumbent firms have more incentives to innovate to exert their market dominance. While industries facing competition innovate, within these industries the firms with high market shares also engage in more innovative activities (Blundell, Griffith, and Van Reenen 1999). Moreover, competition has a heterogeneous effect on innovative activities: the entry of technologically advanced firms will spur incumbent firms close to the technological frontier to innovate and increase their productivity, but discourage efforts to increase productivity in incumbent firms further from the frontier (Aghion and others 2009).[13] The interactions between level of technology, market share, competition, and innovation require more nuanced policies to increase Internet use among firms in ECA.

Governments' Policy Priorities Depend on the Level of Competition

In order to reap the full potential of the Internet for firm productivity and firm entry, ECA countries should focus on improving the business environment to promote competition. There are many approaches to improving the business environment, and each government will need to prioritize reforms based on what is most suitable for its economy. ECA countries can be classified into three groups based on the policy priorities pertinent to their situation: emerging, transitioning, and transforming. A positive relationship is established between use of the Internet by firms and an overall measure of competition, which captures aspects of local and foreign competition (figure 3.17). The countries can be classified into the three groups based on the interaction of these two variables. Countries at the lowest level of competition (below a score of 4) are classified as emerging, countries at a more competitive level (above a score of 4.7 or the 70th percentile) are classified as transforming, and countries in between are classified as transitioning.[14] The countries and the policy priorities in each group are presented in table 3.1.

Reducing tariffs on ICT equipment can improve firms' ability to use the Internet. Many ECA countries have signed the International Technology Agreement (ITA) under the World Trade Organization, where the country commits to remove import tariffs unilaterally on ICT equipment. Eight ECA countries still have import tariffs on ICT products: Armenia and Montenegro, both WTO members, are not in the ITA; and Azerbaijan, Belarus, Bosnia and Herzegovina, Kosovo, Turkmenistan, and Uzbekistan are not WTO members. The tariff rates on basic ICT products such as computers and mobile phones in these eight countries can range from 0.13 to 30 percent (table 3A.2 in annex 3A). These countries have to focus on reducing import tariff rates on ICT products to ensure that their firms do not face high costs to adopt new Internet technologies.

Nevertheless, sectoral restrictions on investment and entrepreneurship remain in several ECA countries. Many ECA countries, even the more developed ones such as France and Italy, have restrictions on foreign direct investment (FDI) that prevent foreign firms from entering certain sectors (figure 3.18). Among the less

FIGURE 3.17
Emerging, transitioning, and transforming digital countries for competition policies

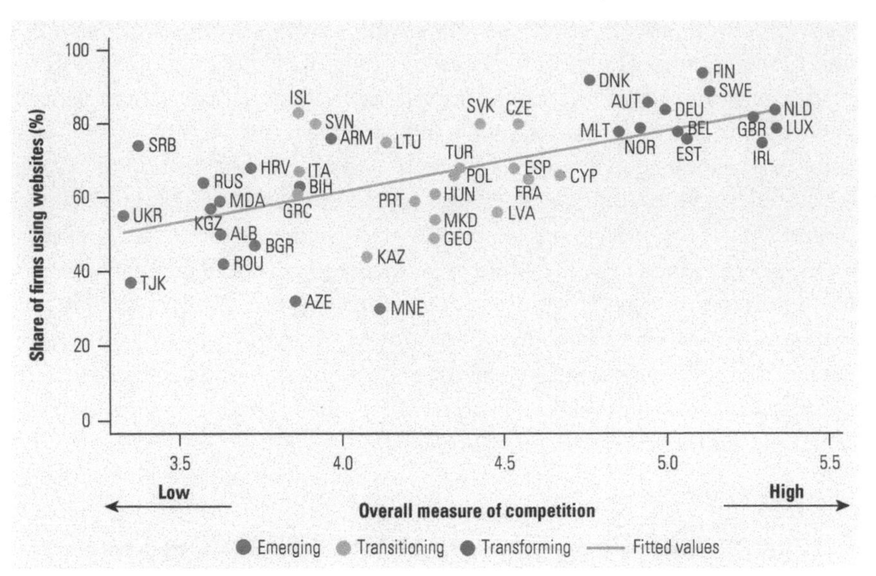

Sources: Calculations based on World Bank Enterprise Survey, Eurostat, and WEF Global Competitiveness Report. All data on firm use of the Internet are from 2013, except for the Russian Federation.
Note: Measure of competition was calculated taking the average scores for eight questions on the local and foreign competition in the WEF Global Competitiveness Report. The four questions on local competition are the (1) intensity of local competition, (2) extent of market dominance, (3) effectiveness of antimonopoly policy, and (4) effect of taxation on incentives to invest. The four questions on foreign competition are the (1) prevalence of trade barriers, (2) prevalence of foreign ownership, (3) business impact of rules on FDI, and (4) burden of customs procedures. All questions are scored from 1 to 7, where 1 is the lowest and 7 is the highest intensity or prevalence. Countries are represented using their ISO 3-alpha code. FDI = foreign direct investment; ISO = International Organization for Standardization.

TABLE 3.1 Policy Priorities for Emerging, Transitioning, and Transforming Digital Countries

Group	Countries	Policy priorities
Emerging	Albania, Armenia, Azerbaijan, Belarus, Bosnia and Herzegovina, Bulgaria, Croatia, Kosovo, Kyrgyz Republic, Moldova, Montenegro, Romania, Russian Federation, Serbia, Tajikistan, Turkmenistan, Ukraine, and Uzbekistan	• Reduce trade and foreign direct investment barriers • Reduce domestic restrictions on sectors • Facilitate entrepreneurship
Transitioning	Cyprus, Czech Republic, Georgia, Greece, Hungary, Iceland, Italy, Kazakhstan, Latvia, Lithuania, FYR Macedonia, Poland, Portugal, Slovak Republic, Slovenia, and Turkey	• Introduce competition policies to prevent antitrust behavior • Enforce competition policies
Transforming	Austria, Belgium, Denmark, Estonia, Finland, France, Germany, Ireland, Luxembourg, Malta, the Netherlands, Norway, Spain, Sweden, and United Kingdom	• Adapt the regulatory environment to deal with digital economy issues, such as multi-sided markets and cybersecurity

Note: Countries are grouped based on their overall level of competition, which is presented in figure 3.17. World Economic Forum data are not available for Belarus, Kosovo, Turkmenistan, and Uzbekistan. These countries are placed in their group by comparing their Economic Bank for Reconstruction and Development transition indicators and the overall distance to frontier score in the World Bank Doing Business indicator.

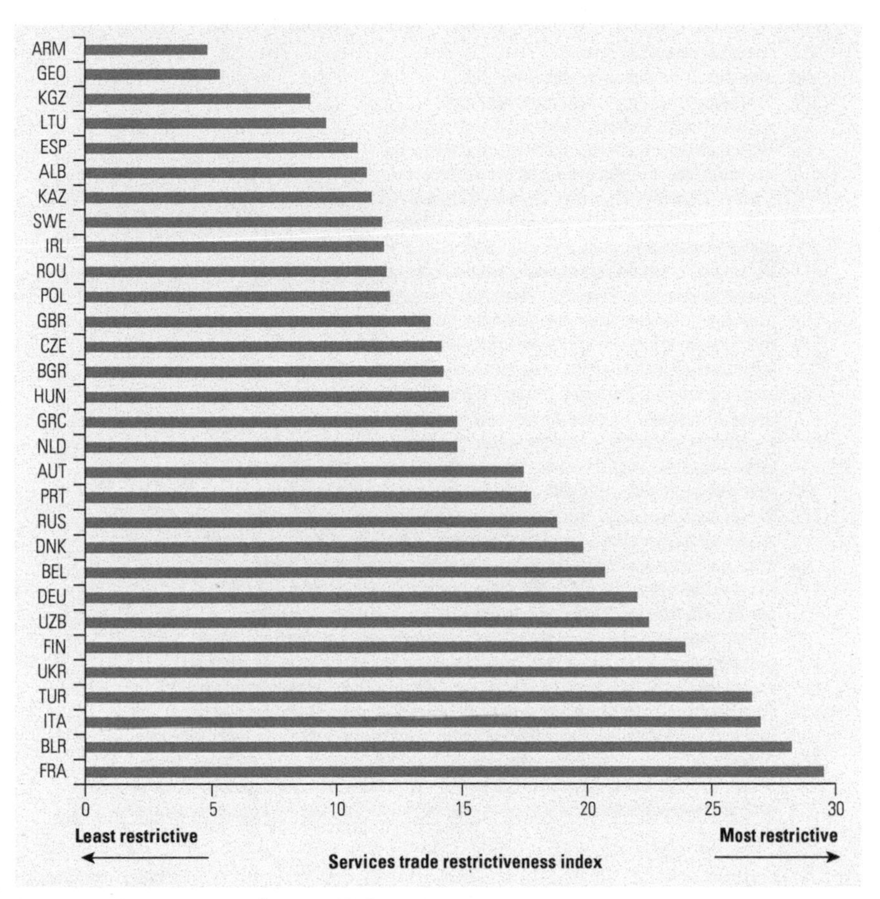

FIGURE 3.18
Restrictions on FDI remain in many ECA countries

Source: Borchert, Gootiiz, and Mattoo 2012.
Note: The index is for Mode 3 services or foreign direct investment.

developed countries in ECA, Belarus, Turkey, Ukraine, and Uzbekistan have the highest levels of restrictions on FDI. Many barriers also remain in the services sector, where efforts to liberalize the sector have stalled. Compared with the goods sector, where liberalization involves reducing tariff rates and easing border controls, the barriers in the services sector are more complex, difficult to quantify, and therefore difficult to remove. Many ECA countries still have high barriers to entrepreneurship in the services sector (figure 3.19). No country has an unrestricted services sector, although Russia comes close, and many countries have a services sector that scores above 3.

Emerging digital countries not only need to introduce regulations and policies to facilitate new businesses, but also need to focus on how these regulations and policies are implemented. Many league tables rank countries on their business environment, and countries compete to improve their rankings. But some of these rankings do not capture the reality facing many firms and may suggest different interpretations of what policy reforms are important (Hallward-Driemeier and Pritchett 2015). There is a large variance between what is captured in the World Bank's Doing Business surveys (estimates by experts such as lawyers and accountants on a typical transaction or the de jure business environment) and what is captured in the Enterprise

FIGURE 3.19
Barriers in the services sector are present in many ECA countries

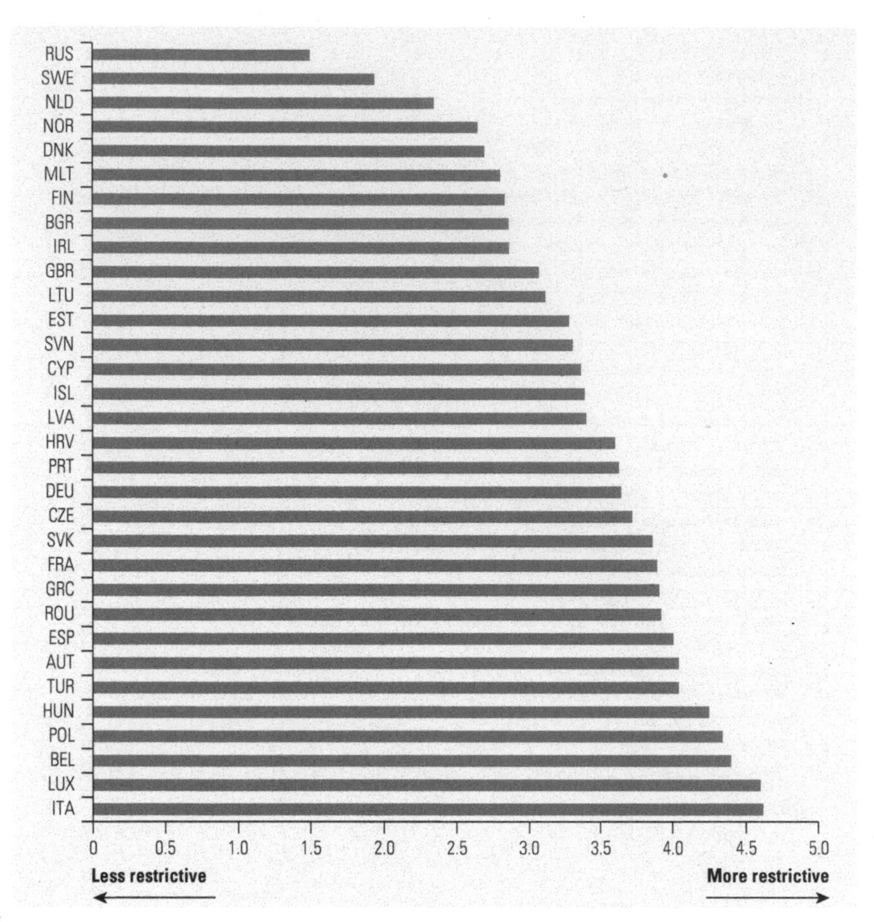

Source: OECD Barriers to Entrepreneurship 2013.

Surveys (surveys of firms on what happened or the de facto business environment). On average, firms spend 10 more days to establish a business than what is estimated by the local experts (figure 3.20).

Transitioning countries have higher levels of competition, as they have started to tackle the fundamental issues faced by many developing countries. Transitioning countries should focus on implementing a strong regulatory framework for competition policy and establishing an agency to enforce these rules. Almost all of the ECA countries have some form of competition policy and a competition agency; Belarus and Turkmenistan are the only two countries without a competition agency. But simply having a competition policy and a competition office does not ensure that the market is regulated properly and the policy is enforced. How competition policy is applied varies widely in the less developed ECA countries (figure 3.21). Only a few countries enforce the competition policy to reduce the market power of incumbents and to promote a competitive business environment—that is, countries with a score of 3 or more in the European Bank for Reconstruction and Development transition indicators for competition policy. None of these countries has a competition policy framework with significant and effective enforcement actions—that is, a score above 4.

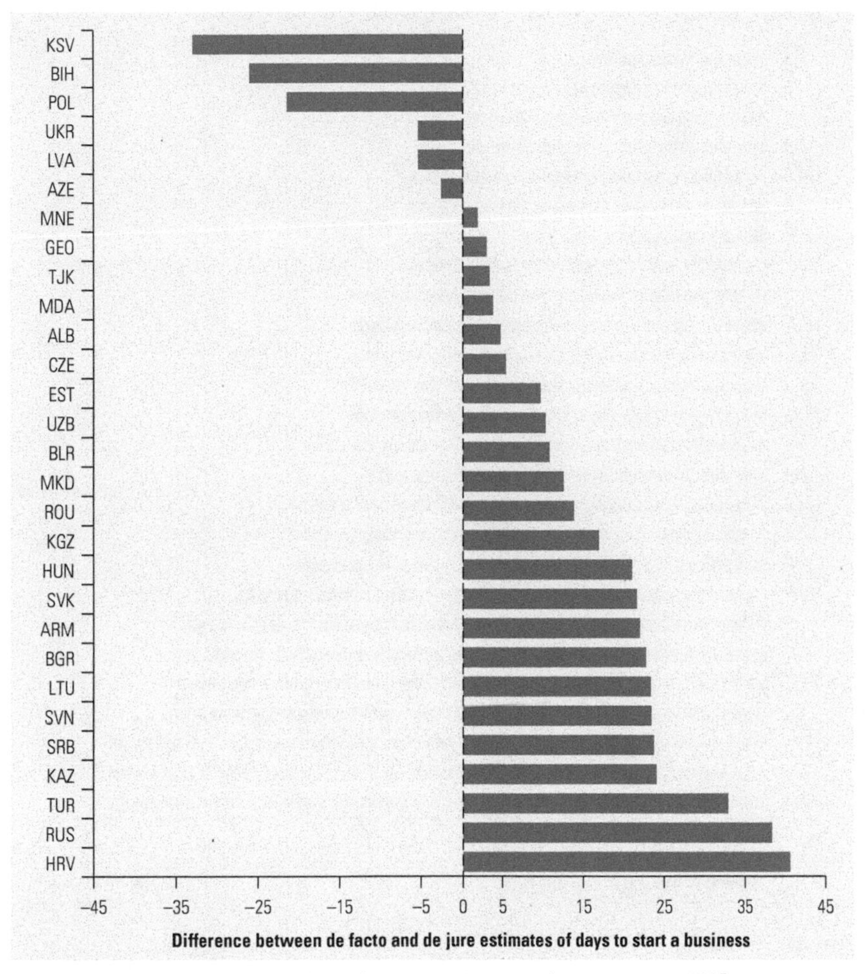

FIGURE 3.20
Firms need more time to start a business than experts estimate

Sources: Calculations based on World Bank Enterprise Survey and Doing Business 2013.
Note: The measure is calculated by taking the difference between the de facto estimate or the "days it takes to obtain an operating license" for the average firm in the Enterprise Survey and the de jure estimate or the "days it takes to start a business" in Doing Business surveys.

While the state of competition policies in the EU is high, the EU countries should continue to strengthen efforts to foster a competitive environment. Competition policies and enforcement are managed at the supranational level by the European Commission, but there is room for interpretation by countries when the policies are translated to national regulations. Thus, there is wide variation in how firms perceive the competition policies across the EU (figure 3.22). Despite having a common competition policy framework, countries in Southern Europe have a weaker competition environment, with an average score of 4, than countries in Western Europe, with an average score of 5.1. None of the countries is viewed as having policies that are at the highest level of effectiveness (score of 6 or 7) in promoting competition.

Transforming digital countries have the highest level of competition among the ECA countries and should focus on responding to new regulatory issues that emerge from the rapidly shifting technology landscape. The Internet has changed

FIGURE 3.21
Many ECA countries have a weak competition policy framework

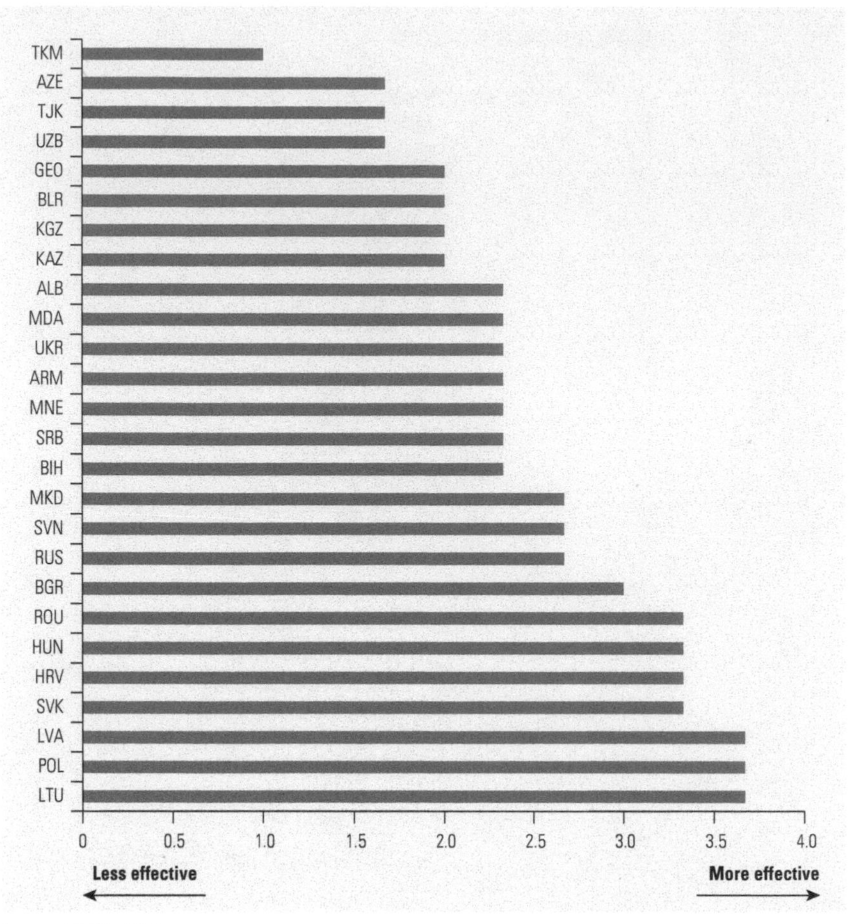

Source: EBRD 2014.
Note: EBRD Transition Indicators on Competition Policy. A larger score means that the country has a more effective competition policy. 1 means no competition legislation and >4 means effective enforcement of competition policy and unrestricted entry to most markets.

business models and how firms operate in markets. Network effects are inherent in the use and deployment of the Internet, and governments need to adapt their competition policies accordingly.

A major outcome of the Internet is the creation of new market structures, called multisided markets, which will affect competition policies. These multisided markets or platforms are different from the usual markets, as there are often two (or more) distinct groups of customers, network effects are present within and across these groups, and an intermediary often brings these customer groups together. Online marketplaces, such as eBay, are a good example that has two groups of customers—buyers and sellers. Two types of network effects are present. First there are usage externalities to the buyers and sellers who benefit from using the marketplace. That is, without an online marketplace, it would be difficult and expensive for these buyers and sellers to get together. Second, there are membership externalities. The system is more valuable to buyers when more different sellers are on the marketplace and to sellers when more different buyers are on the marketplace.

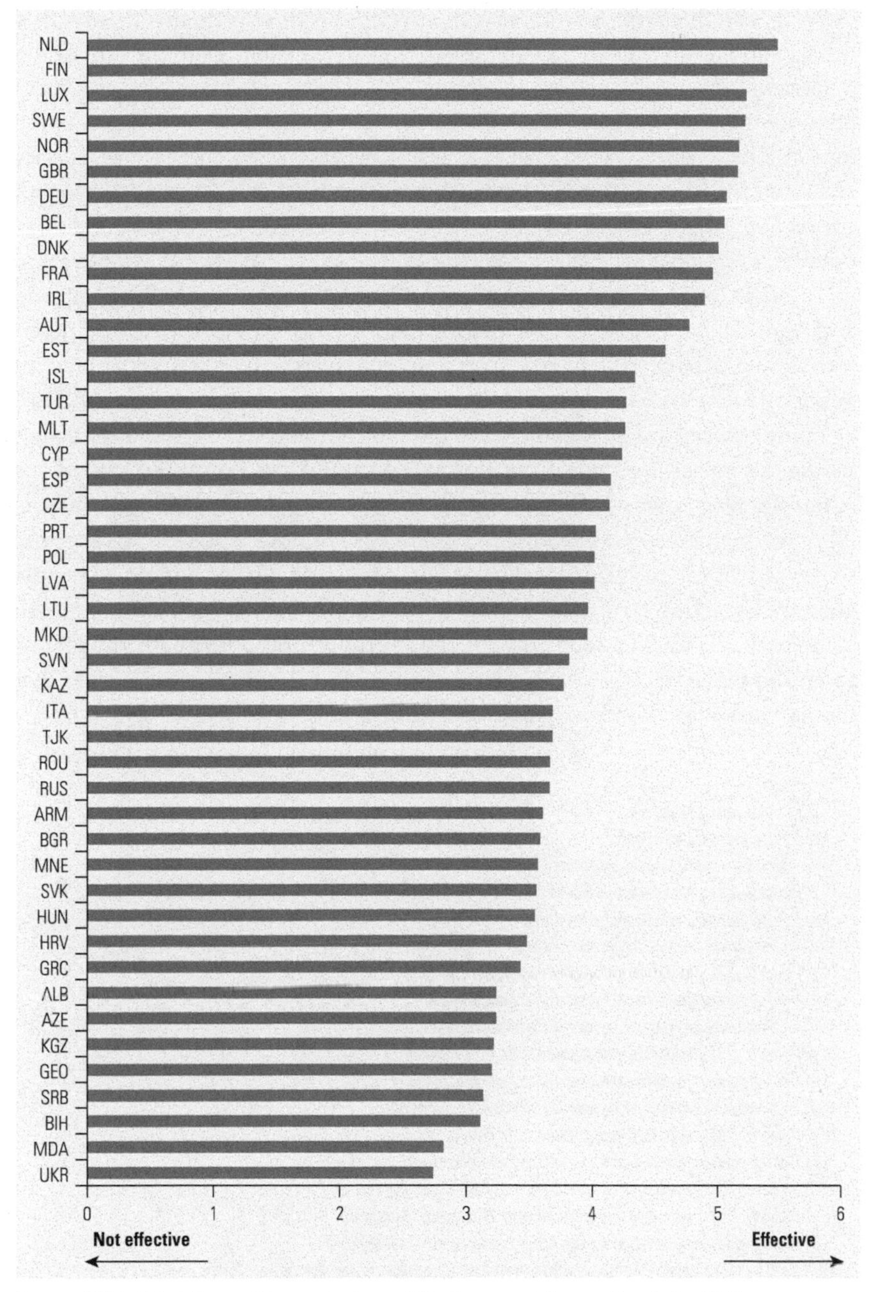

FIGURE 3.22
How firms perceive competition policies varies widely in ECA

Source: WEF Global Competitiveness Index on Effectiveness of Competition Policy 2015.

Competition policies will have to adapt to the emergence of multisided markets. In standard economic models, prices are assumed to be equal to or close to the marginal costs of operations; if prices are higher than marginal costs, the firm has some form of market power. With a multisided market, prices will not necessarily equal marginal costs, as it may be efficient (to maximize the size of the market) for the intermediary or platform to charge certain groups of customers higher prices and others lower prices or even to offer the product for free.[15] Competition agencies cannot assume that the optimal price is equal to the marginal cost that firms face in these multisided markets. Since the intermediary is bringing many customer groups together, a price change in one side of the market will affect participation in another side. The competition agency cannot consider the price increase without considering the effects on other parts of the market (OECD 2009). When market power is excessive, the competition agency should step in to foster a more competitive environment. But it may be difficult for competition agencies to determine the market boundaries and enforce the competition policy in a multisided market.[16]

Another issue that transforming countries, and increasingly emerging and transitioning countries, will need to deal with is cybersecurity. Compared with simpler uses of the Internet, firms may face higher risks of security breaches in adopting the newer Internet technologies such as cloud computing and IoT, as many devices are connected over the Internet and data are shared across different business functions. Concerns over security have prevented many firms in transforming countries from using cloud computing (figure 3.23): on average, 24 percent of firms in transforming countries do not use cloud computing due to concerns over security breaches. While it is mostly up to the firm to handle security breaches by implementing IT security policies, governments can increase their efforts to educate firms on IT security methods, create regulations to ensure that most firms have a minimum level of IT security, and ensure that the national Internet infrastructure is not compromised.

FIGURE 3.23
Many firms across ECA do not use cloud computing because of security concerns

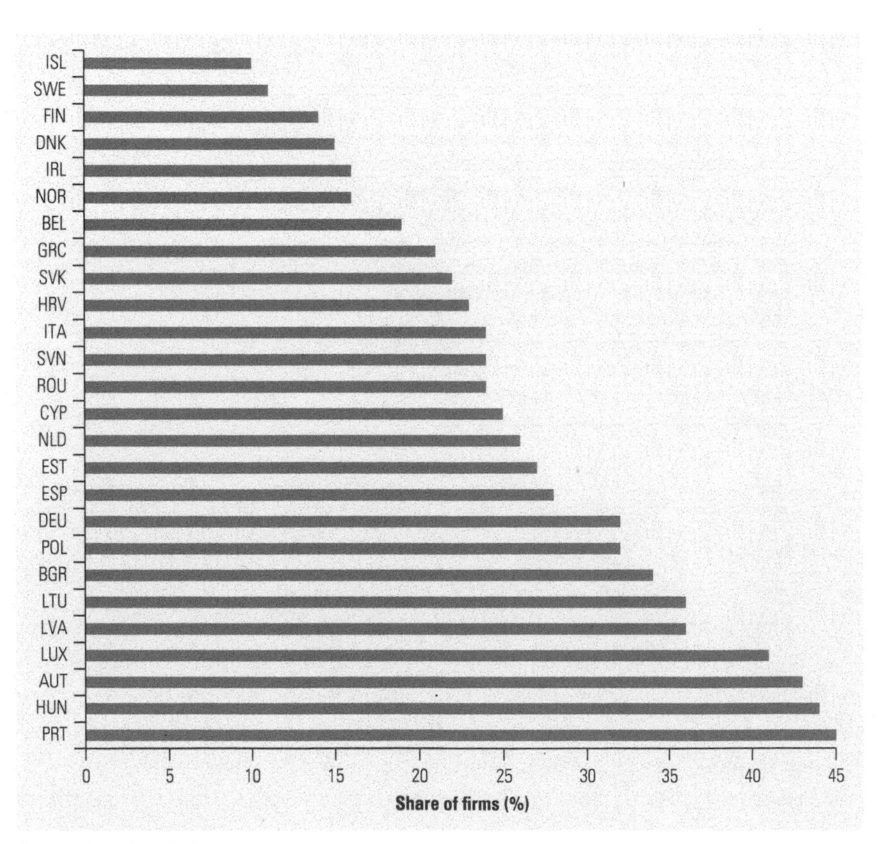

Source: Eurostat 2014.
Note: Data for France, Malta, and the United Kingdom are unavailable.

The Internet Is Generating New Challenges for Policy

The Internet can present new challenges to governments in fostering a pro-competitive environment. The network nature of the Internet and the embedded switching costs can promote concentration, allowing some firms to grow large and dominate a sector. Governments should, however, be less concerned about concentration as long as competition policies are effective and competition agencies enforce them. In multisided markets, intermediaries are often concentrated due to the presence of network effects. Competition agencies need to recognize that conditions are different in these multisided markets than in standard markets and that competition among platforms and intermediaries may not necessarily improve the welfare of consumers and firms (Monopolkomission 2015). For example, platform competition can lower prices and make it easier for firms to use the platforms to enter new markets. But customers can be split across many different platforms, resulting in a small customer base on each platform, and sellers will have to participate in many platforms to generate enough revenue.

Competition can motivate firms to adopt the Internet and use it more intensively, but collective action issues exist that cannot be solved by competition alone. Many productive uses of the Internet may not be implemented because of the inherent network effects in the technology. A firm may think that it is useful and efficient to adopt a technology, but if there are no corresponding users, then the benefits of implementing the technology will be negligible, as in the case of e-invoicing systems. In these situations, governments may introduce standards or norms that can encourage more firms to use a technology. For example, requiring a standard data format is likely to increase the use of data exchanges.

Governments also need to recognize that data are currency in the digital economy and that firms can use this information to reduce consumer welfare. First, when data are not portable, consumer choices are limited. Firms can use proprietary standards, such as the smart lighting system sold by Philips that is incompatible with light bulbs sold by other companies (Schneier 2015), which prevents consumers from switching products. Second, data can be used to create highly customized products, which makes it difficult for consumers to find a substitute product. Buyer power is reduced, and firms can take advantage of consumers by charging high prices. Third, data can allow firms to coordinate their market responses with "robo-sellers," computer systems that use data and pricing algorithms to set market prices. These computer systems can help firms to form oligopolies and coordinate their prices. The first antitrust case was brought by the U.S. competition authorities against an online seller that used robo-sellers to collude with other sellers.[17] In order to ensure that firms do not use data to disadvantage consumers, countries should ensure that data standards are proper and that data interoperability gives users more control over their information and makes them more vigilant about unscrupulous uses of data.

In addition, the data collected by firms raises concerns about online privacy. Many firms collect information through websites and connected devices and use Internet firms (such as Google and Facebook) to target their customers.

Consumers are often unaware of the information they reveal to companies online or the information they provide through many free online services and mobile phone apps. The choice to give up some online privacy is an individual's decision in "payment" for the service. Providing the data is not necessarily bad: the information is useful to provide targeted online searches and suggestions for consumers and can help firms to improve the quality of services. It becomes a concern when the data are used for malicious purposes, such as to target the poor by offering loans with nontransparent or misleading terms[18] or the vulnerable by stealing their online identity to access sensitive information and financial accounts. These concerns about online privacy can prevent consumers from participating in the digital economy.

Faced with these new challenges, governments will have to balance the need for regulation with the desire for innovation. New uses of technology to circumvent competition among firms and harm consumers can foster the need to regulate these new sectors. This need may be further amplified in ECA, as consumers are worried about privacy and legal rights and may be more reluctant to use the Internet if proper regulations are not in place. However, firms will find it difficult to innovate and respond to fast-changing trends if there are too many regulations. Countries will need to determine how much weight to place on regulation versus innovation in formulating their policies.

The Internet holds a lot of potential to increase firm productivity and firm entry. Many ECA countries need to ensure that there is a pro-competitive business environment with fewer entry barriers, proper regulations, and enforcement to harness the full potential of the Internet. Each country has different policy priorities based on the level of competition in the local market; only by ensuring that firms use digital technology more productively can a country reap the full benefits to firm productivity and entry.

Annex 3A. Selected Firm-Level Studies

TABLE 3A.1 Selected Firm-Level Studies Using European Data

Study	Data and method	Findings
Bloom, Sadun, and Van Reenen 2005	• Unbalanced panel of 7,000 U.K. establishments from 1995 to 2004 • ICT measure constructed from ICT capital measured in surveys • Estimation of panel production functions and TFP regressions	ICT capital has a significant impact on productivity. The effect is greater for U.S. than for non-U.S. multinationals or domestic firms. The effect among U.S. multinationals is stronger in IT-intensive industries.
Bloom and others 2005	• Panel of 3,000 U.K. firms from 1994 to 2000 • ICT measure constructed using the value of IT hardware and the number of personal computers per employee • Estimation of production function and tests of heterogeneity of IT impact across different firm characteristics such as size, sector, and time period; tests for spillover effects at the regional and industry level	ICT has a significant effect on productivity, but there is no evidence of ICT spillover at the industry or regional level.

(Continued)

TABLE 3A.1 Selected Firm-Level Studies Using European Data *(continued)*

Study	Data and method	Findings
Forth and Mason 2003	• Panel of 308 U.K. firms from 1997 to 1999 • ICT measure constructed using categorical indicators of different types of ICT • Estimation of relationship using OLS and instrumental variable methods	There is a general positive impact and an interaction with skill shortages.
Hempell 2005	• Cross section of 995 firms in Germany and 972 firms in the Netherlands in 1998 • ICT measure constructed using ICT expenditure • Estimation using a GMM system	The effect of ICT is significant.
Matteucci and others 2005	• Panel of 3,000 firms in Germany from 1997 to 2000, 3,900 firms in Italy from 1995 to 2000, and 2,400 firms in the United Kingdom in 2001 • Construction of various ICT measures, depending on the data: lagged ICT investment for German and Italian firms and duration of Internet access and proportion of workers using a personal computer for British firms • For each data set, use of different estimation methods: OLS regressions with firm fixed effects for the German data; TFP equations for the Italian data; and a cross-sectional Cobb-Douglas production function for the British data	ICT has a significant effect on productivity in manufacturing firms, but not in services firms in Germany and a significant effect on productivity in British services firms. The effect of ICT is weak (but significant) for Italian firms.

Source: Draca, Sadun, and Van Reenen 2007.
Note: GMM = generalized method of moments; ICT = information and communication technology; IT = information technology; OLS = ordinary least squares; TFP = total factor productivity.

TABLE 3A.2 Most Favored Nation Tariff Rates for Non–International Technology Agreement Countries in ECA
Tariff rate (%)

Country	Computers	Mobile phones
Armenia	0.13	3.30
Azerbaijan	3.00	15.00
Belarus	0.13	3.30
Bosnia and Herzegovina	3.44	0.00
Kosovo	2.22	10.00
Montenegro	0.00	0.00
Turkmenistan	0.00	10.00
Uzbekistan	5.00	30.00

Source: Data from the World Integrated Trade Solution TRAINS database, except Kosovo. Tariff rates for Kosovo are from the Kosovo customs website (https://dogana.rks-gov.net/tarik/).
Note: The tariff rates for computers are average tariff rates for all products in Harmonized System code 8471, "automatic data processing machines and units." The tariff rates for mobiles phones are for products in Harmonized System code 8517.11, "line telephone sets with cordless handsets."

Notes

1. Based on the latest data available for these countries from the United Nations Conference on Trade and Development (UNCTAD) business use of ICT data. http://unctad.org/en/Pages/statistics.aspx.
2. Directive 2014/55/EU of the European Parliament and of the Council of 16 April 2014 on electronic invoicing in public procurement.

3. That is, the hydrocarbon sector and sectors subjected to a special consumption tax (mainly tobacco products and alcoholic and soft beverages).
4. Based on a 2014 Forrester survey on mobile banking functionality of 32 large retail banks in 11 countries (Australia, Canada, France, Germany, Italy, the Netherlands, Poland, Spain, Turkey, the United Kingdom, and the United States). The Garanti Bank of Turkey had the highest score.
5. These UNCTAD data are not directly comparable with the data on tax filings. Because of data limitations, the share of firms filing taxes online cannot be compared directly with the share of firms in countries in Central Asia, South Caucasus, and Russia.
6. Broadband Internet does not seem to affect manual tasks.
7. Bloom, Sadun, and Van Reenen (2012) examine 1,600 firms in France, Germany, Italy, Poland, Portugal, Sweden, and the United Kingdom.
8. Commander, Harrison, and Menezes-Filho (2011) find that Indian firms in states with poor labor market policies are associated with lower levels of ICT adoption.
9. See, for example, Aghion and others (2009), Blundell, Griffith, and Van Reenen (1999), and Oliveira and Martins (2010). See Holmes and Schmitz (2010) for a survey of literature examining how local competition can improve firm productivity.
10. Bugamelli, Schivardi, and Zizza (2008) describe case studies of Italian manufacturers that innovate when faced with competitive pressures; Iacovone, Pereira-Lopez, and Schiffbauer (2015) show that Mexican firms use more ICT when faced with import competition from China. See De Loecker and Goldberg (2014) for a survey of literature on how foreign competition improves firm productivity.
11. Bloom, Draca, and Van Reenen (2015) examine the effects of import competition in 12 Western European countries from 1996 to 2007: Austria, Denmark, Finland, France, Germany, Ireland, Italy, Norway, Spain, Sweden, Switzerland, and the United Kingdom.
12. Bank Association of Turkey.
13. Aghion and others (2005) also find that firms at the same high level of technology will innovate more when faced with competition, but competition will discourage firms at low levels of technology.
14. Some modifications were made to some countries in the groups. Iceland and Slovenia were moved to the "transitioning" category because the number of firms using websites is above the 90th percentile. Greece and Italy were moved to the "transitioning" category because, as EU countries, their trade and investment barriers are generally low.
15. Evans and Schmalensee (2014) use the example of Opentable, a restaurant reservation system. Customers are charged nothing (and enjoy a marginal rebate) to attract greater participation, while restaurants are charged both a usage and a membership fee.
16. Competition agencies need to draw the market boundaries to determine who the market players are so that they can assess if there is competition. A multisided market complicates this process. For example, the Apple iPhone is not just a mobile phone, but also an app store, a web browser, a minicomputer, a media player, and a game device.
17. *U.S.* v. *Topkins*, U.S. District Court, Northern District of California, no. 15-cr-00201.
18. Financial companies may offer attractive low initial interest rates, but high subsequent interest rates.

References

Accenture Consulting. n.d. "Garanti Bank: World's First Socially Integrated Banking Service." Client Case Study, Accenture Consulting. https://www.accenture.com/us-en/success-garanti-bank-creating-socially-integrated-mobile-banking-services.aspx.

Acemoglu, D., P. Aghion, C. Lelarge, J. Van Reenen, and F. Zilibotti. 2007. "Technology, Information, and the Decentralization of the Firm." *Quarterly Journal of Economics* 122 (4): 1759–99.

Aghion, P., N. Bloom, R. Blundell, R. Griffith, and P. Howitt. 2005. "Competition and Innovation: An Inverted-U Relationship." *Quarterly Journal of Economics* 120 (2): 701–28.

Aghion, P., R. Blundell, R. Griffith, P. Howitt, and S. Prantl. 2009. "The Effects of Entry on Incumbent Innovation and Productivity." *Review of Economics and Statistics* 91 (1): 20–32.

Akerman, A., I. Gaarder, and M. Mogstad. 2015. "The Skill Complementarity of Broadband Internet." *Quarterly Journal of Economics* 130 (4): 1781–824.

Barseghyan, L. 2008. "Entry Costs and Cross-Country Differences in Productivity and Output." *Journal of Economic Growth* 13 (2): 145–67.

Bartel, A., C. Ichniowski, and K. Shaw. 2007. "How Does Information Technology Affect Productivity? Plant-Level Comparisons of Product Innovation, Process Improvement, and Worker Skills." *Quarterly Journal of Economics* 122 (4): 1721–58.

Bloom, N., M. Draca, T. Kretschmer, and J. Van Reenen. 2005. "IT Productivity, Spillovers, and Investment: Evidence from a Panel of UK Firms." Unpublished mss., Centre for Economic Performance, London School of Economics.

Bloom, N., M. Draca, and J. Van Reenen. 2015. "Trade Induced Technical Change? The Impact of Chinese Imports on Innovation, IT, and Productivity." Oxford University Press on behalf of the *Review of Economic Studies*.

Bloom, N., B. Eifert, A. Mahajan, D. McKenzie, and J. Roberts. 2013. "Does Management Matter? Evidence from India." *Quarterly Journal of Economics* 128 (1): 1–51.

Bloom, N., T. Kretschmer, and J. Van Reenen. 2009. "Work-Life Balance, Management Practices, and Productivity." In *International Differences in the Business Practice and Productivity of Firms*, edited by R. Freeman and K. Shaw. Chicago: University of Chicago Press.

Bloom, N., R. Sadun, and J. Van Reenen. 2005. "It Ain't What You Do, It's the Way You Do IT: Testing Explanations of Productivity Growth Using U.S. Affiliates." Unpublished mss., Centre for Economic Performance, London School of Economics.

———. 2012. "Americans Do IT Better: U.S. Multinationals and the Productivity Miracle." *American Economic Review* 102 (1): 167–201.

———. 2013. "Management as a Technology." Unpublished mss., London School of Economics.

Bloom, N., and J. Van Reenen, 2007. "Measuring and Explaining Management Practices across Firms and Countries." *Quarterly Journal of Economics* 122 (4): 1341–408.

Blundell, R., R. Griffith, and J. Van Reenen. 1999. "Market Share, Market Value, and Innovation in a Panel of British Manufacturing Firms." *Review of Economic Studies* 66 (3): 529–54.

Bocquet, R., O. Brossard, and M. Sabatier. 2007. "Complementarities in Organizational Design and the Diffusion of Information Technologies: An Empirical Analysis." *Research Policy* 36 (3): 367–86.

Borchert, Ingo, B. Gootiiz, and A. Mattoo. 2012. "Guide to the Services Trade Restrictions Database." World Bank Policy Research Working Paper 6109, World Bank, Washington, DC.

Bugamelli, M., and P. Pagano. 2004. "Barriers to Investment in ICT." *Applied Economics* 36 (20): 2275–86.

Bugamelli, M., F. Schivardi, and R. Zizza. 2008. "The Euro and Firm Restructuring." NBER Working Paper 14454, National Bureau of Economic Research, Cambridge, MA.

Capgemini Consulting. 2007. "SEPA: Potential Benefits at Stake; Researching the Impact of SEPA on the Payment Market and Its Stakeholders." Report prepared for the European Commission, Brussels.

Cardona, M., T. Kretschmer, and T. Strobel. 2013. "ICT and Productivity: Conclusions from the Empirical Literature." *Information Economics and Policy* 25 (3): 109–25.

Commander, S., R. Harrison, and N. Menezes-Filho. 2011. "ICT and Productivity in Developing Countries: New Firm-Level Evidence from Brazil and India." *Review of Economics and Statistics* 93 (2): 528–41.

Correa, P. G., A. M. Fernandes, and C. J. Uregian. 2010. "Technology Adoption and the Investment Climate: Firm-Level Evidence for Eastern Europe and Central Asia." *World Bank Economic Review* 24 (1): 121–47.

Crespi, G., C. Criscuolo, and J. Haskel. 2007. "Information Technology, Organisational Change, and Productivity Growth: Evidence from UK Firms." CEP Discussion Paper 783, Centre for Economic Performance, London School of Economics, March.

De Loecker, J., and P. K. Goldberg. 2014. "Firm Performance in a Global Market." *Annual Review of Economics* 6 (1): 201–27.

Draca, M., R. Sadun, and J. Van Reenen. 2007. "Productivity and ICTs: A Review of the Evidence." In *The Oxford Handbook of Information and Communication Technologies*, edited by R. Mansell, C. Avgerou, D. Quah, and R. Silverstone, 100–47. Oxford: Oxford University Press.

eBay. 2015. *Empowering People and Creating Opportunity in the Digital Single Market.* https://www.ebaymainstreet.com/sites/default/files/ebay_europe_dsm_report_10-13-15_1.pdf.

Ensor, B. 2012. "Digital Banking Innovation in Turkey." Forrester blog, September 18. http://blogs.forrester.com/benjamin_ensor/12-09-18-digital_banking_innovation_in_turkey.

EBRD (European Bank for Reconstruction and Development). 2014. *Transition Report: Innovation in Transition.* http://www.ebrd.com/news/publications/transition-report/transition-report-2014.html.

Eurostat. Multiple years. Eurostat Database.

———. 2012. "Final Report: ESSnet on Linking of Microdata on ICT Usage." Eurostat, Luxembourg.

———. 2013. "The Multifaceted Nature of ICT: Final Report of the ESSnet of Linking of Microdata to Analyze ICT Import." Eurostat, Luxembourg.

Evaluation Society in Bosnia and Herzegovina. 2015. "An Independent Impact Assessment of the Effects of Support to Sub-National Regulation Reform in Bosnia and Herzegovina." Report prepared for the International Finance Corporation, Washington, DC.

Evans, D. S., and R. Schmalensee. 2014. "The Antitrust Analysis of Multi-sided Platform Business." In *Oxford Handbook of International Antitrust Economics*, edited by R. Blair and D. Sokol. Oxford: Oxford University Press.

Forth, J., and G. Mason. 2003. "Persistence of Skill Deficiencies across Sectors, 1999–2001." In *Employers Skill Survey: New Analyses and Lessons Learned*, edited by G. Mason and R. Wilson, 71–89. Nottingham: Department for Education and Skills.

Gordon, R.J. 2004. "Why Was Europe Left at the Station When America's Productivity Locomotive Departed?" NBER Working Paper 10661, National Bureau of Economic Research, Cambridge, MA.

Haller, S. A., and I. Siedschlag. 2011. "Determinants of ICT Adoption: Evidence from Firm-Level Data." *Applied Economics* 43 (26): 3775–88.

Hallward-Driemeier, M., and L. Pritchett. 2015. "How Business Is Done in the Developing World: Deals versus Rules." *Journal of Economic Perspectives* 29 (3): 121–40.

Haltiwanger, J., R. S. Jarmin, and J. Miranda. 2013. "Who Creates Jobs? Small versus Large versus Young." *Review of Economics and Statistics* 95 (2): 347–61.

Hempell, T. 2005. "What's Spurious? What's Real? Measuring the Productivity Impacts of ICT at the Firm Level." *Empirical Economics* 30 (2): 427–64.

Holmes, T., and J. Schmitz. 2010. "Competition and Productivity: A Review of Evidence." *Annual Review of Economics* 2 (1): 619–42.

Iacovone, L., M. Pereira-Lopez, and M. Schiffbauer. 2015. "The Complementarity between ICT Use and Competition in Mexico." Background paper for the *World Development Report 2016*, World Bank, Washington, DC.

Kim, Y., and P. Orazem. 2016. "Broadband Internet and New Firm Location Decisions in Rural Areas." Unpublished mss., University of Nebraska-Lincoln.

Marr, B. 2015. "How Big Data Drives Success at Rolls-Royce." *Forbes News*, June 1. http://www.forbes.com/sites/bernardmarr/2015/06/01/how-big-data-drives-success-at-rolls-royce/#1c49437a3ac0.

Matteucci, N., M. O'Mahoney, C. Robinson, and T. Zwick. 2005. "Productivity, Workplace Performance, and ICT: Evidence for Europe and the U.S." *Scottish Journal of Political Economy* 52 (3): 359–86.

Monopolkomission. 2015. "Competition Policy: The Challenge of Digital Markets." Special Report 68, Monopolies Commission, Berlin.

Oliveira, T., and M. F. Martins. 2010. "Understanding e-Business Adoption across Industries in European Countries." *Industrial Management and Data Systems* 110 (9): 1337–54.

OECD (Organisation for Economic Co-operation and Development). 2009. *Two-Sided Markets*. Policy Roundtable. Paris: OECD.

———. 2011. "ICT Value Added Indicator." https://data.oecd.org/ict/ict-value-added.htm.

Paunov, C., and V. Rollo. 2015. "Overcoming Obstacles: The Internet's Contribution to Firm Development." *World Bank Economic Review* 29 (Suppl. 1): S192–204.

Schneier, B. 2015. "How the Internet of Things Limits Consumer Choices." *The Atlantic*, December 24. http://www.theatlantic.com/technology/archive/2015/12/Internet-of-things-philips-hue-lightbulbs/421884/.

World Bank. 2013. Enterprise Surveys (http://www.enterprisesurveys.org), World Bank.

———. 2015. "Doing Business 2016: Measuring Regulatory Quality and Efficiency." Washington, DC: World Bank.

———. *World Development Indicators*. World Bank.

WEF (World Economic Forum). 2013. *The Global Competitiveness Report 2013–2014*. World Economic Forum, Geneva.

———. 2015. *The Global Competitiveness Report 2015–2016*. World Economic Forum, Geneva.

WITS (World Integrated Trade Solution). WITS Database.

4

Digital Trade

After graduating from university, Emina Terzić joined her father, Mr. Terzić, in their textile manufacturing firm. Ms. Terzić is eager to employ what she learned in university to modernize and expand the business. In particular, she wants to take advantage of Bosnia and Herzegovina's growing integration with the European Union and move into the retail business of selling dresses and blouses directly to European customers.

Ms. Terzić wants to use e-commerce to sell their clothes directly to the customers in Europe so that she can remove the middleman. Setting up an e-commerce website entails lower investments and risks than opening up physical stores. It is relatively easy for the company to hire a web designer to set up an e-commerce website. But there are larger, more difficult issues to tackle: how to receive and send payments to customers, how to set up a supply chain to ensure that orders are fulfilled, how to send orders out to customers, and how to deal with export and customs issues. Some of these issues will depend on the quality of the trade infrastructure and systems and are not within the control of the company.

This continuation of the story from chapter 3 highlights the main focus of this chapter: firms need more than the Internet to conduct digital trade, such as logistics infrastructure, trade facilitation, and payment systems, and to lower trade barriers.

Introduction and Main Messages

The Internet greatly increases firms' potential to produce new goods and services, and serve new markets. The Internet reduces transaction costs—communication, information, and coordination—through the use of e-mail, websites, and dedicated platforms and online marketplaces, making it easier for firms to participate in international trade. Online platforms can reduce the matching and information costs that can affect international trade more than domestic trade and provide mechanisms such as feedback and guarantees that improve consumer trust in online sellers. These advantages facilitate the participation of smaller firms in export activities. In combination with innovations in logistics, particularly container shipping, the reduction of transaction costs through the Internet has led to an enormous expansion of global value chains. The Internet is also having a dramatic impact on services, especially in the retail, trade, and finance sectors, by enabling firms to create new digital products, such as music, videos, and books, and to digitize their services and deliver them over long distances. This ability has led to a remarkable global expansion of business, professional, and technical services exports.

Despite the enormous potential benefits, use of the Internet is relatively limited in Europe and Central Asia (ECA). Fewer firms use the Internet to sell their products and services in the ECA region than in other regions. The countries in Western and Northern Europe have smaller shares of e-commerce sales in gross domestic product (GDP) than Japan and the United States. ECA firms that use the Internet to sell products tend to sell more to domestic than to foreign markets, missing opportunities to expand their markets. And the export of digitally enabled services is particularly low in many ECA countries.

Establishing the appropriate enabling environment is critical to facilitating more digital trade and attracting and retaining firms, particularly as the Internet has greatly expanded firms' mobility. A firm's choice of location is no longer determined solely by where the customers are, but also by factors such as business environment, good source of labor, and proximity to amenities. First, firms need a good logistics infrastructure to limit the costs of exporting goods and a developed payment system to ensure that they can conduct their financial transactions. Second, firms need a well-oiled trade system that is not impeded by unnecessary paperwork or complicated customs procedures. Third, firms need to have access to international markets for their exports of goods and services. While tariffs are low in most countries, many other trade barriers impede exports, for example, nontariff measures and services restrictions that can hinder the flow of digital trade.

A framework for classifying countries by the relationship between the share of firms engaging in e-commerce and the quality of infrastructure and payment systems can be used to choose priorities for improvements in the enabling environment. Emerging digital countries face significant difficulties in delivering goods and effecting payments and should focus on improving their logistics infrastructure and allowing for easier and more secure online transactions. Transitioning digital countries have these fundamentals in place and should focus on simplifying and streamlining trade procedures, while increasing the incentives to adopt new technology through more competitive pressures. Finally, transforming digital countries have the foundations of a good digital trade system, but need to negotiate with trade partners to reduce barriers to their exports.

This chapter considers how the Internet can increase exports, discusses how e-commerce is transforming sales to both domestic and foreign consumers, and considers how the Internet is making firms more mobile. The policy priorities for governments are then reviewed, followed by the new challenges for policy.

The Internet Can Increase Exports by Reducing Trade Costs

Higher Internet use is related to more international trade. A wide variety of studies shows a positive correlation between Internet use and international trade, using various measures of Internet use (number of Internet users, number of web hosts, and communication costs) and trade (export flows, export growth, and openness of trade). One of the earliest studies finds that a 10 percentage point increase in the growth of Internet use (measured by the number of web hosts) was associated with a 0.2 percentage point increase in the growth of bilateral merchandise exports from 1995 to 1999 (Freund and Weinhold 2004). Similarly, an increase in Internet penetration is associated with the growth of services imports and exports (Freund and Weinhold 2002; Choi 2010). Another study shows that an increase in the number of broadband users by 10 percent is associated with a rise of the ratio of total trade to GDP of 1.94 percentage points (Riker 2014). Using the regression estimates to perform some simple calculations, the projected rise in broadband users increased trade openness by an average of 6.88 percentage points in high-income countries and by 1.67 percentage points in developing countries (figure 4.1).[1]

Examining the effects of the Internet on international trade is complicated by the endogenous relationship between Internet adoption and international trade. Higher Internet adoption can increase international trade, but more trade can lead to higher Internet adoption through development effects (higher income levels can lead to more Internet infrastructure development) and firm selection effects (firms that do not adopt the Internet are less competitive and exit the market). While establishing causality in the relationship between Internet use and trade

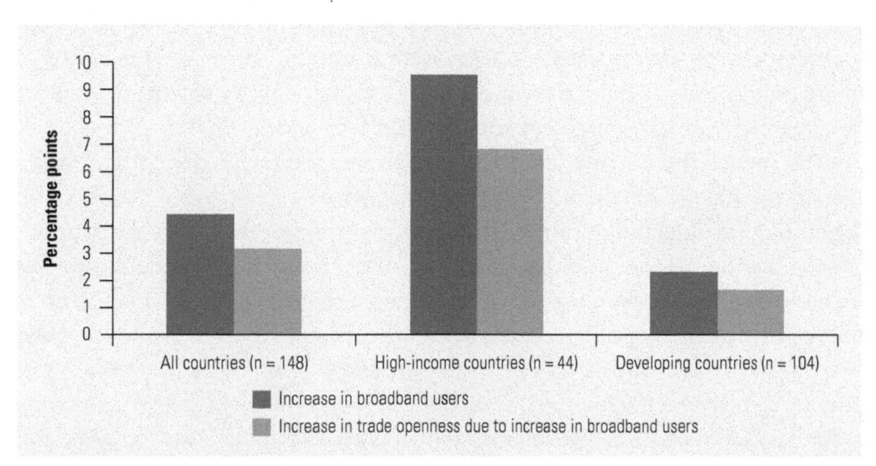

FIGURE 4.1
Trade openness can increase by 1.7 to 6.9 percentage points due to projected increase in broadband users

Source: Riker 2014.
Note: The increase in trade openness resulting from increases in broadband is calculated by multiplying the estimated regression coefficient with the projected increase in the broadband users from 2012 to 2017. The numbers in the parentheses next to the axis categories are the countries in each category.

may be challenging, evidence suggests that trade costs play an important role in explaining this relationship.

The Internet helps firms to export more by reducing trade costs. Despite the decline in tariffs in recent years, trade costs remain high due to nontariff measures, such as sanitary and phytosanitary (SPS) regulations and technical barriers to trade (TBTs), as well as communication and information barriers. These costs are high even in high-income countries, reaching as much as 170 percent ad valorem tariff equivalent (Anderson and van Wincoop 2004). This is one reason that few firms participate in exporting, even in high-income countries. For example, in the United States only 4 percent of firms exported in 2000 (Bernard and others 2007). Trade costs are even higher in lower-income countries. The ability to incur these costs can only be met by firms that are large or productive enough, and small firms are often restricted from exporting (Chaney 2008; Melitz 2003). In Europe, for example, firms with more than 249 workers are 15–40 percentage points more likely to export than firms with 10–19 workers.[2]

The Internet can significantly reduce the fixed costs of trade, for example, the up-front costs of product design, marketing and customer service, logistics and distribution, and activities related to meeting standards and regulations and obtaining import licenses. The Internet reduces these costs by facilitating connections between firms and customers. Firms can use the Internet to learn about export markets and use communication technologies, such as e-mail and Voice over Internet Protocol (VoIP), to liaise with their suppliers and customers. The costs to match buyers and sellers can be reduced when firms use online platforms to provide a marketplace for customers and firms. These platforms can also reduce capital costs for firms, as they do not need to establish a physical storefront to reach their customers.

Lower fixed costs mean that firms with Internet access are more likely to export. A study of firms in ECA finds that, controlling for firm and country characteristics, manufacturing firms with Internet access are 27 percentage points more likely to export and service firms with Internet access are 15 percentage points more likely to export (Clarke 2008).[3] The study also shows that, when firms are already exporting, Internet access does not increase the amount of exports. Thus, the Internet is more likely to increase the number of goods (the extensive margin) than the average value of goods (the intensive margin) exported. While higher Internet adoption in the exporting country increases both exporters' intensive and extensive margins, the impact on the extensive margin is higher by 1.1 percentage point for a 10 percent increase in Internet adoption (Osnago and Tan 2016).

The impact of the Internet on trade costs differs by the type of goods. A reduction in communication costs has a greater impact on differentiated goods, which often require detailed information exchanges, than on homogeneous goods (Fink, Mattoo, and Neagu 2005; Tang 2006). The Internet greatly reduces the cost of trade by transforming books or recorded media containing music, video, or computer software into digital goods, where the cost of reproduction is virtually zero and the cost of transport is minimal. Due to network effects, the Internet also has a higher impact on costs for trading pairs when both countries have a high level of Internet adoption. Countries with high Internet penetration can communicate more and find more matches between each other, than can trading pairs where only one or neither partner has a high level of Internet adoption (Osnago and Tan 2016). Country pairs where both have a high level of Internet adoption have 29.6 percentage points more bilateral exports, 21.7 percentage points more trade on the intensive margin, and 6.5 percentage points more trade on the extensive margin, compared with other country pairs (figure 4.2).

Although the impact is smaller, the Internet can also reduce the variable costs of trade. Firms can reduce communications costs by using e-mail or VoIP to reach their customers or streamline their logistics and distributing processes by using e-commerce websites and online platforms. Firms can practice "drop-shipping," a distribution model where the retailer does not have the product in stock but passes the order to a wholesaler who fulfills the order. This allows firms to reduce their distribution costs and, more important, their capital costs tied to inventory. With this operations model, firms also can improve product customization, as the product can be made at the time of ordering. Firms can then cater to the "long tail"—the market for niche products—while still reaching a larger consumer base.

One approach to measuring the impact of the Internet on trade costs, which implicitly encompass both fixed and variable costs, concerns how the Internet can reduce the negative effects of distance and borders on trade. There is a rich literature on how distance and borders create frictions that reduce international trade (for a summary of the literature, see Anderson and van Wincoop 2004). Even within the European Union (EU)—a market free of trade barriers for the flow of goods—distance reduces the value of trade and the number of shipments, especially at very short distances. The distance effects are stark: when trading regions are more than 250 kilometers apart, the total value of trade and the number of shipments decline rapidly and remain flat thereafter (figure 4.3).

The Internet reduces the effects of distance; the impact of distance on trade is about 65 percent smaller for online than for offline trade flows in a sample of 62 countries (Lendle and others 2012). Distance still has some negative impact on online transactions: the size of trade flows between two countries is negatively related to the distance between them, even after controlling for the extent of Internet adoption in them (Freund and Weinhold 2004; Osnago and Tan 2016). The lingering effect of distance may be capturing some effects of tastes and preferences. E-commerce websites selling differentiated and taste-dependent

FIGURE 4.2
Impacts of the Internet are higher if both countries have high adoption levels

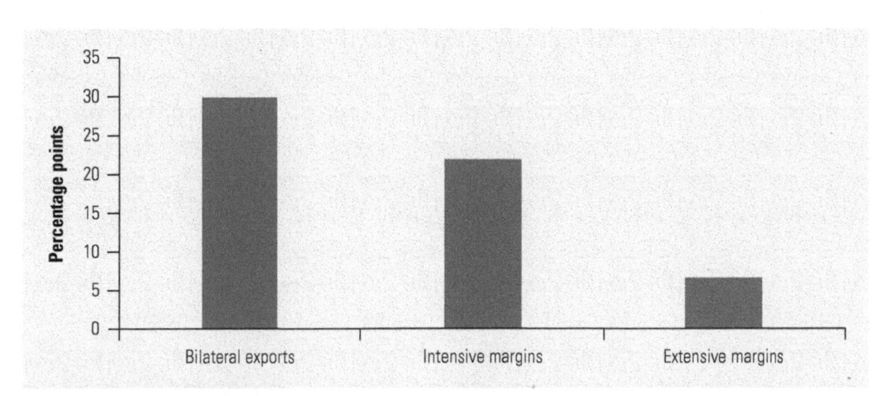

Source: Osnago and Tan 2016.
Note: The effects of the Internet is for country pairs with both high levels of Internet adoption compared to other bilateral pairs.

FIGURE 4.3 Distances between EU regions reduce trade flows and extensive margins of trade

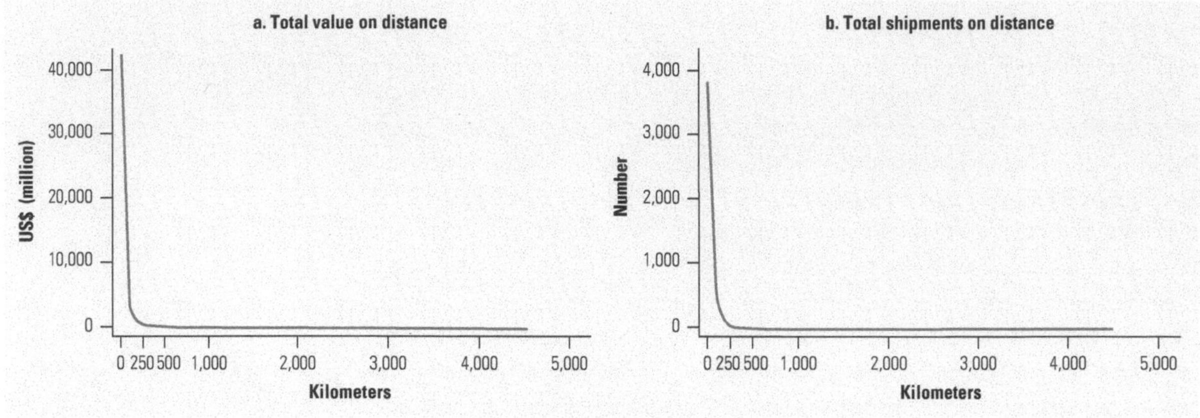

Source: Tan 2016.
Note: The analyses use a Gaussian kernel estimator in STATA to estimate the relationship between trade flows and distance. The data is bilateral freight flows between 278 EU NUTS-2 level regions collected by Eurostat. Distance is calculated as a weighted average of actual distance travelled by the freight trucks.

goods, such as music and games, are more affected by distance than websites selling more general products and services such as technology information and financial information (Blum and Goldfarb 2006).

The Internet-driven reduction in communication and coordination costs also boosts trade by increasing firms' ability to manage the outsourcing of intermediate goods to the lowest-cost locations, thus significantly increasing the growth of global value chains.[4] The entry of multinational companies is positively correlated with the level of Internet use of firms, and more foreign affiliates are located in countries with higher use of Internet of firms (Alfaro and Chen 2015). The positive relationship between Internet use of firms and multinational entry is stronger when the multinational firm is in an industry that uses communication technologies more intensively and has fewer routine tasks. The Internet also enables countries to develop new sectors that export services and digital products. As digital technologies develop, more physical products can be digitized and consumed over the Internet. Development of new sectors that can deliver services and products online is important for many ECA countries, in particular, the landlocked countries in ECA.

The Internet may be increasing global trade in services, but digitally enabled services exports from ECA countries are small. Exports of information and communication technology (ICT) services average 4.4 percent of GDP in Western Europe, but only 0.8 percent in other parts of ECA (figure 4.4). The exports of ICT services are even lower in the Russian Federation, Turkey, South Caucasus, and Central Asia, where exports of ICT services average below 0.3 percent of GDP in all countries except Armenia (1.1 percent) and Tajikistan (1.3 percent). Conversely, Central, Northern, and Eastern Europe have larger shares of ICT services exports.

The available data do not provide a complete picture of ICT services exports. Services are difficult to measure, as they are intangible and leave little administrative trail when crossing borders. Services delivered through the Internet are particularly difficult to capture in statistics.[5] In addition, official statistics do not

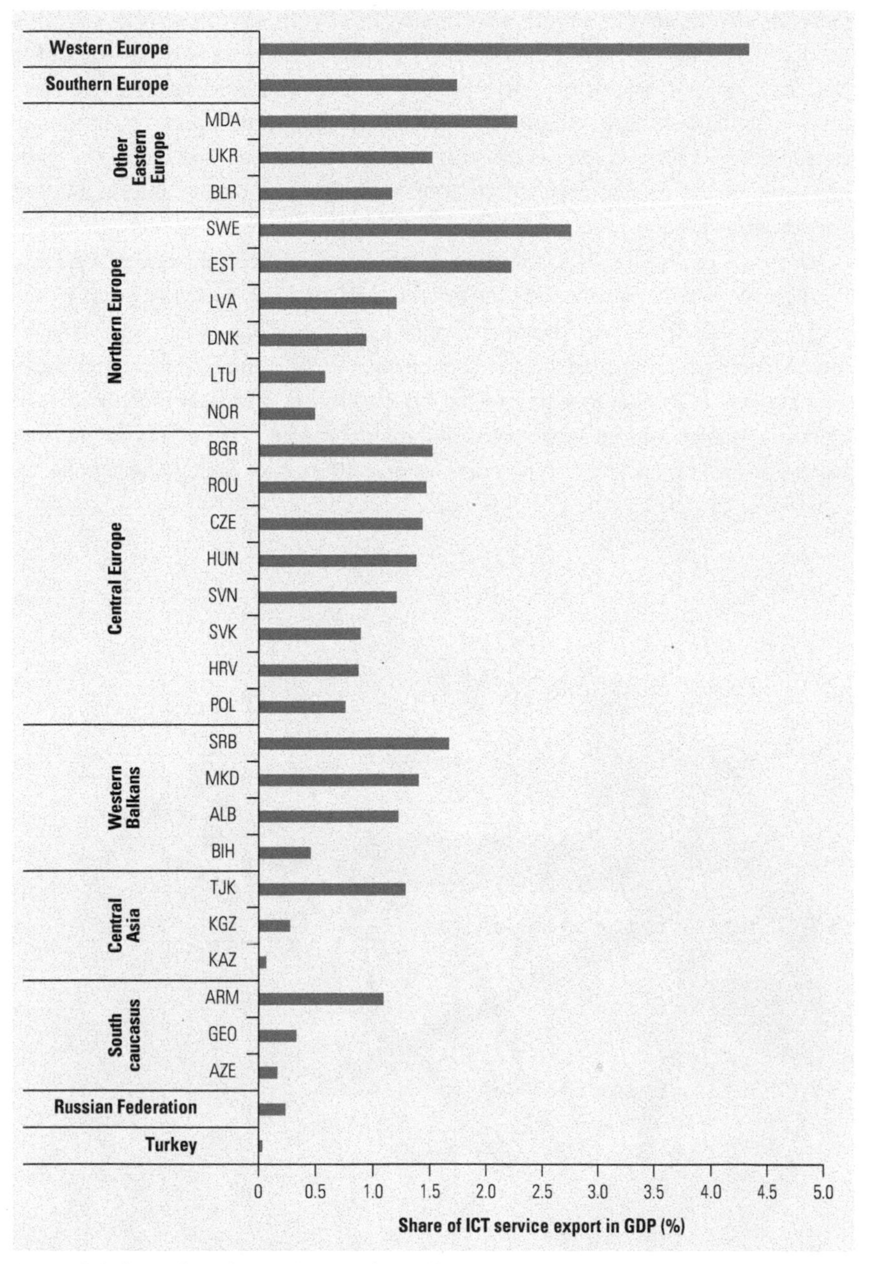

FIGURE 4.4
Small share of ICT services exports in the GDP of ECA countries

Source: Calculations based on UNCTAD and WDI 2014.
Note: The ICT services are the exports of the "telecommunications, computer, and information services" subsector. GDP = gross domestic product; ICT = information and communication technology.

record free services provided on the Internet. For example, services in the form of e-mail, video, search services, and social media websites are usually provided for free and paid for through advertising revenues. These ad-supported business models make it difficult to assess the economic value of these services.

An alternative measure of a country's service exports can be gleaned from the amount of Internet communication each region receives. Many offices use VoIP to communicate with their foreign affiliates and subsidiaries—for example, to instruct factories, to request information technology (IT) help, or to provide customer service. The Internet voice traffic entering a country can indicate the amount of business and professional services a country exports. The VoIP data are not a perfect indication of services exports, as the data are not separated into business and personal VoIP calls and are based on surveys from national regulators. The data, however, are the only comprehensive source of voice traffic that captures bilateral flows between many countries.

The ECA region has 20 percent of global voice traffic, behind East Asia and Pacific (EAP) with 28 percent and Latin America and the Caribbean (LAC) with 24 percent (figure 4.5).[6] In 2014, 78 percent of voice traffic in the ECA region was within the region, compared with almost 50 percent of voice traffic in the EAP region. This may indicate that ECA countries are not taking advantage of digital technologies to market and export more services to international markets, especially to the large North American market, which accounts for only 13 percent of ECA's VoIP traffic.

FIGURE 4.5
ECA has a sizeable amount of incoming voice traffic, but most is from other ECA countries

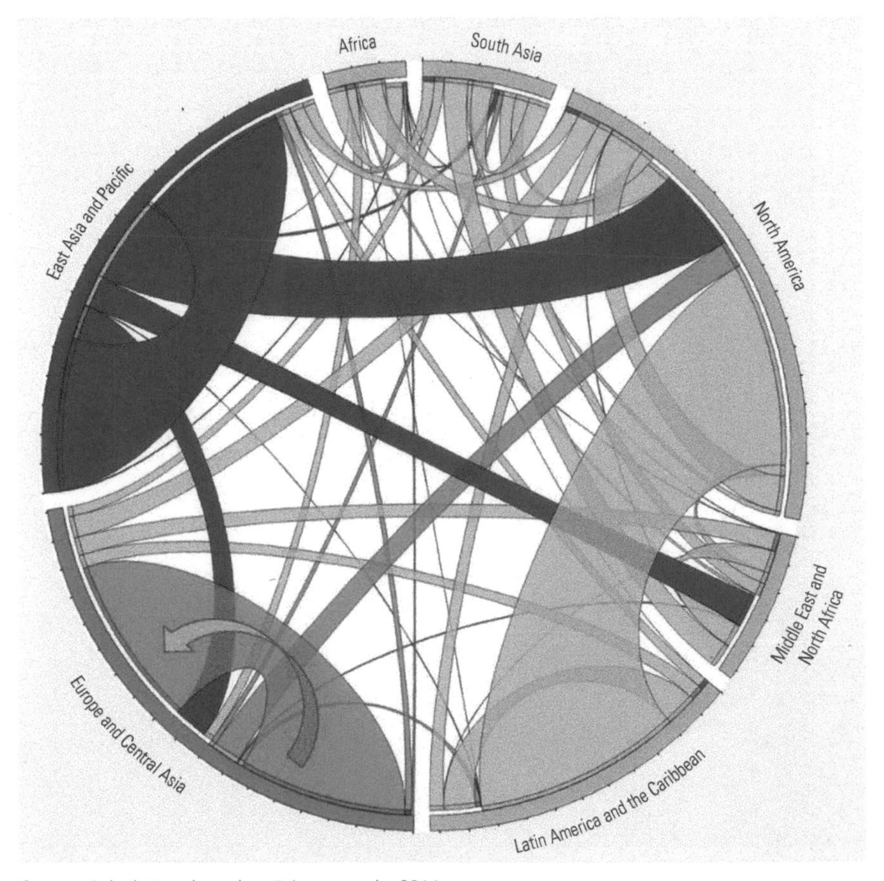

Source: Calculations based on Telegeography 2014.
Note: The incoming flows for each region are calculated by aggregating the amount of VoIP calls for the latest available year (either 2013 or 2014) entering each country in that region. The size of the ribbons represents the amount of flows between the regions, measured in millions of minutes. The flows for the Europe and Central Asia region are shaded blue and the flows for the East Asia and Pacific region are shaded red. For example, the blue ribbon with the yellow arrow represents the flows from the Europe and Central Asia region to itself. Regions are geographical classifications that include developed and developing countries.

E-Commerce Is Transforming Sales to Both Domestic and Foreign Consumers

Firms can conduct e-commerce either through online platforms that serve many buyers and sellers or through their own website. Online platforms, such as Alibaba, Amazon, and eBay, can provide a marketplace for customers to find sellers and firms to market their products to customers. These platforms also remove uncertainties about products by providing consumers with information through photographs and rating systems and increasing consumer trust through feedback and contractual enforcement mechanisms. These services reduce the risks of asymmetric information inherent in many foreign transactions, especially for firms in developing countries. For example, many platforms act as a trusted intermediary between buyers and sellers, as they can handle product complaints and protection against fraudulent sellers. Alibaba even guarantees foreign buyers a refund if the product is not delivered on time or does not fit the description.

The role of these platforms in reducing costs by increasing information is shown in a study of the sale of used cars on eBay, which finds that sellers with better information (photos and text) are able to sell at higher prices (Lewis 2011). Taobao, an Alibaba website in China, also indicates the online status of sellers and allows buyers to communicate instantly with sellers to verify product details. Ratings and feedback mechanisms can also provide the prospective buyer with independent information about the seller.[7] Customers pay nothing to shop on these platforms, and firms pay relatively low fees to register.

Online platforms facilitate exports by smaller firms, which otherwise can face very high fixed costs in exporting. Many firms on eBay have fewer than 10 employees, but are still able to engage in international trade. Online platforms also facilitate exporting to diverse markets. On average, firms on eBay reached 27 export destinations in 2014, many of which covered many countries (figure 4.6).

Firms in Western Europe are reaching fewer export destinations through online platforms than firms in the less developed subregions of ECA. This may reflect the high costs of starting a business and trading across borders in the ECA countries, causing more firms to use online platforms to export. For example, Ukrainian firms face very high costs in starting a business (Ukraine was ranked 112th in the world in 2012 in the World Bank Doing Business surveys) and trading across borders (ranked 140th), and many Ukrainian firms use online platforms to export: Ukrainian firms reached an average of 37 export markets and a total of 152 export markets on eBay (eBay 2015). Firms on these online platforms reached 8.4 destinations on average, while offline firms reached 2.8 destinations (Lendle and Vézina 2015).[8]

The Internet is particularly important for firms in small countries, which may face difficulties in achieving efficient scale if serving only domestic markets. For example, eBay sales from German and U.K. firms to other EU countries constitute less than 10 percent of total eBay sales, while 80 percent of eBay sales from firms in Luxembourg and Slovenia go to other EU countries (figure 4.7). More generally, the share of sales to other EU countries in total eBay sales is negatively correlated with the size of GDP. Thus, the benefits of e-commerce are highest for the smaller and less developed countries in the ECA region.

FIGURE 4.6
Commercial sellers on eBay in ECA are able to reach an average of 27 export destinations

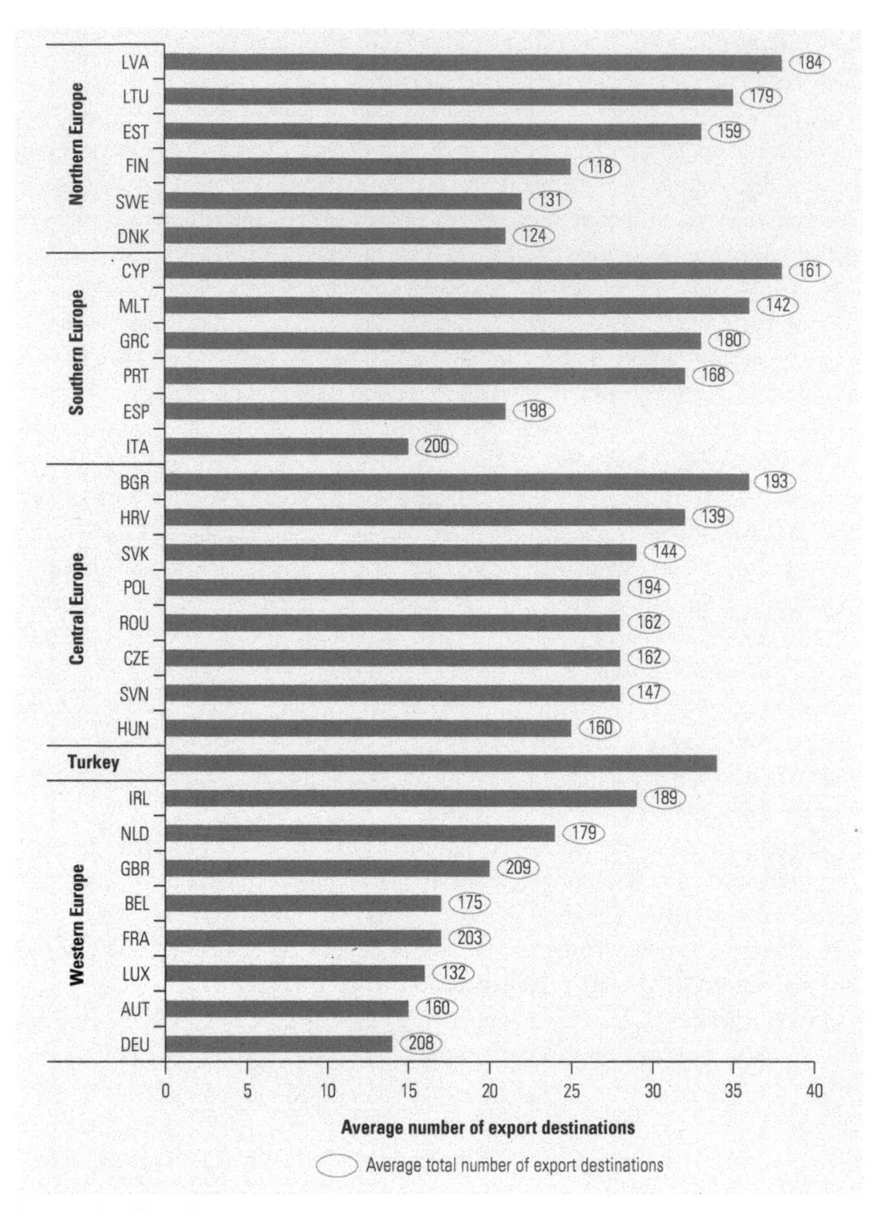

Source: eBay 2014, 2016.
Note: Commercial sellers on eBay are firms that operate on eBay with transactions of more than $10,000 per year. Data for Turkey are obtained from eBay 2014: the data are from 2013 and do not have the average total number of export destinations.

Use of the Internet to sell products is limited in ECA. With the most recent available data, only about 10 percent of firms in many ECA countries used the Internet to sell their products and services in 2009, a smaller share than in all regions other than EAP (figure 4.8).[9] Despite the lower level of economic development and Internet penetration, a higher percentage of firms in Africa sell online than in these ECA countries. With the exception of Croatia, the share of firms selling online increased between 2009 and 2015 (for the limited number of countries with data) by an average of 104 percent (figure 4.9).

FIGURE 4.7 E-commerce allows firms to reach a larger consumer base and achieve economies of scale

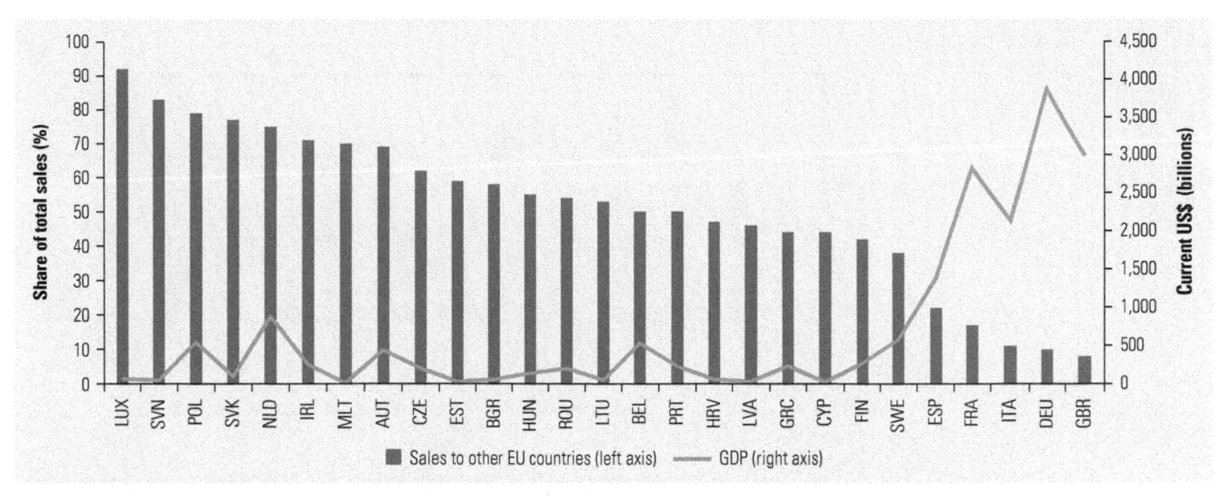

Sources: Calculation based on eBay 2015 and World Development Indicators 2014.
Note: eBay sales to other EU countries is a percentage of total sales by the firms in the country. EU = European Union; GDP = gross domestic product.

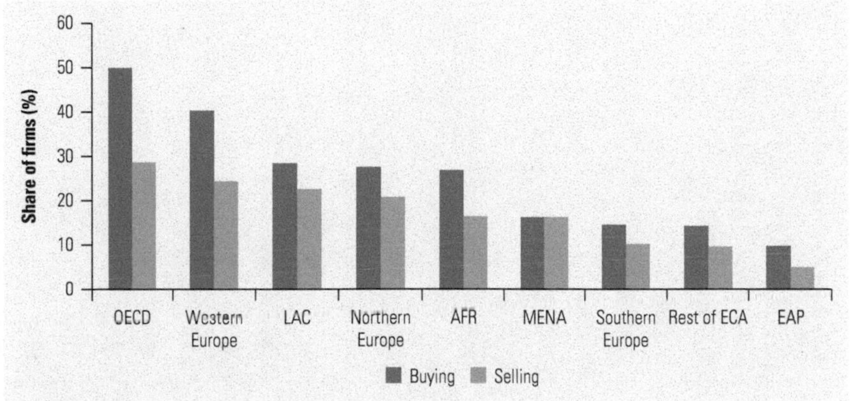

FIGURE 4.8
Firms in many ECA countries are not participating in e-commerce

Source: Calculations based on UNCTAD Business of ICT, circa 2009.
Note: The most recent data were used for each country and the average percentage is taken for the available data in each region. The high-income countries are Australia, Canada, Japan, New Zealand, and Singapore. AFR = Africa; EAP = East Asia and Pacific; ECA = Europe and Central Asia; LAC = Latin America and the Caribbean; MENA = Middle East and North Africa; OECD = Organisation for Economic Co-Operation and Development.

However, the share of firms selling online in other parts of ECA remained well below the 35 percent share in Western Europe. The share of business-to-consumer (B2C) e-commerce sales in ECA averages 1.6 percent of GDP, ranging from 0.4 to 5.7 percent in 2014 (figure 4.10).[10] The developed countries in ECA have smaller shares of e-commerce sales than do Japan and the United States, and the difference is larger when the average does not include the United Kingdom, which has the highest share of e-commerce sales in ECA. It is revealing that more firms buy online than sell online in all regions, because it is more difficult to set up an e-commerce website for sales. The firm needs to link its e-commerce website to its supply chain, manufacturing, logistics, and payment systems before it can start selling its products, while there are fewer costs to purchasing online, which can often be conducted over e-mail.

FIGURE 4.9
The share of ECA firms selling online rose between 2009 and 2015

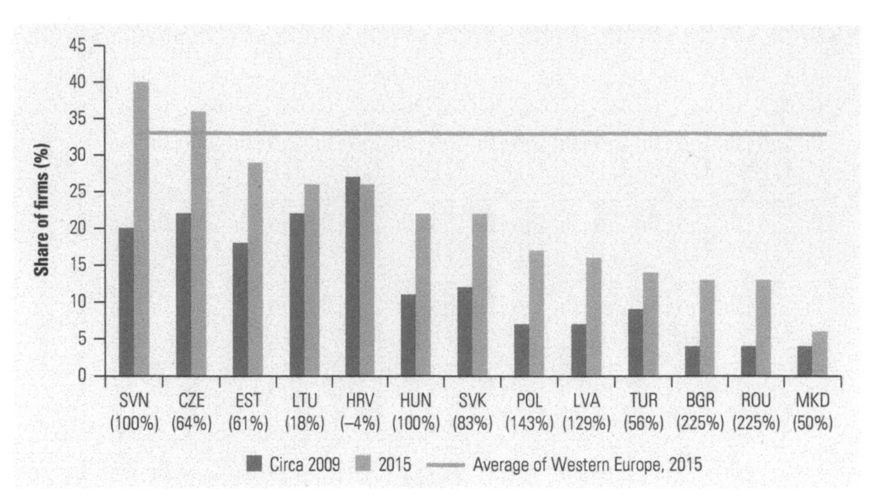

Source: Calculations based on Eurostat 2009 and 2015.
Note: The latest data were taken for the countries when available. The green line is the average share of wholesale and retail firms that have received an order through computer mediated networks in 2015 for countries in Western Europe. The percentage change between the shares of firms in 2009 to 2015 is provided in the parentheses under each country code.

FIGURE 4.10
ECA countries have small shares of e-commerce sales compared with other countries

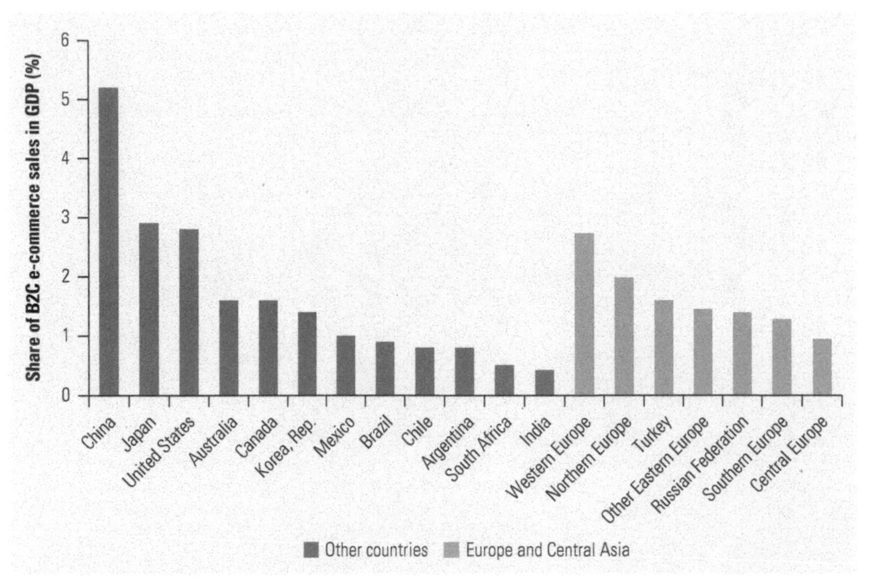

Source: Calculations are based on 2014 data taken from various reports by the E-commerce Foundation 2014 and 2015.
Note: The averages are calculated for different ECA regions. The business-to-consumer (B2C) e-commerce shares in GDP are calculated as the total sales of goods and services in sectors that are more likely to be conducted through e-commerce. These sectors are event tickets, fashion, food and health, sports and recreation, toys, electronics, insurance, travel, media and entertainment, and telecom. GDP = gross domestic product.

The largest firms tend to sell more through e-commerce than do small firms. In Central Europe and the Western Balkans, firms with more than 250 employees, on average, receive 18 percent of turnover from e-commerce, compared with about 11 percent for firms with 50–249 employees and 6 percent for firms with 10–49 employees (figure 4.11). The difference between the percentages of turnover received from e-commerce by large versus small firms can be as large

FIGURE 4.11
Large firms benefit most from e-commerce

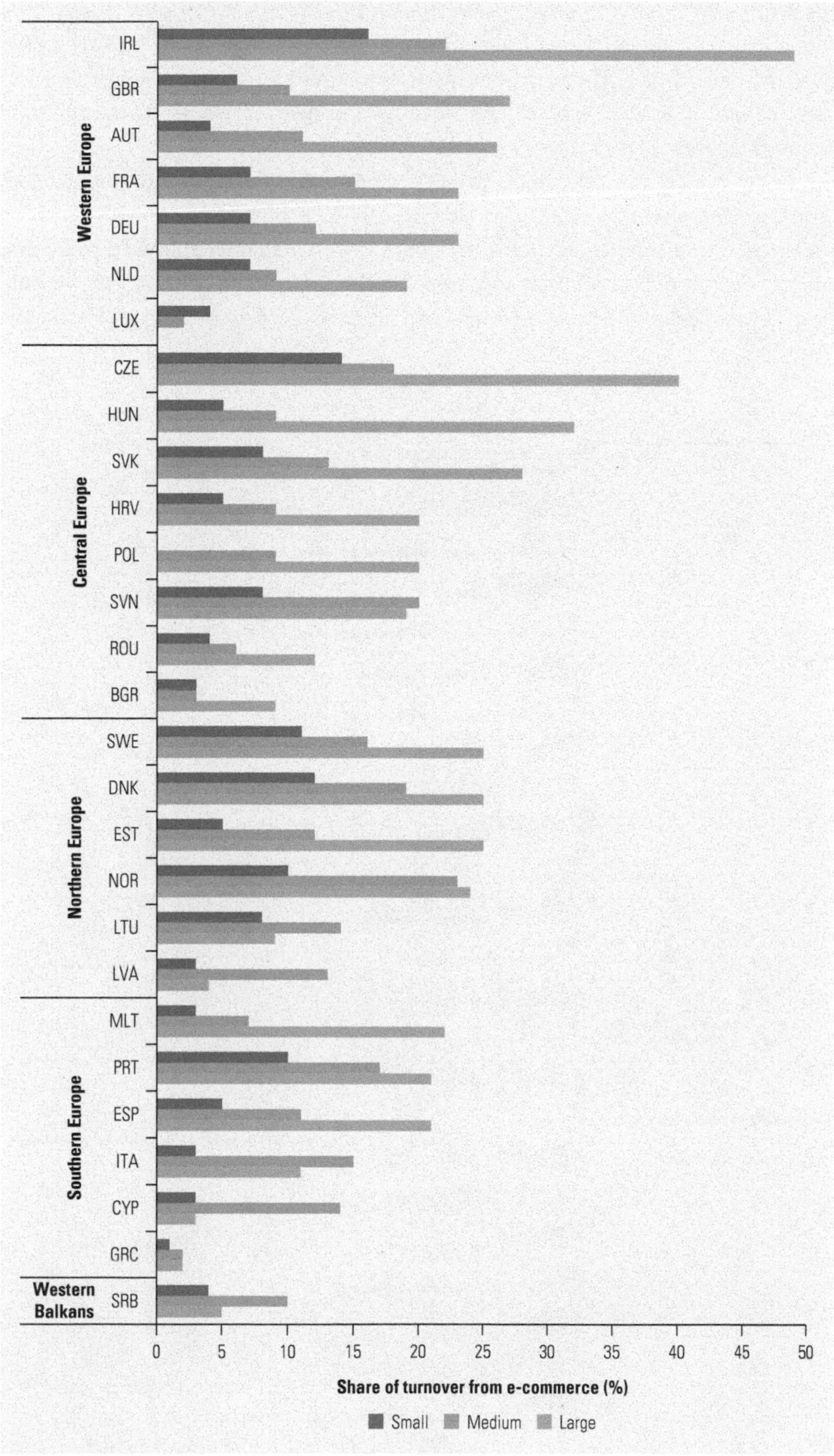

Source: Eurostat 2015.
Note: Data are from 2015 for all countries except Serbia and Latvia, which are from 2014. Data for Belgium, Finland, Iceland, and FYR Macedonia are not available. The size of the firms is defined by Eurostat where small-sized firms have 10–49 employees, medium-sized firms have 50–249 employees and large-sized firms have more than 250 employees.

as 4 times in Croatia, 5 times in Estonia, and 6.4 times in Hungary. Large firms are more likely to sell through e-commerce given the high costs, including specialized skills and ICT equipment. As noted, small firms often sell goods and services through online platforms because the fixed costs are spread among numerous users.

Firms in ECA are more likely to sell online to domestic than to foreign consumers. The share of ECA firms selling online to domestic consumers is, on average, more than three times the share of firms selling online to the world or to other non-EU countries (figure 4.12). The difference between firms selling domestically and to the world is as large as 23 percentage points in the

FIGURE 4.12
Overseas sales by firms in ECA are limited

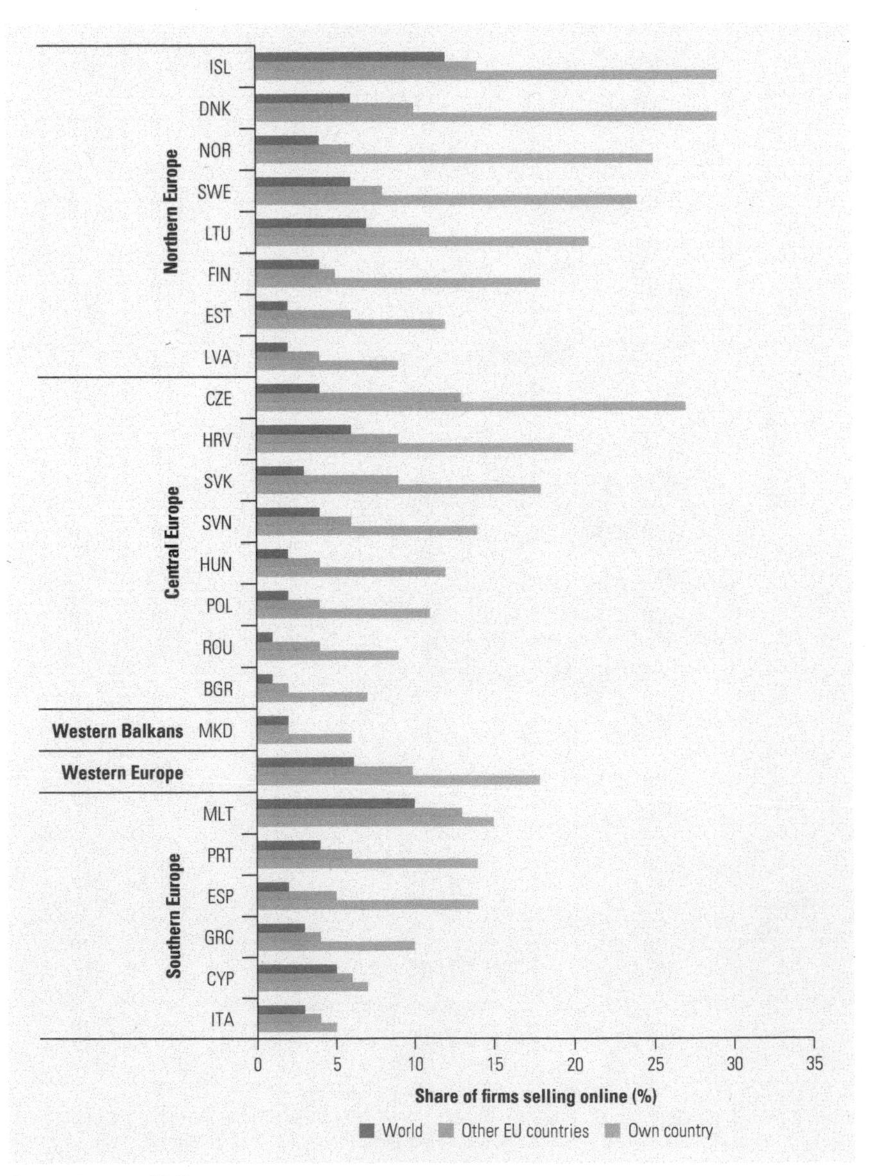

Source: Eurostat 2013.
Note: Countries in the "world" classification exclude other EU countries.

Czech Republic and Denmark. Despite the lack of trade barriers or borders within the EU, a smaller share of EU firms sell to consumers in other EU countries than to domestic consumers (the Czech Republic again has the largest difference: 14 percentage points). This may reflect the fact that local products tend to be more suited to consumer preferences or the unwillingness to purchase from foreign firms. Consumers usually have an ordered preference in where they purchase from—domestically, then from neighboring countries, then from foreign countries farther afield. This pattern is observed among consumers in the EU.[11]

The Internet Makes Firms More Mobile

The Internet makes firms more mobile by allowing them to outsource many activities, communicate easily from a distance, and deliver services from any location. Firms can choose their location based on the available resources and are no longer as tied to locations near large consumer markets. Knowledge spillovers may be less important for firms to obtain new information, as scientific knowledge can be exchanged over the Internet. Moreover, the Internet enables firms to rely on remote workers rather than solely on the local job market. This is more feasible for routine tasks that can be codified (Leamer and Storper 2014). In short, many firms are no longer forced to locate near either their customers or their workers. Firms' increased mobility means that countries have greater scope to attract production and jobs by creating the right enabling environment, but also greater potential to lose firms to competing countries.

Greater mobility enables firms to respond either to the attractions of agglomeration in cities or to the reduced costs in the periphery. On the one hand, firms often locate in dense urban environments to reap the benefits from agglomeration: a large supply of workers with a variety of skills, proximity to consumers, availability of specialized inputs, and learning by interacting with and observing the technological improvements of neighboring firms (referred to as knowledge spillovers). On the other hand, firms may choose to locate in the periphery, because urban locations impose congestion costs, such as higher land rents, wages, and cost of services.

An example of firms' flexibility in choice of location is that ECA e-commerce firms do not necessarily locate in population centers. Even within a country, the region with the highest digital density—the number of firms that sell on eBay and the amount of sales—is not always the region with the highest GDP or population (map 4.1). For example, in the United Kingdom, the Greater Manchester region has the highest number of eBay firms and eBay sales per 100,000 inhabitants, but the Inner London region has the highest GDP and the Surrey and East and West Sussex region has the highest population. Moreover, new technologies such as 3D printing are further changing how firms are located and can increase the dispersion of firms.

It is difficult to determine the impact of the Internet on firms' choice of location in principle. The tendency to gain from agglomeration is likely strongest for firms providing services that can be delivered over the Internet. These services firms

MAP 4.1 A high digital density does not need to correspond with high GDP or population density

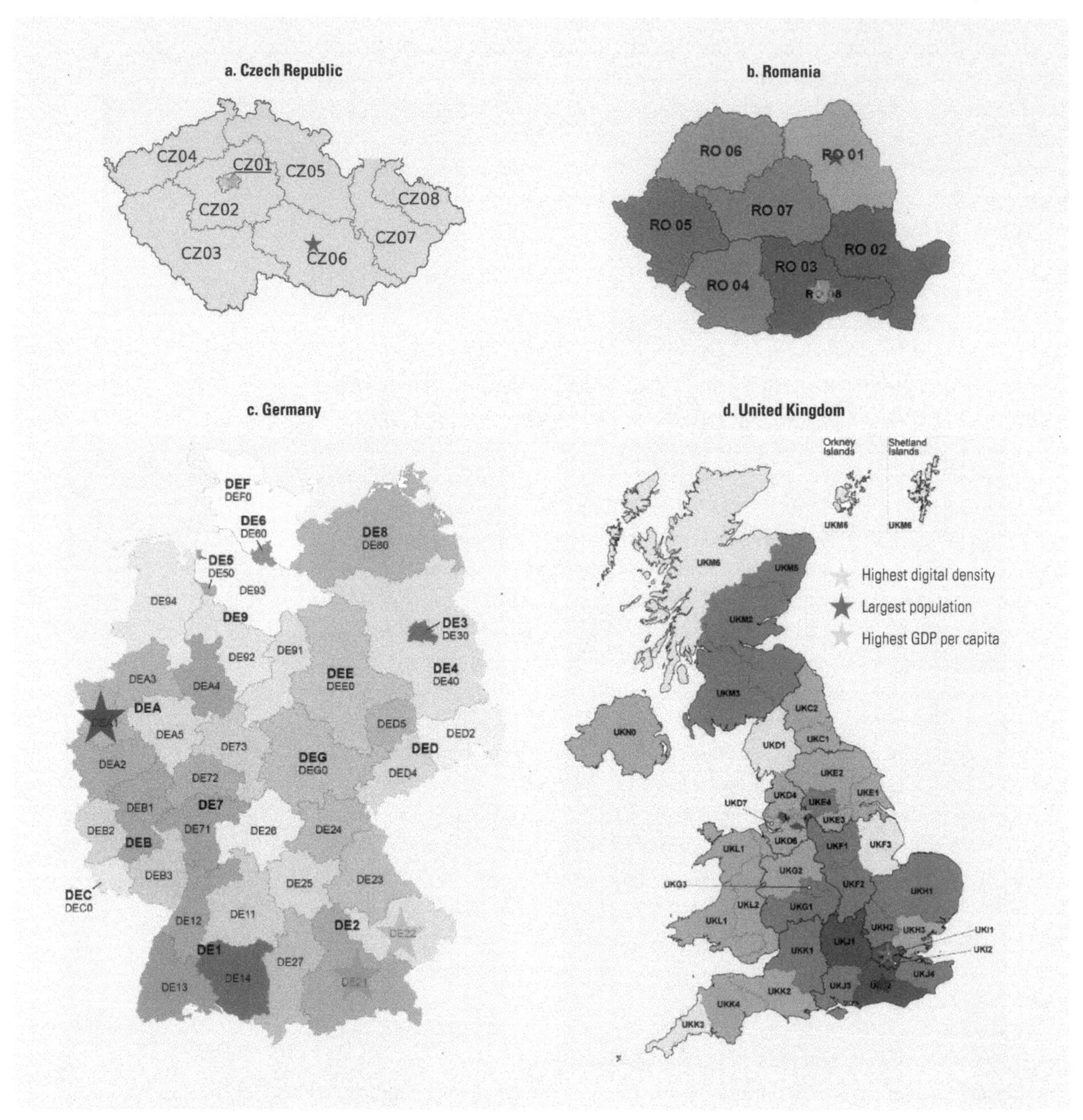

Sources: eBay 2015 and Eurostat.
Note: Digital density is obtained from eBay 2015 and measures the number of eBay firms per 100,000 inhabitants and the sales by eBay firms per 100,000 inhabitants at the NUTS 2 region level. Population and regional production data are obtained from Eurostat. All data are for 2014.

tend to perform complex tasks that benefit from face-to-face interaction among skilled, often specialized, workers, whose supply is greater in cities. By contrast, many manufacturing firms are more likely to perform routine tasks and to require less face-to-face interaction, so they can use ICT to communicate and can locate in peripheral areas with lower costs.

The agglomeration of firms can also have a positive effect on the Internet use of firms in that location. As the density and concentration in a location increase, the costs of adopting the Internet and other new technologies fall. A larger, denser location will have lower costs of adopting Internet technologies, as these locations are more likely to have more Internet infrastructure and a larger labor pool to provide the specialized skills needed to implement the new technology. Moreover, the presence of more firms can create more competition among firms, which increases the incentives for firms to adopt new technology (chapter 3).

Countries need to adopt broad horizontal policies designed to develop sectors and to get the fundamentals—education, business environment, and infrastructure—right. While these policies can be termed as industrial policies, they are different from the more interventionist policies adopted by countries in the mid-twentieth century. Industrial policy, per se, is not bad, but it depends on the issues or problems to which the industrial policies are applied.[12] Industrial policies, or transformative productivity policies, can be beneficial when they provide public goods on issues that affect the economy. The egregious example is market interventions that deal with a specific sector, such as tax exemptions for the chemicals sector. Industrial policies have to be partnered with competition, as competitive forces act as a discipline on the firms in that sector. When industrial policies are targeted at competitive sectors, they can increase the productivity and innovation of firms in that sector (Aghion and others 2015). Moreover, tax policies aimed at attracting certain firms can create unnecessary competition among countries and may be counterproductive, as the mobility afforded by the Internet allows firms to move their profits easily to the lowest tax regime. Ultimately, firm location will be determined by access to skills and talent, entrepreneurial activities, and capital.

Countries should not try to pursue narrow industrial policies that predict the next growth sector. For example, many ECA countries are motivated by the growth of ICT services exports in Bulgaria, Poland, and Romania[13] and use the business process outsourcing (BPO) sector as a popular reference point when thinking about developing sectors that can export more digitally enabled services. But with the fast-evolving technology, the BPO sectors may not be the next engine of growth, as call centers may be replaced by robots. With better technology to recognize voice and textual answers, it may not be necessary for call centers to perform repetitive work such as updating and changing customer information. The market failures in the economy may be too complex, and industrial policies do not address them (Duranton 2011). There is no clear and unambiguous evidence that policies that promote an industrial cluster improve the productivity, employment, and innovation of firms in that cluster (Uyarra and Ramlogan 2012). There is also evidence that policies to promote industrial clusters do not have long-term benefits for firms. The BioRegio policies were designed to develop biotechnology sectors in German regions, and a study finds that, while the winners of public research and development (R&D) grants had more patents and collaboration, the effects of these grants were temporary and did not have significant outcomes in the period after winning the grant (Engel and others 2013).

Policy Priorities for the Government

Countries need to create the right enabling environment to encourage greater e-commerce, which can involve a wide range of policies. Priorities for reform should be based on what is suitable for the economy and firms. The countries in ECA can be classified into three groups—emerging, transitioning, and transforming—based on the relationship between the share of firms engaging in e-commerce and the quality of logistics infrastructure and payment systems, which captures the basic enabling environment for digital trade (figure 4.13). The categorization is based loosely on the following criteria: countries with weaker logistics infrastructure and payment systems (below a score of 4) are classified as emerging, countries with intermediate-quality infrastructure and systems (between 4 and 5) are classified as transitioning, and countries with a high level (above 5) are classified as transforming.[14]

Emerging digital countries need to strengthen their logistics infrastructure and develop their payment systems to allow for easier and more secure online transactions, so that firms can deliver their goods and receive payments for them. Transitioning digital countries have these fundamentals in place and should focus on making it easier for firms to export by reducing the amount of paperwork and simplifying and streamlining trade procedures. Some of the transitioning

FIGURE 4.13
Policies for creating the enabling environment for digital trade for emerging, transitioning, and transforming digital countries

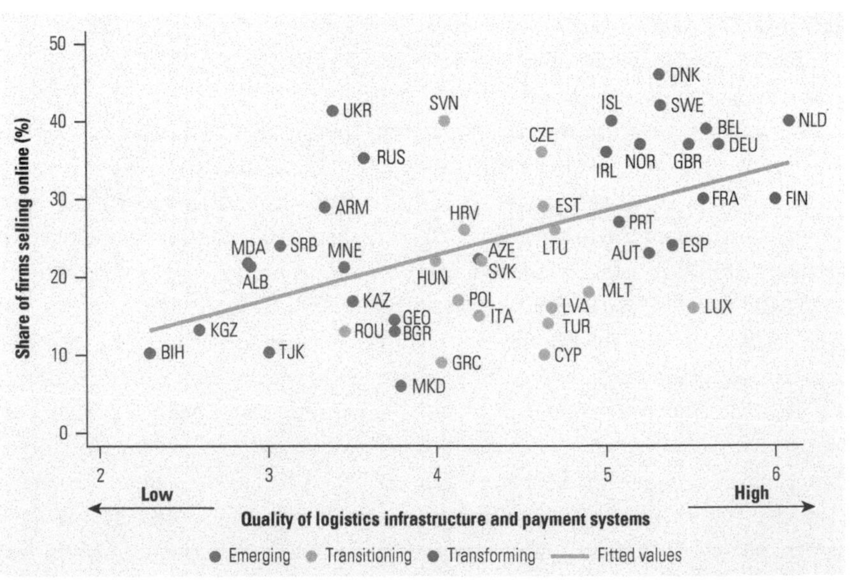

Sources: Calculations based on World Bank Enterprise Survey, Eurostat, and WEF Competitiveness Report.
Note: Data for firms selling were calculated from the Enterprise Survey and Eurostat and obtained for the latest available years. The quality of logistic infrastructure and payment systems is calculated as an average of the response from executives in the WEF Competitiveness Survey. The questions are the quality of roads, railroad infrastructure, port infrastructure and air transport infrastructure, and the affordability and availability of financial services, where the answers are ranged from 1 (worst) to 7 (best). The year of the data from the WEF Competitiveness Survey corresponds to the year for the data of firms selling online for each country.

countries already have low costs of exporting, but they still have low percentages of firms engaging in e-commerce. In these cases, transitioning countries should focus on increasing the incentives to adopt new technology by encouraging a more competitive environment. Lastly, transforming countries have the foundations of a good digital trade system, but will need to negotiate with their trade partners to reduce tariff and nontariff barriers on goods and restrictions on services. The countries and policy priorities for each group are presented in table 4.1. The three groups of policy priorities provide a guide for ECA countries, but a country may also need to pursue policies beyond those recommended for its group, as the policies to create an enabling environment are difficult to sequence and are often complementary.

Emerging digital countries can facilitate more e-commerce by ensuring that the infrastructure is in place to ship goods. A solid logistics infrastructure is necessary for efficient e-commerce. While the Internet can help firms to export more, high-quality logistics infrastructure can strengthen the effect. A study examining the entry of foreign products into the U.S. market finds that the probability of product entry increases 0.65 percent when there are 10 additional Internet users per 100 people, but the probability increases 1.18 percent when these 10 additional Internet users are in a country with highly efficient export logistics (Riker 2015).

The logistics system encompasses many components—freight transportation, warehousing, border clearance, and domestic postal system—and inefficiencies in the system will increase the trade costs for firms. Azerbaijan, Georgia, the Kyrgyz Republic, FYR Macedonia, Turkmenistan, and Uzbekistan have scores at or below the midpoint of 2.5 of the logistics performance index (figure 4.14). These countries have additional challenges, as they have difficulty accessing major shipping routes or are landlocked. In addition, countries lacking a good postal system will increase the costs for e-commerce companies that rely on the domestic postal system to make the last-mile parcel delivery to their consumers. Countries with

TABLE 4.1 Policy Priorities for Emerging, Transitioning, and Transforming Countries

Group	Countries	Policy priorities
Emerging	Albania, Armenia, Azerbaijan, Belarus, Bosnia and Herzegovina, Bulgaria, Georgia, Kazakhstan, Kosovo, Kyrgyz Republic, FYR Macedonia, Moldova, Montenegro, Russian Federation, Serbia, Tajikistan, Ukraine, and Uzbekistan	• Improve logistics infrastructure • Develop online payment systems
Transitioning	Croatia, Cyprus, Czech Republic, Estonia, Greece, Hungary, Italy, Latvia, Lithuania, Luxembourg, Malta, Poland, Romania, Slovak Republic, Slovenia, and Turkey	• Enhance trade facilitation measures • Improve competition (Chapter 3)
Transforming	Austria, Belgium, Denmark, Finland, France, Germany, Iceland, Ireland, the Netherlands, Norway, Portugal, Spain, Sweden, and United Kingdom	• Reduce trade barriers in partner countries (nontariff measures, service restrictions)

Note: Countries are grouped based on the relationship between the number of firms selling online and the quality of infrastructure and payment systems, which is presented in figure 4.13. The World Economic Forum data are not available for Belarus, Kosovo, Turkmenistan, and Uzbekistan. These countries are classified based on their scores in the Logistics Performance Indicators and the World Bank's Findex database.

FIGURE 4.14
The logistics sector among some ECA countries is weak

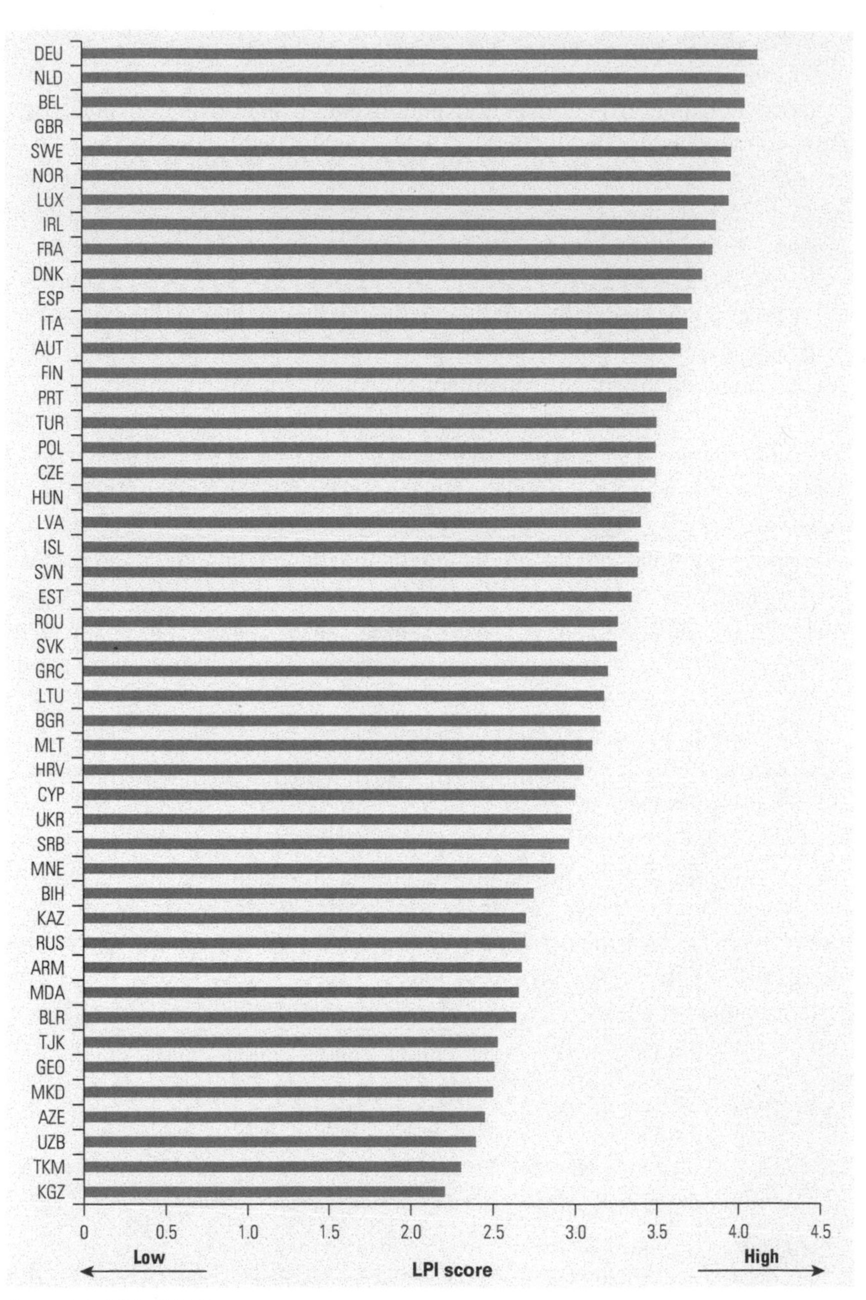

Source: World Bank 2014.
Note: The overall LPI score is a weighted average of six dimensions that includes efficiency of clearance process, quality of trade and transport, and competence and quality of logistics services. LPI = Logistics Performance Indicators.

weak logistics performance can improve their logistics sector by encouraging more competitiveness within the sector, allowing the development of third-party logistics services, and incorporating technology to improve the traceability of shipments.

In addition to logistics infrastructure, emerging countries need to ensure that there is a developed payment system that will facilitate online transactions.

Online transactions are usually carried out with a credit card or through online payment methods such as PayPal. As online payment systems become more widespread, measured by the market share of PayPal in the country, cross-border online trade in the EU also increases (Gomez-Herrera, Martens, and Turlea 2013). Moreover, online payment systems may offer more payment security over credit and debit cards, and concerns over payment security prevent many consumers from purchasing products and services online. These concerns may rise as countries engage more in e-commerce and as consumer education about payment fraud increases. The share of consumers who are worried about payment security tends to be higher in the more developed ECA countries (figure 4.15).

If individuals do not have a credit card, they are less likely to purchase anything online. On average, only 15 percent of individuals in many parts of ECA countries have credit cards, compared with about 50 percent in Western Europe (figure 4.16).

FIGURE 4.15
Consumers in ECA worry about payment security and more so in developed ECA countries

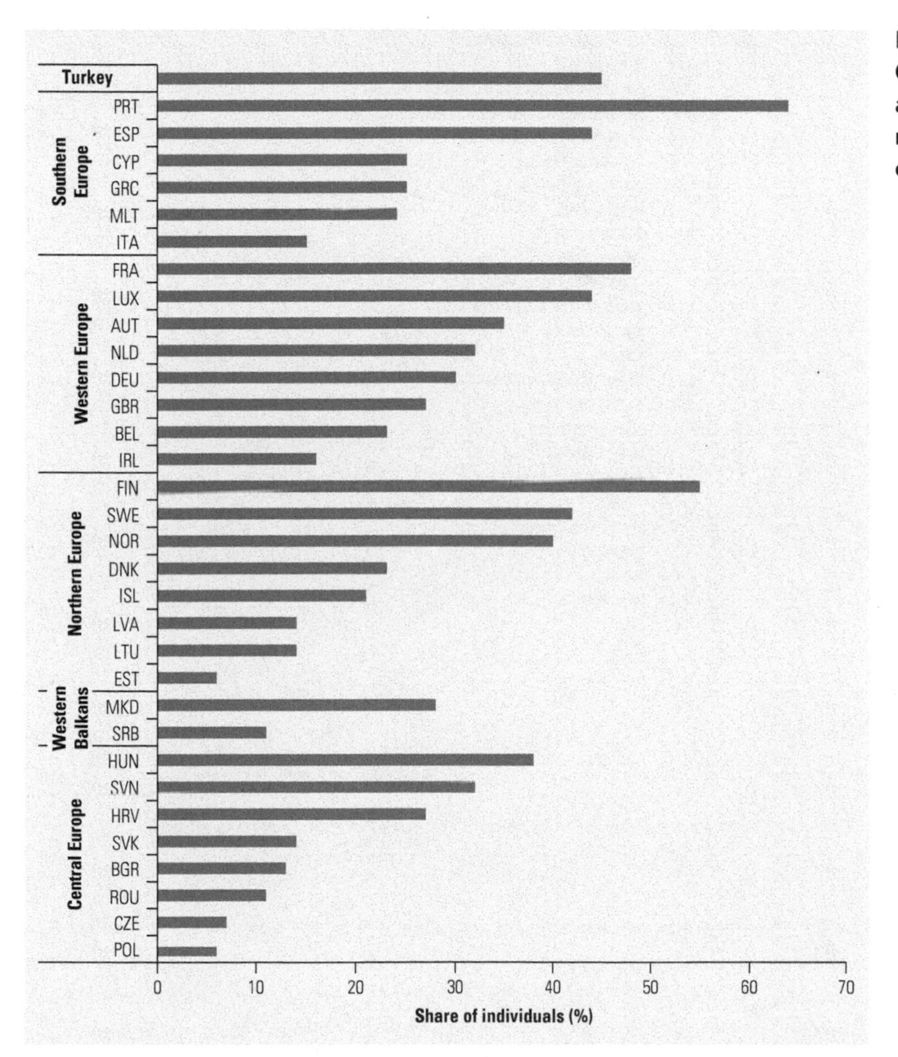

Source: Eurostat.
Note: The percentage of individuals who did not purchase something online in the last 12 months because of payment security concerns. The data for all countries are for 2015, except for Iceland and Serbia, which have data for 2009.

FIGURE 4.16
Many individuals in ECA countries have no credit cards

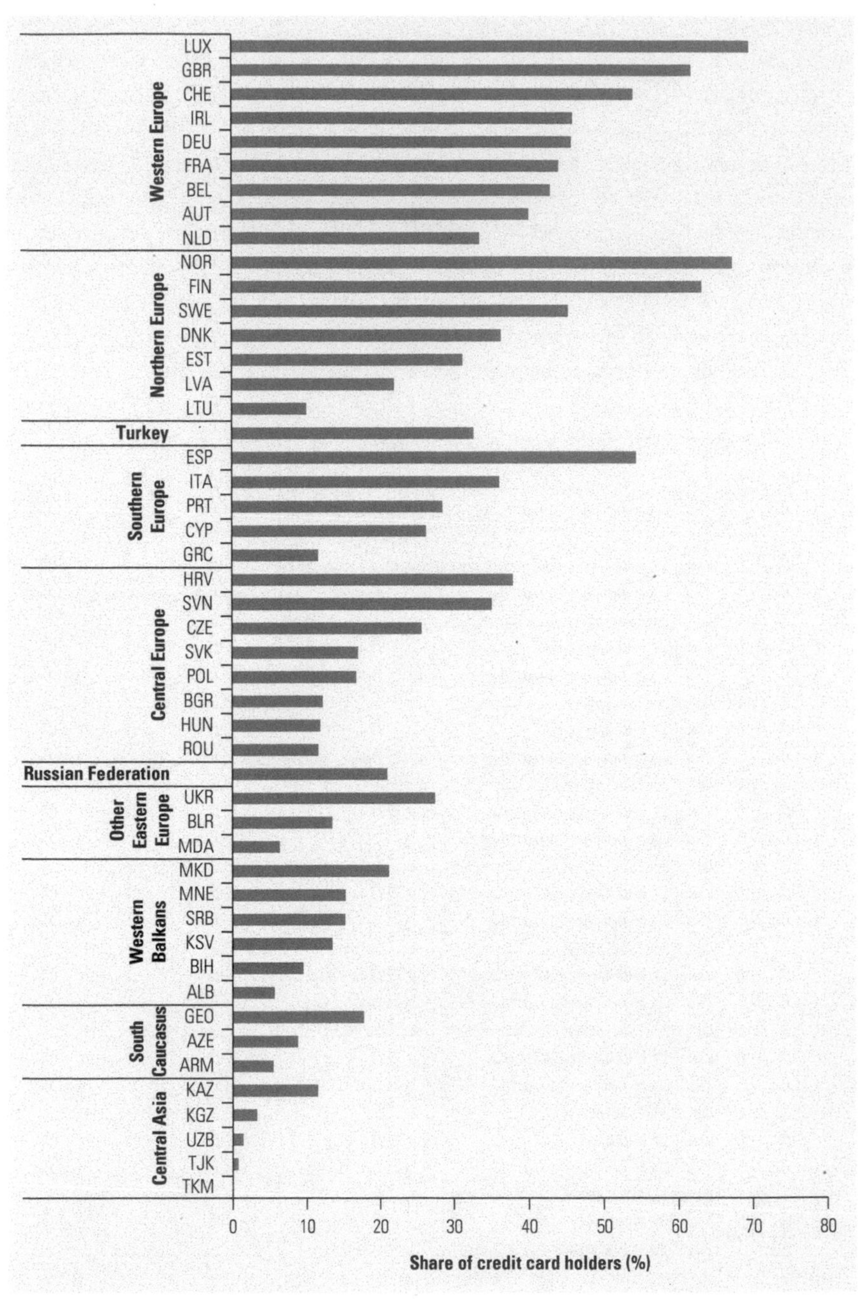

Source: 2014 data from the World Bank's Findex database (Demirguc-Kunt and others 2015).

Similar to the situation in logistics, financial access is less developed in Central Asia and South Caucasus; 5 percent or less of individuals have credit cards in Armenia, Tajikistan, Turkmenistan, and Uzbekistan. Firms in many countries in Central Asia, Eastern Europe, and South Caucasus also do not have access to online payment services, making it doubly difficult to sell things online.[15]

While the lack of a good logistics infrastructure and payment systems is important, more firms find that a weak logistics infrastructure is an obstacle to e-commerce sales. On average, 21 percent of firms in ECA that are not selling online find that logistics are an obstacle to e-commerce and 15 percent find that payment systems are an obstacle (figure 4.17). In addition, more small and medium firms list both of these issues as obstacles than large firms.

After establishing logistics infrastructure and payment systems, transitioning countries have to focus on trade facilitation measures, which refer to the administrative requirements surrounding exports and imports. Border and

FIGURE 4.17
Poor logistics and payment systems are obstacles to e-commerce sales

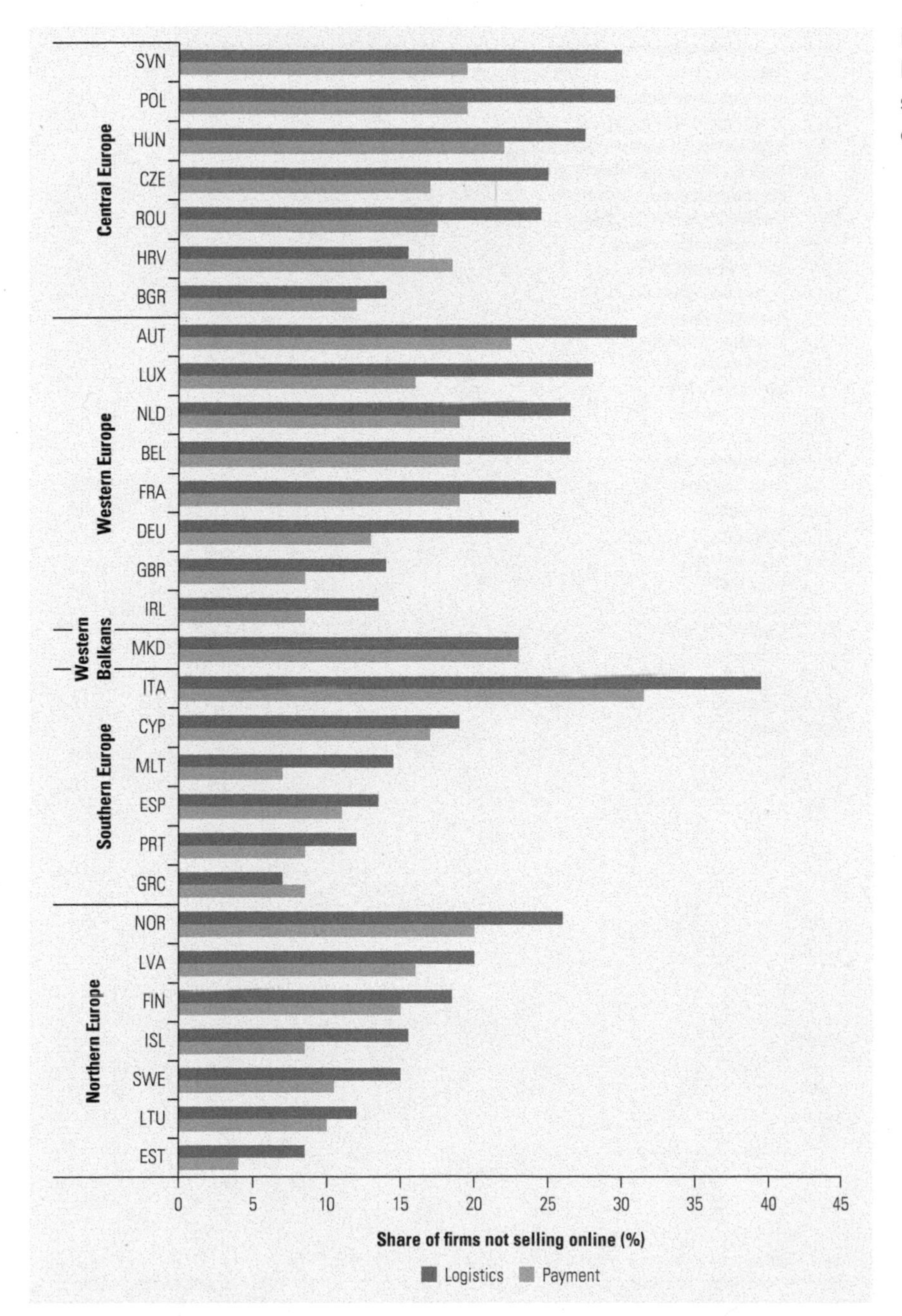

Source: Eurostat 2013.

documentation requirements associated with exporting a container are lower in transitioning countries than in emerging economies (for example, US$1,004 in Kazakhstan and US$1,625 in Russia), but not insignificant. For example, these costs total US$330 in Estonia and Greece, US$350 in Cyprus, and US$443 in Turkey (figure 4.18). By contrast, these costs are close to zero in many Western European countries. For ECA countries, streamlining procedures is the trade facilitation measure that could have the largest impact on trade

FIGURE 4.18
High costs of exporting in some ECA countries

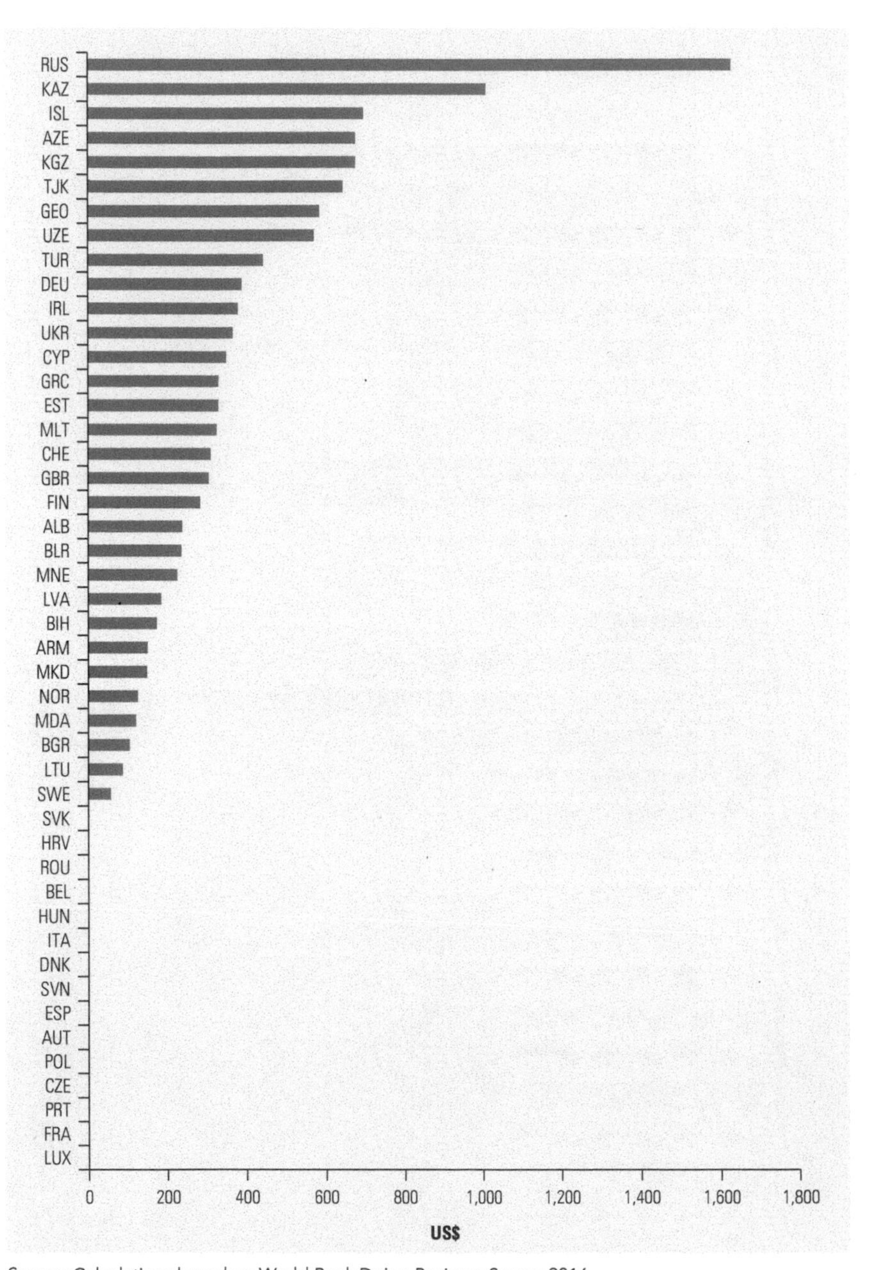

Source: Calculations based on World Bank Doing Business Survey 2016.
Note: The total costs to export includes the costs of border and documentation compliance. The Doing Business survey assumes that the costs are based on shipping one 15 metric tons of container of auto parts (HS 8708) for each country.

flows, potentially reducing trade costs by 2.2 to 2.8 percent (figure 4.19) (Moïsé and Sorescu 2013). Trade facilitation improvements can also increase online trade. eBay introduced a global shipping program that handles the shipping and customs clearance for eBay sellers. Sellers selected for the program had 2.7 percent more exports than sellers not selected, and the extension of the program to all sellers increased exports 1.27 percent and product variety 1 percent (Hui 2015).[16]

A transitioning digital country can undertake many areas of trade facilitation: providing more information about customs procedures and requirements to firms, simplifying regulations, increasing institutional capacity in the customs agency, simplifying customs procedures to reduce paperwork, and improving risk assessment procedures to reduce inspection times. Countries should examine the costs of trading from a supply chain perspective—at what stage of purchase, export, or transport do firms encounter higher costs—to determine which trade facilitation measure they should tackle. Countries can leverage ICT to modernize customs agencies and procedures. One example is Albania, which employed the Automated System for Customs Data (ASYCUDA) to improve its risk management and inspection processes between 2007 and 2012. While this example is about improving import processes, it highlights how trade facilitation and ICT can improve trade outcomes. Before 2006, the Albania customs agency subjected all shipments to physical inspections, but after using ASYCUDA, the inspection rates of imports dropped from 43 to 12 percent and the share of imports taking more than one day to clear customs dropped by half to 7 percent. The reduction in customs clearance time increased the value of imports by 7 percent.[17]

FIGURE 4.19
Enhancing trade facilitation in ECA countries can reduce trade costs

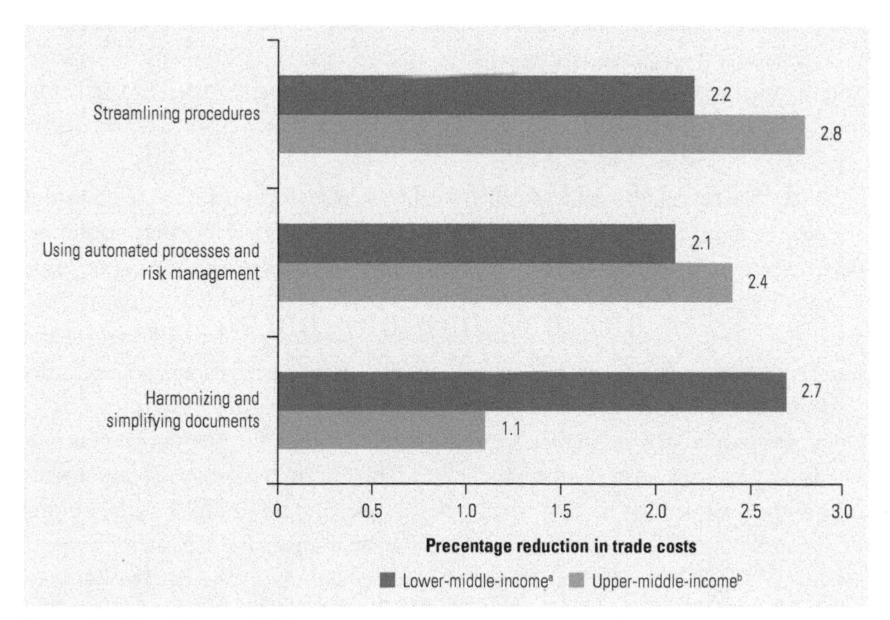

Source: Moïsé and Sorescu 2013.
Note: Percentage reduction in trade costs was calculated after enhancing trade facilitation.
a. Includes Armenia, Georgia, Moldova, and Ukraine.
b. Includes Albania, Azerbaijan, Belarus, Bosnia and Herzegovina, Bulgaria, Kazakhstan, Latvia, Lithuania, FYR Macedonia, Montenegro, Romania, Russian Federation, and Serbia.

Some transitioning countries have low costs of exporting but low levels of e-commerce activities. These countries can reduce the focus on having trade facilitation measures and increase the focus on having a pro-competitive business environment to encourage more firms to engage in e-commerce. The transitioning EU member countries (Croatia, Hungary, Italy, Luxemburg, Poland, and the Slovak Republic) have negligible costs of exporting, but less than 25 percent of their firms sold goods or services online in 2015. These countries could improve their business environment by introducing more pro-competitive policies and easing entry barriers to allow more local and foreign entrants. As discussed in chapter 3, competitive pressures can motivate firms to adopt new technologies and processes. The entry barriers in the service sectors of these countries are among the highest in ECA: in particular, Italy, Luxembourg, and Poland have the highest score in the Organisation for Economic Co-operation and Development (OECD) measure of barriers to entrepreneurship in the services sectors (see figure 3.19 in chapter 3). Lower foreign barriers in the services sectors will force incumbents to compete with foreign entrants, who are more likely to bring new technology and operational processes.

A key issue with the enabling environment is that many e-commerce firms face complicated regulations when they enter the market. Some e-commerce firms may sell services and products such as health services and cosmetics that are regulated by government agencies for public health and safety reasons. Government agencies, however, may not be familiar with the operations of online companies. In particular, some e-commerce companies are intermediaries between merchants and customers and sell a variety of products and services on their websites. In these cases, a government unfamiliar with the e-commerce business model may subject the company to many different regulations, imposing an unnecessary and possibly prohibitive administrative burden on the company. One of the first e-commerce companies in FYR Macedonia, Grouper.mk, faced these issues when it started (box 4.1). Similarly, a Swedish online travel agency needed to fulfill onerous establishment requirements set by the Irish authorities before it could market its travel services online to Irish consumers (Kommerskollegium 2011).

After facilitating the exports of their firms, transforming digital countries should focus on the trade barriers their firms face in the importing countries. Firms can face many trade barriers: while high tariff rates are no longer an issue in many countries, firms exporting physical goods face nontariff measures (NTMs) such as technical regulations and conformity assessment. Small firms are the most negatively affected by NTMs, as they often lack the resources and capacity to deal with them. One group of NTMs involves technical barriers to trade (TBTs), which are regulations and standards that establish specific product characteristics (size, functions, and performance) and labeling or packaging requirements before the product can be imported. Intra-EU exports do not face TBTs, but the ECA countries that are not within the EU will have to meet the EU technical regulations. Within ECA in 2015, the Czech Republic, Denmark, and the Netherlands imposed the highest number of TBTs (figure 4.20). An ECA firm is also likely to face many TBTs if it chooses to export to major markets outside of ECA. For example, the United States imposed more than 1,200 TBTs, and China imposed more than 1,100 TBTs in 2015.

BOX 4.1 The Challenges of Setting Up an e-Commerce Company in FYR Macedonia

The Grouper.mk website was launched in January 2011 by Ms. Nina Angelovska. The website acts as an intermediary between end users and merchants, providing a wide range of goods and services from household appliances to vacation rental apartments to online education courses. The company has more than 100,000 registered users and 2,000 merchants. The website is the leading e-commerce website in FYR Macedonia.

The website was launched at a time when only 2 percent of individuals were ordering or purchasing goods and services on the Internet in FYR Macedonia, according to Eurostat data. As one of the first movers in the industry, Grouper.mk faced many challenges when entering the Macedonian market:

> Many challenges and obstacles were in the way: lack of habit for online buying, small share of active payment cards in circulation, safety concerns and lack of trust in buying online, lack of skills and adoption of new trends among merchants, legal issues, lack of skills and knowledge of law makers and affected institutions, and administrative burden.
>
> *Ms. Angelovska, chief executive officer and cofounder of Grouper.mk website*

The regulatory environment and legal framework in FYR Macedonia was not ready for e-commerce activities. The law for e-commerce was basic and did not go into details about particular issues of what rules and regulations applied to the sale of products and services. When a specific product or service is sold online, the rules concerning that product or service are applied. This, however, created ambiguities for Grouper.mk, which is a company acting as an intermediary for merchants instead of selling the products and services, and for the government inspectors. At that time, the government lacked a full understanding of the emerging field of e-commerce, and many different interpretations of the laws were possible.

The company was able to resolve these issues through consultations with the government and support from other aid agencies. The government had a project named "We Learn from the Business Community," and Grouper.mk was invited to outline the obstacles and challenges in the e-commerce market. The Ministry for Information Society also worked on a report with the U.S. Agency for International Development and created a report examining the general challenges in the e-commerce market in FYR Macedonia and providing policy recommendations.

Source: Based on an e-mail interview with Nina Angelovska.

ECA firms exporting online services also face restrictions in entering markets, as countries may not recognize the foreign firm or may subject the firm to onerous domestic regulations. Countries may have offered access to certain service sectors under its commitments in the World Trade Organization (WTO) General Agreement on Trade in Services (GATS), but the GATS was negotiated before the Internet became widespread (pre-1995).[18] There are WTO cases where a country denied access to a foreign online services provider even though it committed to open up that services sector in the GATS.[19] A study of French firms finds that they are less likely to export services to highly regulated markets and that regulations are still a deterrence in EU countries where there is no discrimination between EU firms (Crozet, Milet, and Mirza 2016). EU countries have tried to ban the online sale of services from other EU countries: Germany tried to ban the online sale of over-the-counter pharmaceuticals, and

FIGURE 4.20
ECA firms will face many trade barriers in major trading partners

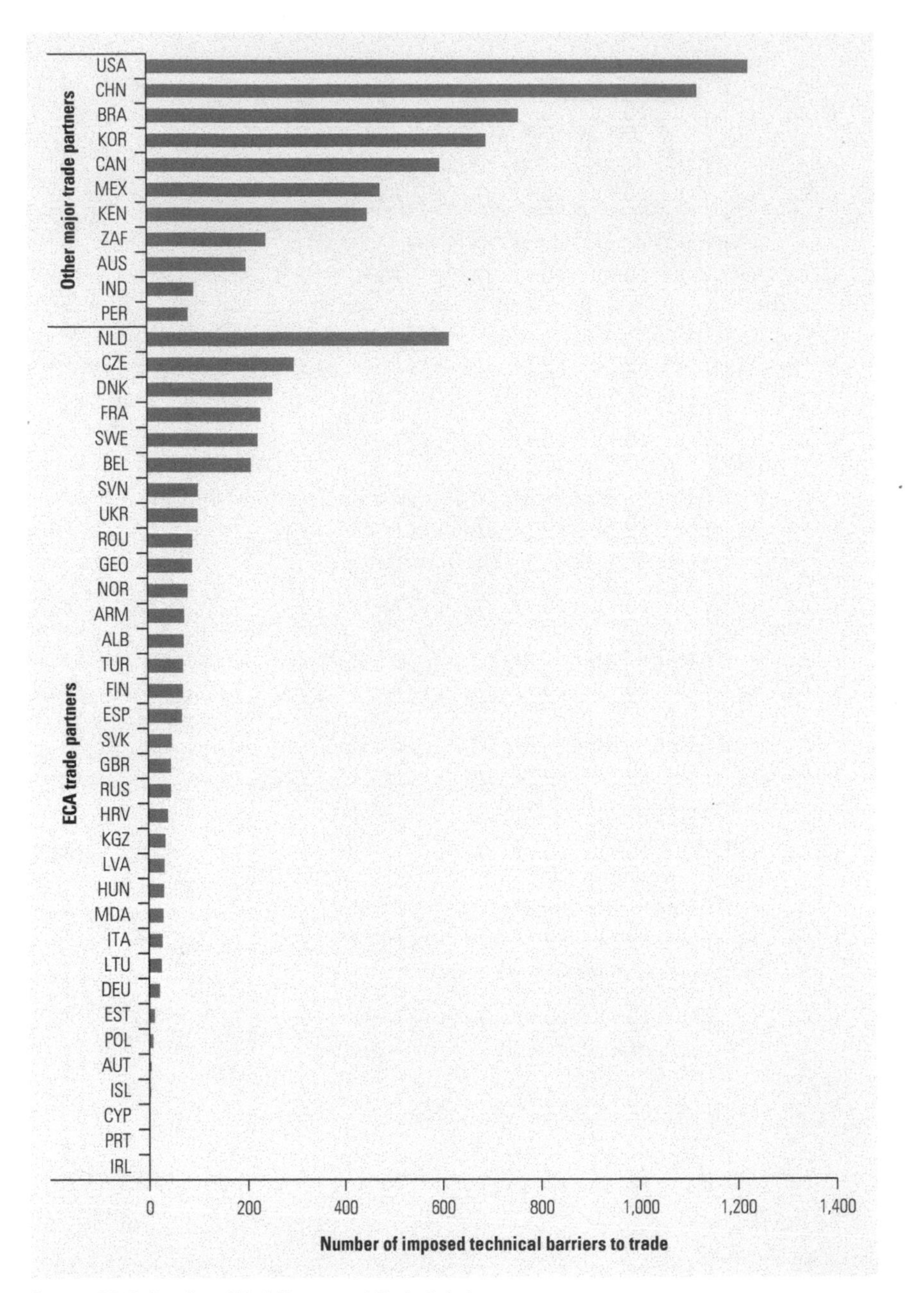

Source: 2015 data from World Integrated Trade Solution.

Hungary tried to ban the online sale of contact lenses by requiring the sale of contact lenses to take place in a physical store with a minimum of 18 square meters (Kommerskollegium 2011). These countries argued that the requirement was applied equally to all firms, but the bans went against the EU principle of free movement of goods and services. These countries were unsuccessful, and it took a legal case to settle this issue, which creates uncertainty and costs for exporting firms. Finally, export of services over the Internet can be complicated by issues of intellectual property, where countries cannot access content from other countries as they are under different intellectual property regimes. Countries have to be cognizant of the issues firms may face when exporting services online and either handle these issues preemptively in trade agreements or provide support to firms facing such barriers in the importing country.

Transforming countries may face a difficult task of reducing the barriers to digital trade in the importing countries, depending on a country's bargaining power and the trade partner's interests. These issues can be handled bilaterally between countries, but for many ECA countries, it is better to approach the issue at the international or regional level. The WTO has been working on issues of NTMs and has created a lot of momentum behind the trade facilitation agenda, especially with the conclusion of the WTO Trade Facilitation Agreement in 2013. This agreement raises the issue of good trade facilitation practices and will help to reduce the costs that firms face from NTMs. Unfortunately, international efforts to improve access to services sectors have been slower at the international level. Since the transforming countries are all within the EU, they can use the EU working groups and discussion forums to address the barriers to services trade. If the issue concerns an EU firm accessing other non-EU markets, these countries can rely on the regional grouping to provide bargaining power, especially with larger trading partners like the United States.

Even though the focus has been on firms and the supply-side issues of digital trade, countries should also examine whether there are demand-side issues. Consumers may lack the trust to purchase products and services online either because they do not trust the online merchants or they do not trust the payment systems. Online marketplaces such as eBay have instituted rating systems and other programs to increase the consumers' trust of online sellers.[20] The lack of consumer protection legislation can make consumers particularly reluctant to purchase from foreign sellers, as it may be difficult to obtain refunds or protection against fraud. The European Commission recognizes that these demand-side issues are also important and are being tackled in the EU digital single market agenda.

New Challenges in Digital Trade

The increase in data flows that accompany digital trade flows presents new challenges for firms. Service exports that are delivered over the Internet are essentially data flows on the Internet infrastructure. Goods exports also often require significant data flows: firms use consumer data to manufacture more customized goods, provide customer service, and complete online transactions. Even physical shipments rely on data flows, as these shipments have data packets that contain information about the good and the customer and that provide traceability of goods through the Internet of Things (IoT). However, barriers to these data flows can impede the free flow of digital trade. Countries are erecting new barriers to data flows amid concerns about privacy and security. These concerns may be legitimate, but some countries are using these barriers to protect local firms.

Many countries are erecting new barriers to cross-border data flows, even as data flows increase with more digital trade. Firms need a free flow of data to operate across national borders, especially as production processes become more fragmented and goods and services become more digitized. Large multinational firms will have to transfer data within the firm to control and coordinate their international operations: manage their workers and production processes, transfer technical and marketing data, supervise and maintain an efficient supply chain, and control financial payments and transactions. The emerging exports of digitally enabled services and digital goods such as music, books, and media content will increase the need for a free flow of data. There are, however, new concerns that countries are imposing barriers on cross-border data flows and preventing sensitive information from leaving the national borders.[21] Some barriers are legitimate. For example, EU citizens have concerns about privacy and data misuse, and the European Commission has implemented strict regulations concerning how and where data are handled. To meet privacy concerns, the EU has recently concluded the new Privacy Shield Framework with the United States to govern transatlantic data transfers. But other countries may impose barriers because of a need to protect local industries. A study that simulated the effects of these data barriers on the local economy estimates that data barriers can reduce GDP and exports by 1.7 percent each (Bauer and others 2014).

The cross-border nature of data flows presents new challenges for ECA countries that are interested in promoting more digital trade. Within their own borders, ECA countries have to balance their citizens' concerns about privacy and security with the private sector's need for free flow of data to participate in digital trade and global value chains. Storing the data locally may not be more secure than storing the data in dedicated secure servers overseas. When faced with barriers to data flows imposed by other countries, ECA countries can ensure that they have strong regulations that protect data following internationally recognized principles, such as the OECD's Privacy Guidelines and its Declaration on Cross-Border Data Flows and the Asia-Pacific Economic Cooperation's Privacy Framework.[22] Barriers to data flows can also be approached in international and regional agreements, such as the Trans-Pacific Partnership Agreement, which includes provisions that protect the flow of data.

Notes

1. See Riker (2014) on how the growth of broadband users is forecast. The effects of broadband users on trade openness should be taken with some caution, as it is a partial equilibrium analysis.
2. Barba Navaretti and others (2010) examine the probability of exporting of firms in seven European countries: Austria, France, Germany, Hungary, Italy, Spain, and the United Kingdom.
3. Falk and Hagsten (2015) also find that exporting firms have higher levels of Internet sets than non-exporters, and this stylized fact is true across the manufacturing and services sector.
4. Many studies have confirmed this relationship for firms in different countries: Canada (Baldwin and Gu 2008), Germany (Rasel 2013), Ireland (Murphy and Siedschlag 2013),

Italy (Benfratello, Razzolini, and Sembenelli 2009), the United Kingdom (Abramovsky and Griffith 2006), and globally based on a large data set of multinational firms and their subsidiaries (Alfaro and Chen 2015).

5. A technical note by UNCTAD (2015) examines the possible ways of measuring ICT-enabled services.
6. In the data, the countries grouped under ECA include the developed and developing countries in the region. These countries are Albania, Andorra, Austria, Belarus, Belgium, Bosnia and Herzegovina, Bulgaria, Croatia, Cyprus, the Czech Republic, Denmark, Estonia, Finland, France, Georgia, Germany, Gibraltar, Greece, Greenland, Hungary, Iceland, Ireland, Italy, Latvia, Lithuania, Luxembourg, the former Yugoslav Republic of Macedonia, Malta, Moldova, Monaco, the Netherlands, Norway, Poland, Portugal, Romania, Russia, Serbia, the Slovak Republic, Slovenia, Spain, Sweden, Switzerland, Ukraine, and the United Kingdom.
7. Some issues arise when evaluating the impact of feedback mechanisms on online sales. For example, there is an omitted variable bias since people are less reluctant to give negative feedback. See Bajari and Hortaçsu (2004) for more details on these issues and a summary of the literature.
8. The authors compare firm-level exports in the World Bank's Exporter Dynamics database and eBay data.
9. The information provided in figure 4.8 is the latest available for e-commerce activities among firms across a wide number of countries. The average percentage of firms in EAP region may be underestimated. The latest available data for China are for 2005, when only 6 percent of Chinese firms were selling online, but e-commerce has exploded in China, fueled by the Alibaba.com website in the last five years.
10. The share of B2C e-commerce sales is measured by calculating the sales of goods and services in sectors that are more likely to be conducted through e-commerce.
11. Cowgill, Dorobantu, and Martens (2013) examine the data collected by Google through AdWords, its online advertising software, which tracks shoppers through their clicks and converts transactions to sales.
12. See Inter-American Development Bank (2014) for a discussion on this issue.
13. In the 2016 A.T. Kearney global service location index, which ranks ICT offshoring locations, Romania is ranked 13th, just behind Poland, ranked 10th, and Bulgaria, ranked 12th.
14. Belarus, Kosovo, Turkmenistan, and Uzbekistan are not covered in the World Economic Forum competitiveness survey. The quality of their infrastructure and payment systems can be approximated using the Logistics Performance Indicators (LPI) and the Findex database. Turkmenistan and Uzbekistan have scores below the passing grade of 2.5 on the LPI, and Belarus just passes with a score of 2.64. Less than 1 percent of individuals use the Internet to purchase goods in Turkmenistan and Uzbekistan, 5.2 percent in Kosovo, and 21.7 percent in Belarus. As their infrastructure and payment systems are not developed, they are classified as emerging countries.
15. The absence of PayPal in Central Asia and Caucasus can be attributed to three reasons: a business decision by PayPal (size of market, ease of entry, competition), a regulatory issue (payment and banking regulations), or infrastructure issues (broadband and mobile connectivity, access to banking "rails," or systems to conduct banking services).
16. The eBay sellers were selected randomly, so that this is likely the exogenous impact of the program rather than reflecting a selection bias.
17. Fernandes, Hillberry, and Mendoza Alcantara (2015) describe the IT improvements in Albania using the updated ASYCUDA package, which has a risk management module allowing customs authorities to improve their risk management systems and inspection processes.
18. The definition of online services is unclear in the access commitments, as online services can be mode 1 services (services delivered to a country from another country) or mode 2 services (services consumed outside the individual's country).

19. See the WTO dispute settlement case (no. DS285) regarding the provision of online gambling services by Antigua and Barbuda to the United States.
20. Lendle and others (2012) find that eBay reduces the information frictions and increases trust for online buyers.
21. Chander and Le (2015) provide a good summary of data nationalism in France, Germany, Russia, and other countries.
22. See UNCTAD (2016) for a discussion. The report also has a set of policy guidelines for countries in this area.

References

Abramovsky, L., and R. Griffith. 2006. "Outsourcing and Offshoring of Business Services: How Important Is ICT?" *Journal of the European Economic Association* 4 (2-3): 594–601.

Aghion, P., J. Cai, M. Dewatripont, L. Du, A. Harrison, and P. Legros. 2015. "Industrial Policy and Competition." *American Economic Journal: Macroeconomics* 7 (4): 1–32.

Alfaro, L., and M. Chen. 2015. "ICT and Multinational Activity." Background paper for *World Development Report 2016*, World Bank, Washington, DC.

Anderson, J. E., and E. van Wincoop. 2004. "Trade Costs." *Journal of Economic Literature* 42 (3): 691–751.

Bajari, P., and Hortaçsu, A., 2004. "Economic Insights from Internet Auctions." *Journal of Economic Literature* 42 (2): 457–486.

Baldwin, J. R., and W. Gu. 2008. *Outsourcing and Offshoring in Canada*. Statistics Canada.

Barba Navaretti, G., M. Bugamelli, G. Ottaviano, and F. Schivardi. 2010. "The Global Operations of European Firms." Policy Brief 2010/05, Bruegel, Brussels.

Bauer, M., H. Lee-Makiyama, E. van der Marel, and B. Verschelde. 2014. "The Costs of Data Localisation: Friendly Fire on Economic Recovery." ECIPE Occasional Paper 3/2014, European Centre for International Political Economy, Brussels.

Benfratello, L., T. Razzolini, and A. Sembenelli. 2009. "ICT Investment and Offshoring: Empirical Evidence from Microdata." Unpublished mss., Department of Economics and Financial Sciences G. Prato, University of Turin.

Bernard, A. B., J. B. Jensen, S. J. Redding, and P. K. Schott. 2007. "Firms in International Trade." *Journal of Economic Perspectives* 21 (3): 105–30.

Blum, B. S., and A. Goldfarb. 2006. "Does the Internet Defy the Law of Gravity?" *Journal of International Economics* 70 (2): 384–405.

Chander, A., and U. P. Le. 2015. "Data Nationalism." *Emory Law Journal* 64 (3): 677–739.

Chaney, T. 2008. "Distorted Gravity: The Intensive and Extensive Margins of International Trade." *American Economic Review* 98 (4): 1707–21.

Choi, C. 2010. "The Effect of the Internet on Service Trade." *Economics Letters* 109 (2): 102–04.

Clarke, G. R. 2008. "Has the Internet Increased Exports for Firms from Low- and Middle-Income Countries?" *Information Economics and Policy* 20 (1): 16–37.

Cowgill, B., C. L. Dorobantu, and B. Martens. 2013. "Does Online Trade Live Up to the Promise of a Borderless World? Evidence from the EU Digital Single Market." Digital Economy Working Paper 2013/08, Institute for Prospective Technological Studies, Seville.

Crozet, M., E. Milet, and D. Mirza. 2016. "The Impact of Domestic Regulations on International Trade in Services: Evidence from Firm-Level Data." *Journal of Comparative Economics* 44 (3): 585–607.

Demirguc-Kunt, Asli, Leora Klapper, Dorothe Singer, and Peter Van Oudheusden. 2015. "The Global Findex Database 2014: Measuring Financial Inclusion around the World." Policy Research Working Paper 7255.

Duranton, G. 2011. "California Dreamin': The Feeble Case for Cluster Policies." *Review of Economic Analysis* 3 (1): 3–45.

eBay. 2014. "Commerce 3.0: A Springboard for Turkey's Small Business to the Global Economy." https://www.ebaymainstreet.com/sites/default/files/eBay_Commerce-3_Empowering-Indian-Businesses-Entrepreneurs.pdf.

———. 2015. "Empowering People and Creating Opportunity in the Digital Single Market." https://www.ebaymainstreet.com/sites/default/files/ebay_europe_dsm_report_10-13-15_1.pdf.

———. 2016. "Small Online Business Growth Report: Towards an Inclusive Global Economy." https://www.ebaymainstreet.com/sites/default/files/ebay_global-report_2016-4_0.pdf.

eCommerce Foundation. 2014. *BRICS B2C E-commerce Report 2014 (Light Version).*

———. 2015a. *Central Europe B2C E-commerce Report 2015 (Light Version).*

———. 2015b. *Eastern Europe B2C E-commerce Report 2015 (Light version).*

———. 2015c. *Latin America B2C E-commerce Report 2015 (Light Version).*

———. 2015d. *North America B2C E-commerce Report 2015 (Light Version).*

———. 2015e. *Northern Europe B2C E-commerce Report 2015 (Light Version).*

———. 2015f. *Southern Europe B2C E-commerce Report 2015 (Light Version).*

———. 2015g. *Western Europe B2C E-commerce Report 2015 (Light Version).*

Engel, D., T. Mitze, R. Patuelli, and J. Reinkowski. 2013. "Does Cluster Policy Trigger R&D Activity? Evidence from German Biotech Contests." *European Planning Studies* 21 (11): 1735–59.

Eurostat. Multiple years. Eurostat Database.

Falk, M., and E. Hagsten. 2015. "E-commerce Trends and Impacts across Europe." UNCTAD Discussion Paper 220, United Nations Conference on Trade and Development, Geneva.

Fernandes, A. M., R. H. Hillberry, and A. Mendoza Alcantara. 2015. "Trade Effects of Customs Reform: Evidence from Albania." Policy Research Working Paper 7210, World Bank, Washington, DC.

Fink, C., A. Mattoo, and I. C. Neagu. 2005. "Assessing the Impact of Communication Costs on International Trade." *Journal of International Economics* 67 (2): 428–45.

Freund, C., and D. Weinhold. 2002. "The Internet and International Trade in Services." *American Economic Review* 92 (2): 236–40.

———. 2004. "The Effect of the Internet on International Trade." *Journal of International Economics* 62 (1): 171–89.

Gomez-Herrrera, E., B. Martens, and G. Turlea. 2013. "The Drivers and Impediments for Cross-Border e-Commerce in the EU." Digital Economy Working Paper 2013/02, Institute for Prospective Technological Studies, Seville.

Hui, X. 2015. "Entry Costs and SMEs' Exports: A Randomized Field Experiment." Unpublished mss., Ohio State University.

Inter-American Development Bank. 2014. *Rethinking Productive Development.* New York: Palgrave Macmillan.

Kommerskollegium. 2011. "Survey of E-Commerce Barriers within the EU: 20 Examples of Trade Barriers in the Digital Internal Market." Kommerskollegium 2011-2, National Board of Trade, Stockholm.

Leamer, E. E., and M. Storper. 2014. *The Economic Geography of the Internet Age.* London: Palgrave Macmillan.

Lendle, A., M. Olarreaga, S. Schropp, and P.-L. Vézina. 2012. "There Goes Gravity: How eBay Reduces Trade Costs." World Bank Policy Research Working Paper 6253, World Bank, Washington, DC.

Lendle, A., and P. L. Vézina. 2015. "Internet Technology and the Extensive Margin of Trade: Evidence from eBay in Emerging Economies." *Review of Development Economics* 19 (2): 375–86.

Lewis, G., 2011. "Asymmetric Information, Adverse Selection and Online Disclosure: The Case of eBay Motors." *The American Economic Review* 101 (4): 1535–1546.

Melitz, M. J. 2003. "The Impact of Trade on Intra-Industry Reallocations and Aggregate Industry Productivity." *Econometrica* 71 (6): 1695–725.

Moïsé, E., and S. Sorescu. 2013. "Trade Facilitation Indicators: The Potential Impact of Trade Facilitation on Developing Countries' Trade." Trade Policy Paper 144, OECD, Paris.

Murphy, G., and I. Siedschlag. 2013. "Determinants of Offshoring: Empirical Evidence from Ireland." SERVICEGAP Discussion Paper 38, Economic and Social Research Institute, Dublin.

Osnago, A., and S. W. Tan. 2016. "Disaggregating the Impact of the Internet of International Trade." Policy Research Paper 7785, World Bank, Washington, DC.

Rasel, F. 2013. "Offshoring and ICT—Evidence for German Manufacturing and Service Firms." ZEW Discussion Paper 12-087, Centre for European Economic Research, Mannheim.

Riker, D. 2014. "Internet Use and Openness to Trade." Research Note 2014-12C, Office of Economics, U.S. International Trade Commission, Washington, DC.

———. 2015. "The Internet and Product-level Entry into the U.S. Market." Research Note 2015-05B, Office of Economics, U.S. International Trade Commission, Washington, DC.

Tan, S. W. 2016. "Spatial Frictions and Trade in the European Single Market." Background paper for this book, World Bank, Washington, DC.

Tang, L. 2006. "Communication Costs and Trade of Differentiated Goods." *Review of International Economics* 14 (1): 54–68.

Telegeography. 2014. *Voice Traffic Routes Database.*

UNCTAD (United Nations Conference on Trade and Development). 2015. "International Trade in ICT Services and ICT-Enabled Services: Proposed Indicators from the Partnership on Measuring ICT for Development." *Technical Notes on ICT for Development* 3, UNCTAD, Geneva.

———. 2016. "Data Protection Regulations and International Data Flows: Implications for Trade and Development." UNCTAD, Geneva.

Uyarra, E., and R. Ramlogan. 2012. "The Effects of Cluster Policy on Innovation." In *Compendium of Evidence on the Effectiveness of Innovation Policy Intervention.* Manchester: Manchester Institute of Innovation Research.

WEF (World Economic Forum). 2013. *The Global Competitiveness Report 2013-2014.* World Economic Forum, Geneva.

———. 2015. *The Global Competitiveness Report 2015-2016.* World Economic Forum, Geneva.

———. 2015. *Doing Business 2016: Measuring Regulatory Quality and Efficiency.* Washington, DC: World Bank.

5

The Internet Is Changing the Demand for Skills

Concerns about the effects of technology on jobs are not new. Textile workers in England sought to destroy machines during the nineteenth century to protest the automation of textile production. U.S. presidents John F. Kennedy and Lyndon B. Johnson were explicit in their concerns that more rapid technological change could increase unemployment. Even John Maynard Keynes warned about the rise of technological unemployment during the 1930s.

Despite the vast improvements in the last 50 years, technology has not made human labor obsolete. While new technologies have replaced some workers, they have complemented others by making them more productive. Even though history should allay the concerns about changing technology, anxiety about the job replacement effects of information and communication technology is on the rise. Are these concerns relevant? The answer to this question depends on the time horizon under consideration. While technological change is a crucial driver of productivity and income growth, it could generate disruptions in the short term. This chapter focuses on how countries in the Europe and Central Asia region can minimize these disruptions.

Introduction and Main Messages

The Internet, by complementing and augmenting certain skills and replacing others, is generating dramatic changes in the labor markets of Europe and Central Asia (ECA). On the one hand, the Internet can boost the productivity of workers performing complex cognitive tasks, such as managers, architects, and journalists, by facilitating the coordination of a firm's resources and the collection and processing of vast amounts of information. On the other hand, the Internet can make the skills of workers performing routine tasks, such as low-level administrators, obsolete by increasing automation and outsourcing. Skilled workers performing routine tasks, such as bookkeepers and proofreaders, may also be replaced by online resources (box 5.1).

While previous chapters have shown that the depth of Internet adoption has been slow in ECA, labor market polarization (a decline in the share of workers performing routine tasks accompanied by an increase in the share of both high-skill and low-skill workers performing nonroutine tasks) has been greater in ECA than in other regions. This likely reflects the impact not only of the Internet but also of other structural changes.

The limited availability of skills that are complementary to effective use of the Internet constrains Internet adoption, as shown by the higher return to skills in countries where the Internet intensity of jobs is low and a significant share of skilled labor is absorbed by Internet-intensive sectors. The shortage of advanced computer skills in ECA, which largely reflects levels of education, also constrains Internet adoption. The degree of shortage is shown by the low level of advanced computer skills and high level of vacancies in computer jobs. The shortage of skills could intensify in the older, rapidly aging ECA economies. In many countries, the lower level of skills in older workers appears to be driven by a mixture of cohort effects and a deterioration of skills with age.

The Internet might increase inequality in ECA. The workers performing nonroutine cognitive tasks, who are in greater demand with the Internet, are more likely to be in the top 60 percent of the income distribution, while the workers performing routine or manual tasks are typically at the middle or the bottom of the income distribution. ICT-intensive firms pay higher wages than other firms, and between-firm inequality makes the largest contribution to inequality in the region. In addition, Internet adoption is associated with a rise in the share of income going to the holders of capital, who tend to be richer than workers, whose share of national income is declining.

The most efficient policies for generating benefits from the Internet in ECA differ by the level of Internet adoption. Emerging countries with the lowest levels of Internet use should focus on devoting more resources to early childhood development and improving the quality of primary and secondary education. Transitioning countries, where the higher-order skills required for the Internet economy are limited, should focus on improving the match between higher secondary and tertiary education and demand for skills, reforming curricula to encourage greater innovation and creativity, and establishing a regulatory regime for the labor market that encourages training. Finally, transforming countries, with

BOX 5.1 What Are the Nonroutine Cognitive Tasks Demanded in the Internet Economy?

Skills can be classified by the ease of complementarity and by the ease of automation (table B5.1.1). For example, while the tasks carried out by both street vendors or managers cannot be easily automated, information and communication technology (ICT) can improve the productivity of the latter to a larger extent than that of the former. However, a street vendor's tasks are less likely to be computerized than those carried out by a skilled worker performing routine tasks such as bookkeeping. While both nonroutine manual and cognitive tasks are less likely to be substituted by ICT, only workers performing the latter are likely to experience an earnings boost from using new technologies, as displaced routine workers will likely compete for low-paid jobs in nonroutine manual occupations (Autor 2014). Thus, rising Internet adoption is associated with higher earnings and employment rates among workers who perform nonroutine cognitive tasks.

Throughout this chapter, nonroutine cognitive tasks are defined using the classification of occupations by task content used in studies such as the *World Development Report 2016* (WDR16; World Bank 2016). Skills surveys, which collect detailed information on the skills and activities performed at work, enable a deeper analysis of what nonroutine tasks typically involve. Using these surveys, it is possible to explore the correlation between the likelihood of having an occupation intensive in nonroutine cognitive tasks and a set of specific tasks and skills.

As seen in figures 5.1 and 5.2, the skills of the Internet economy involve a wide range of categories. Workers in occupations whose demand is increasing with the expansion of the Internet are more likely to use literacy and numeracy skills more intensively at work. Interpersonal and organizational skills are also relevant, as well as being autonomous and able to solve problems at work. Nonroutine cognitive occupations are also more likely to involve critical thinking and learning new skills. More important, the characteristics of occupations associated with nonroutine cognitive tasks are, in general, quite similar across economies at different stages of development.

TABLE B5.1.1 ICTs Can Complement Certain Skills, but Substitute Others

	Ease of complementarity (technology is labor augmenting)	
Ease of automation (technology is labor saving)	**High (tasks intensive in cognitive, analytical, and socioemotional skills)**	**Low (tasks intensive in manual skills)**
High (routine tasks)	1. Bookkeepers, proofreaders, clerks	2. Machine operators, cashiers, typists
Low (nonroutine tasks)	3. Researchers, teachers, managers	4. Cleaners, hairdressers, street vendors

Source: World Bank 2016, adapted from Acemoglu and Autor 2011.
Note: Workers in occupations in quadrant 4 can benefit greatly because the majority of their tasks are difficult to automate and the core of their work is in tasks in which digital technologies make them more productive. Occupations in quadrants 1 and 2 are composed of many tasks that can be easily automated. Productivity in occupations in quadrant 3 is by and large not directly affected by digital technologies.

high levels of Internet use, should emphasize developing advanced technical skills (in part by reducing gender gaps in technical fields) and encouraging lifelong learning to ensure that their aging workforces keep up with technological progress.

The chapter opens with a review of labor market polarization in ECA and then discusses the kinds of skills required in the new economy. The next section

FIGURE 5.1
Which are the skills behind nonroutine cognitive tasks? Evidence from STEP Surveys

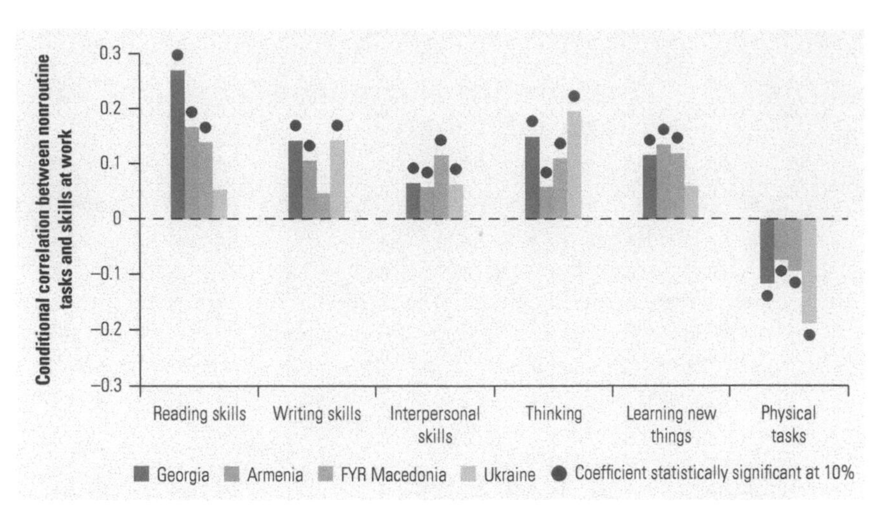

Note: STEP Surveys (urban areas). Each bar is the coefficient from an OLS regression of a binary variable equal to one if the worker carries out nonroutine cognitive tasks on sociodemographic variables (high school education, college, age, age squared, and gender) and cognitive and socio-emotional skills (dummy variables equal to one for high intensity of skills). OLS = ordinary least squares.

FIGURE 5.2 Which are the skills behind nonroutine cognitive tasks? Evidence from PIAAC Surveys

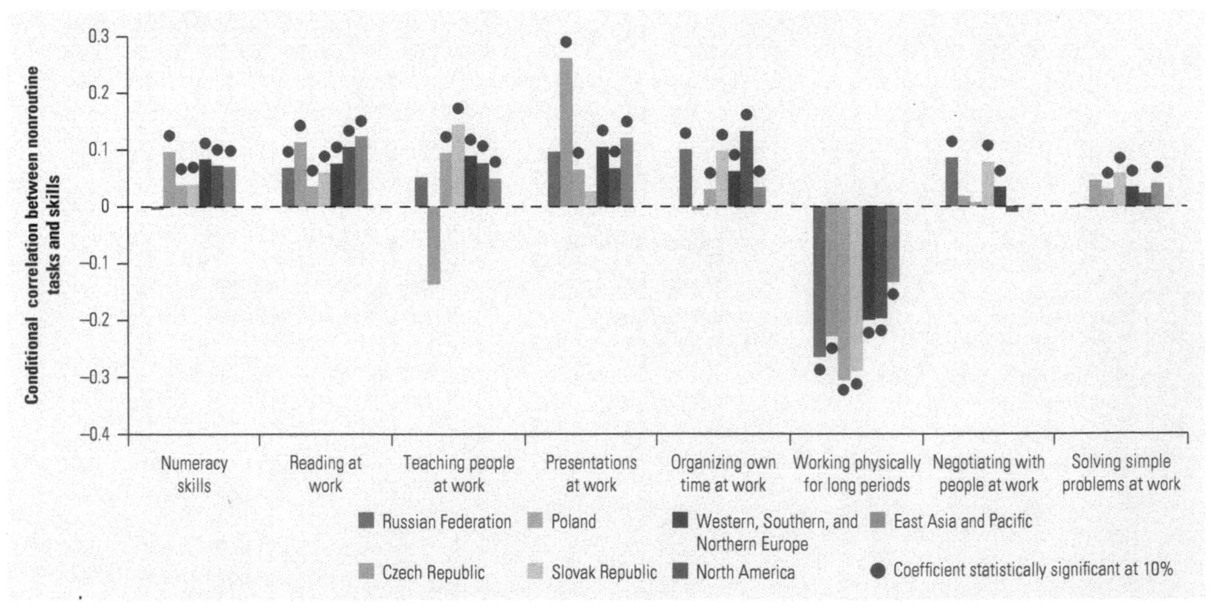

Note: PIAAC Surveys. Each bar is the coefficient from an OLS regression of a binary variable equal to one if the worker carries out nonroutine cognitive tasks on sociodemographic variables (high school education, college, age, age squared, and gender) and cognitive and socioemotional skills (dummy variables equal to one for high intensity of skills). OLS = ordinary least squares.

presents evidence that a lack of skills complementary to the Internet and a shortage of advanced computing skills limit Internet adoption in the region, followed by a discussion of the potential obsolescence of skills in older, rapidly aging economies. A final section shows that most of the efficient policies for reaping the benefits of the Internet depend on the level of Internet adoption.

Is the Internet Increasing Labor Market Polarization in ECA?

Some of the technologies associated with the Internet can help to explain the decline in demand for workers performing routine tasks (see, for example, Acemoglu and Autor 2011; Autor and Dorn 2013; Autor, Katz, and Kearney 2008). Moreover, the Internet has been associated with an increasing polarization of the labor market, where routine middle-skill occupations disappear and both low- and high-skill occupations represent a rising share of the total number of jobs. The ECA region experienced, on average, a larger decline in routine employment than other parts of the world, mirrored by an increase in high- and low-skill occupations (figure 5.3). Within ECA, the share of routine workers has fallen in every subregion. The degree of labor market polarization across ECA countries does not seem to be driven by the stage of development, as Western European countries experienced a decrease similar to the regional average, while poorer countries (such as Central Asia, Turkey, and the Western Balkans) experienced both the largest and the smallest changes.

The more rapid labor market polarization in ECA than elsewhere is not entirely due to the Internet, as the depth of Internet adoption by individuals and firms tends to be lower in ECA than in many other regions (see chapters 1–4).

FIGURE 5.3 Labor markets are becoming more polarized everywhere, especially in ECA

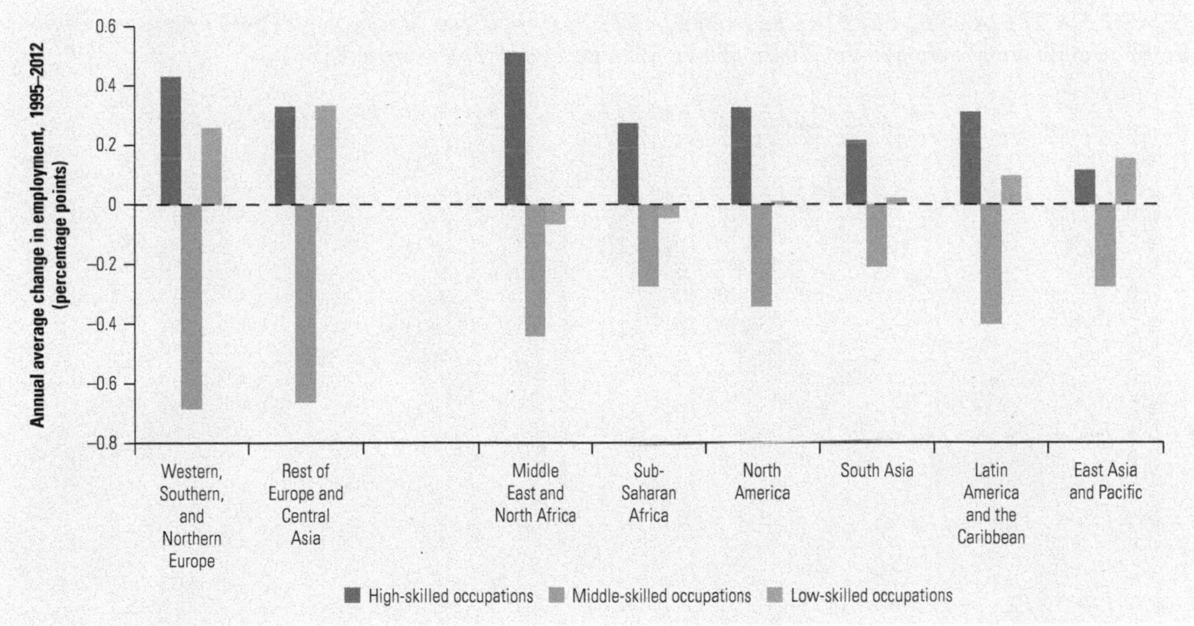

Source: World Bank 2016.
Note: The figure displays changes in employment shares between circa 1995 and circa 2012 for countries with at least seven years of data. The classification follows Autor (2014). High-skilled occupations (intensive in nonroutine cognitive and interpersonal skills) include legislators, senior officials, managers, professionals, and technicians and associate professionals. Middle-skilled occupations (intensive in routine cognitive and manual skills) comprise clerks, craft and related trades workers, plant and machine operators, and assemblers. Low-skilled occupations (intensive in nonroutine manual skills) refer to service and sales workers and elementary occupations. For the United States, comparable data could be accessed only for a short period (2003–08); consistent with Autor (2014), the observed polarization is limited in this period, with most of it having taken place in earlier years.

Labor market polarization is also driven by structural changes in ECA economies, possibly including other aspects of technological change and trade and labor market liberalization. In Central Europe, for example, more detailed data on the task content of occupations show that the decline in the employment share of routine occupations was driven mostly by a decline in routine manual occupations. By contrast, the share of routine cognitive occupations increased or remained stable (Lewandowski, Hardy, and Keister 2016), driven mostly by a movement of labor out of agriculture. However, cyclical influences do not appear to play a major role, as the fall in the share of workers performing routine tasks is roughly constant over time in major ECA subregions, as well as in countries outside the region.[1]

The increase in Internet adoption was higher when countries implemented reforms to the telecommunications sector that aimed to improve competition, increase provision, and lower prices (Howard and Mazaheri 2009; Winkler 2016). Even though the opening of telecommunications markets cannot fully explain the fall in the share of routine labor, these trends are strongly correlated. While the share of routine labor was relatively stable in the years before the reform (that is, not statistically different from that of the year the reform was implemented), this share declined every year thereafter (figure 5.4). Moreover, these results control for other countrywide changes over time, including gross domestic product (GDP) per capita, industry and services as a share of GDP, and working-age population as a share of total population, so they are not affected by general trends in

FIGURE 5.4 The opening up of the telecommunications market was accompanied by a significant decrease in the routine labor employment share, and an increase in Internet users in ECA

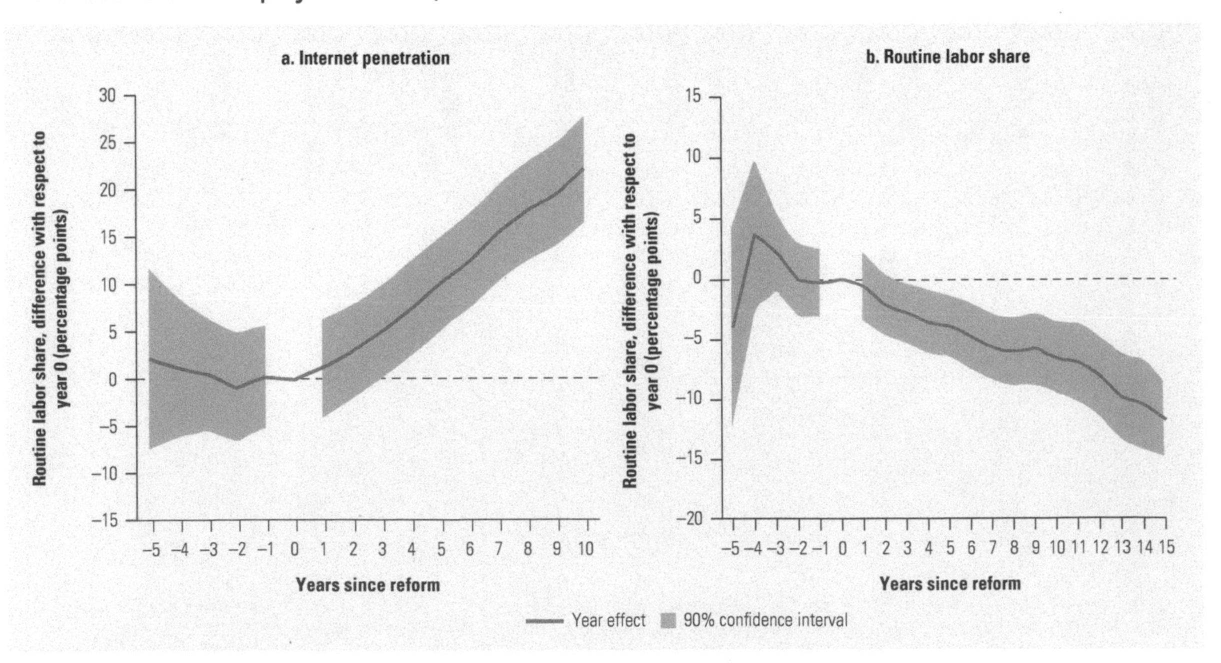

Sources: LFS Eurostat 1995-2013, ILOSTAT 1995-2013, and Regulatory Status Database (Telegeography 2017).
Note: The blue line shows the regression coefficients associated with dummy variables indicating the years since the reform, where the omitted one is the year of the reform (year 0). These results control for calendar year dummy variables, GDP per capita PPP, industry and services GDP shares, and share of working age population. GDP = gross domestic product; PPP = purchasing power parity.

the share of routine employment. An analysis of the impact of four forms of telecommunications liberalization—market liberalization, privatization, establishment of a regulatory body separated from political oversight, and achievement of full regulatory autonomy from the executive branch—using microdata from European economies provides evidence consistent with this hypothesis, showing that liberalization was associated with a larger decline in the share of men in routine occupations in ICT-intensive sectors than in other sectors.

Which Are the New Economy Skills?

While it is clear that the Internet is rapidly changing the demand for skills in the region, it is less clear what types of skills are necessary to carry out the nonroutine cognitive tasks demanded in the new economy. In brief, the skills of the Internet economy are difficult to replicate with the new technologies. These skills are in part acquired in formal education systems. As seen in figure 5.5, college graduates across ECA are at least 40 percentage points more likely to carry out nonroutine cognitive tasks at work than individuals without a university degree. Accordingly, unskilled workers across ECA are more likely to carry out routine or manual tasks.

Nevertheless, the skills associated with the Internet economy are not only those embedded in college degrees. Skills surveys available for many ECA countries can be used to disentangle the impact of more specific skills typically not measured in labor force surveys. The results in figure 5.6 confirm that college graduates are more likely than individuals with primary education or less to carry out nonroutine cognitive tasks. Controlling only for age and gender, in most countries, workers with a college degree are at least 60 percent more likely than workers with a

FIGURE 5.5 Formal education is highly correlated with the task content of jobs

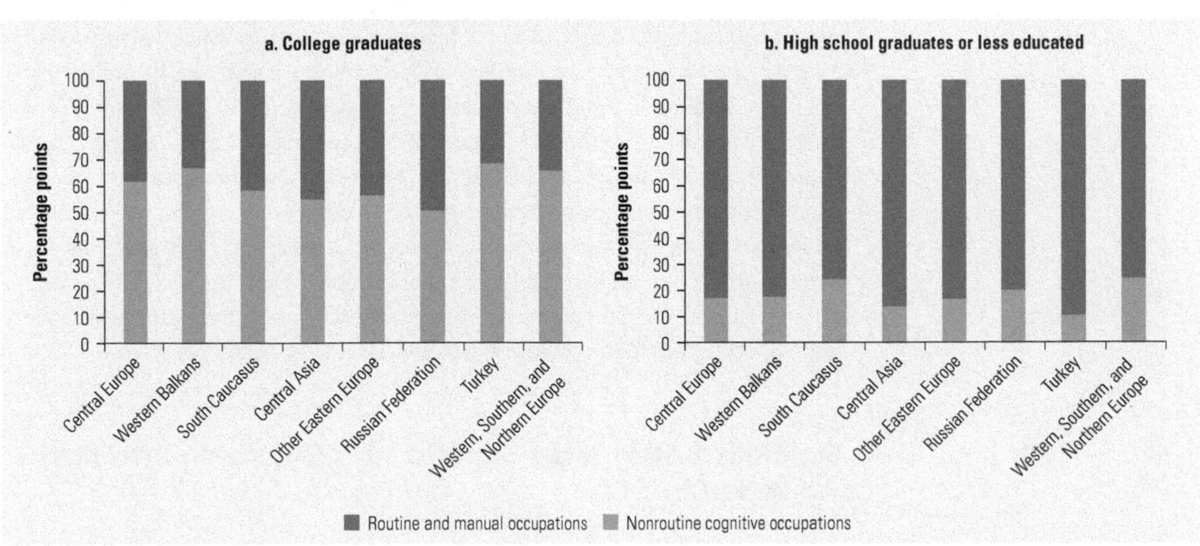

Source: LITS II 2010.
Note: Green bars are the difference between the share of college graduates performing nonroutine cognitive tasks and the share of less educated workers performing nonroutine cognitive tasks.

FIGURE 5.6 The association between education and tasks weakens when controlling for other skills

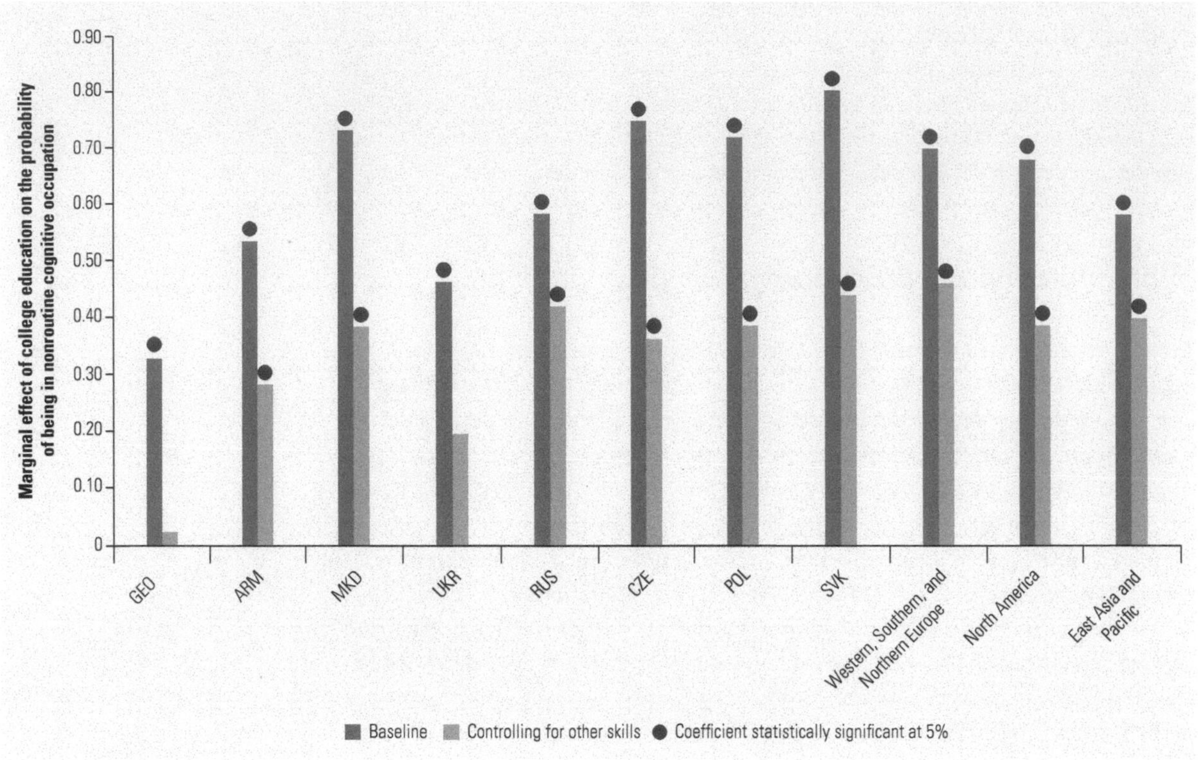

Source: LITS II 2010.
Note: Each bar is the coefficient from an OLS regression of a binary variable equal to one if the worker carries out nonroutine cognitive tasks on a dummy variable equal to one if the worker has a college degree and other control variables (high school education, age, age squared, and gender). The green bars are the coefficients when including other measures of cognitive and noncognitive skills. OLS = ordinary least squares.

primary education or less to carry out higher-order tasks. However, when including a set of cognitive and noncognitive skills as control variables, this relationship becomes weaker (see the description of skills in box 5.1). In Armenia, Central Europe, and the former Yugoslav Republic of Macedonia, the association decreases by about half. In Georgia and Ukraine, when comparing workers with the same level of cognitive and socioemotional skills, the correlation is no longer statistically different from zero. The correlation between education and tasks also weakens in developed economies and the Russian Federation, although to a lesser extent. In other words, while diplomas matter for getting a job in the Internet economy, more specific cognitive and socioemotional skills play an important role.

Do Skills Bottlenecks Significantly Constrain Internet Adoption in ECA?

The low depth of Internet adoption in ECA described in chapters 1–4 may be driven by a lack of skills that are complementary to the Internet—for example, a college education or advanced computer skills. Evidence on the increasing return to skills and on the shortage of advanced computer users in ECA indicate that both are important.

The Shortage of Complementary Skills

As in other regions, the returns to skills are increasing in many ECA countries. Moreover, the supply of skills tends to be lower and the return to skills is higher in countries where jobs are less Internet-intensive (figure 5.7). For example, the return to a college education is higher in the South Caucasus and Central Asia, where Internet intensity is low, than in Finland or Japan, where Internet intensity is high. This may indicate that the supply of college graduates is lower (relative to demand) in the poorer countries, which limits the adoption of the Internet by firms and households.

Within Europe, the skills of the new economy are particularly scarce in Central Europe. Figure 5.8 shows the wage returns to nonroutine cognitive and manual tasks for countries with comparable data. These estimates control for other factors that may be correlated with tasks, wages, and other socioeconomic characteristics of the individual, such as education, and the firm. The results indicate that the returns to nonroutine cognitive tasks have increased in Central European economies, in particular, in Bulgaria and Romania. Moreover, returns to these tasks are already higher in this subregion, which may indicate a shortage of certain skills required to perform nonroutine cognitive tasks. Accordingly, the wage penalty to workers performing nonroutine manual tasks is greater in Central Europe than in other regions, which may suggest that the skills required to perform these tasks are in relatively larger supply in this subregion. The results in figure 5.9 also suggest that skills bottlenecks are more prevalent in Central Europe than in its richer neighbors, since the returns to college education are higher among the former.

FIGURE 5.7
Returns on skills tend to be higher in countries where jobs are less Internet intensive

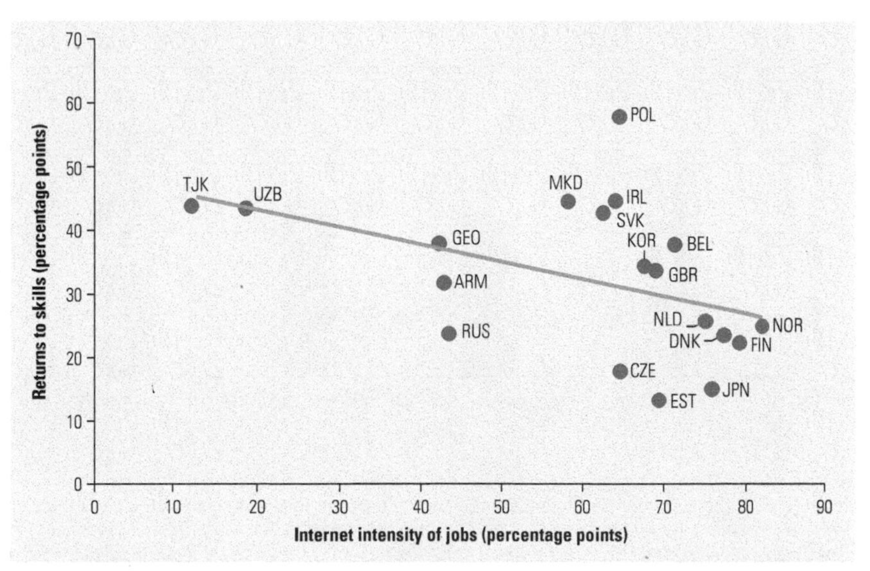

Note: Internet intensity is measured as the share of workers using a computer at work. The microdata come from PIAAC (Western Europe, East Asia, Russia, Central Europe, and Estonia), STEP (urban areas of Armenia, Georgia, Ukraine, and FYR Macedonia), and CALISS (Kyrgyz Republic, Tajikistan, and Uzbekistan). The returns to skills were estimated using an OLS regression of the logarithm of hourly earnings on individual characteristics. These coefficients are associated to a dummy variable equal to one if the worker has a college degree. The omitted category is primary education. OLS = ordinary least squares.

FIGURE 5.8 Central European countries have higher wage return (penalty) to nonroutine cognitive (manual) tasks

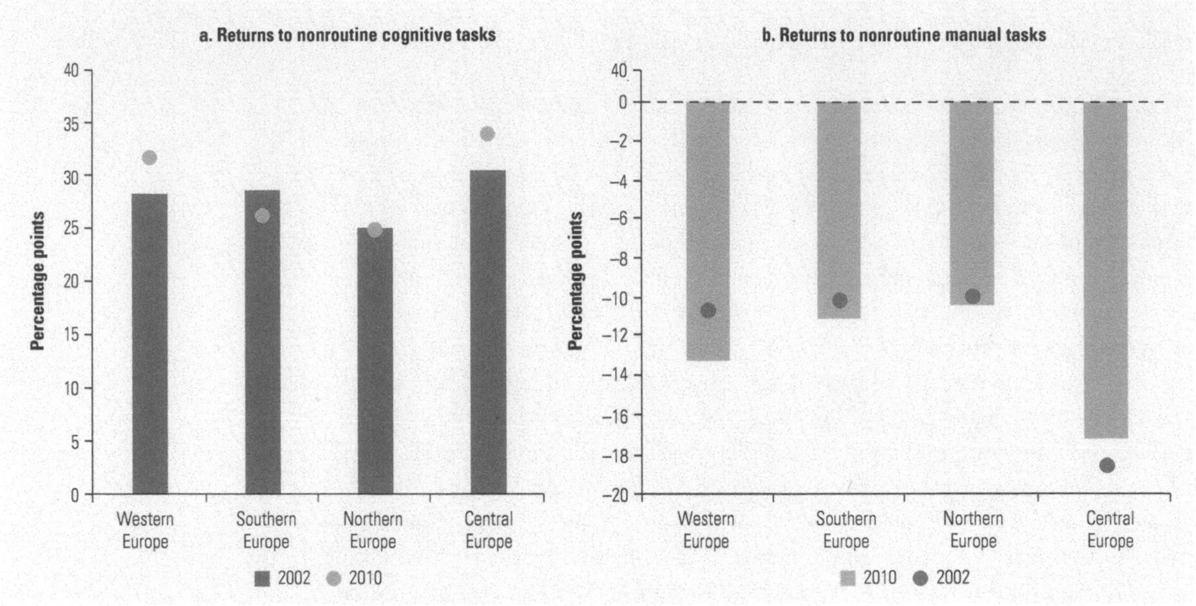

a. Returns to nonroutine cognitive tasks

Percentage points

40 35 30 25 20 15 10 5 0

Western Europe | Southern Europe | Northern Europe | Central Europe

■ 2002 ● 2010

b. Returns to nonroutine manual tasks

Percentage points

40 0 −2 −4 −6 −8 −10 −12 −14 −16 −18 −20

Western Europe | Southern Europe | Northern Europe | Central Europe

■ 2010 ● 2002

Source: Estimates based on data from Structure of Earnings Survey, Eurostat (various years).
Note: Median coefficient for the region. The coefficients were estimated using an ordinary least squares regression of the logarithm of hourly earnings on individual and firm characteristics. These coefficients are associated to dummy variables equal to 1 if the worker carries out nonroutine cognitive or nonroutine manual tasks (averaged at the subregion level). The omitted category is routine tasks. All of them are statistically significant at the 1 percent level.

FIGURE 5.9
Returns on skills are increasing in Central and Southern Europe

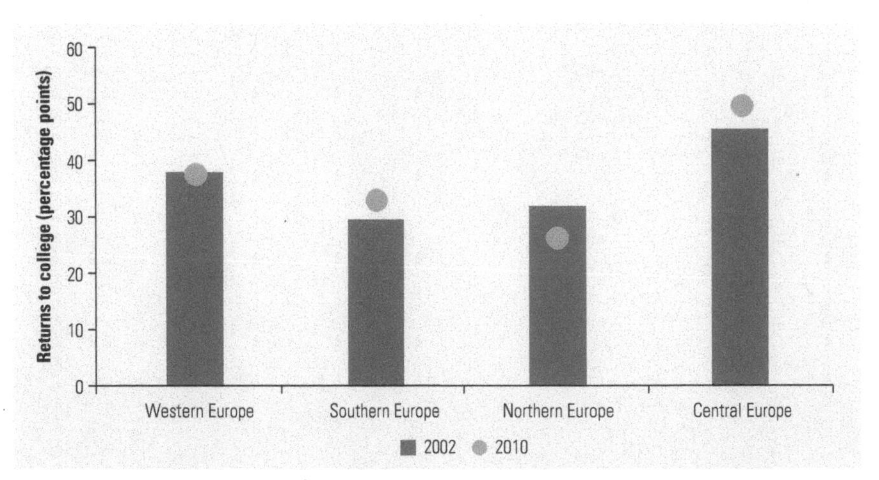

Source: Estimates based on data from Structure of Earnings Survey, Eurostat (various years).
Note: Median coefficient for the region. The coefficients were estimated using an OLS regression of the logarithm of hourly earnings on individual and firm characteristics. These coefficients are associated to a dummy variable equal to 1 if the worker has a college degree (averaged at the subregion level). The omitted category is primary education. All of them are statistically significant at the 1 percent level. OLS = ordinary least squares.

Another piece of evidence that skills bottlenecks limit the depth of Internet adoption in developing ECA is that Internet-intensive sectors are absorbing a significant share of the skilled labor force. Skilled workers are significantly more likely than unskilled workers to work in Internet-intensive sectors (figure 5.10). In Central Asia, Poland, and the Slovak Republic, skilled workers are at least 15 times more likely to work in Internet-intensive sectors than their unskilled peers. In contrast, skilled workers in developed economies are only 3–4 times more likely to work in high Internet-intensive sectors than unskilled ones. In the South Caucasus and Ukraine, skilled workers are also more likely than unskilled workers to work in high Internet-intensive sectors, but the ratios are small and not statistically significant. This may reflect the fact that the samples in the South Caucasus and Ukraine are representative of urban areas only and thereby fail to capture geographic disparities.

FIGURE 5.10 Skilled workers are more likely to work in Internet-intensive sectors than unskilled ones, especially in developing Europe and Central Asia

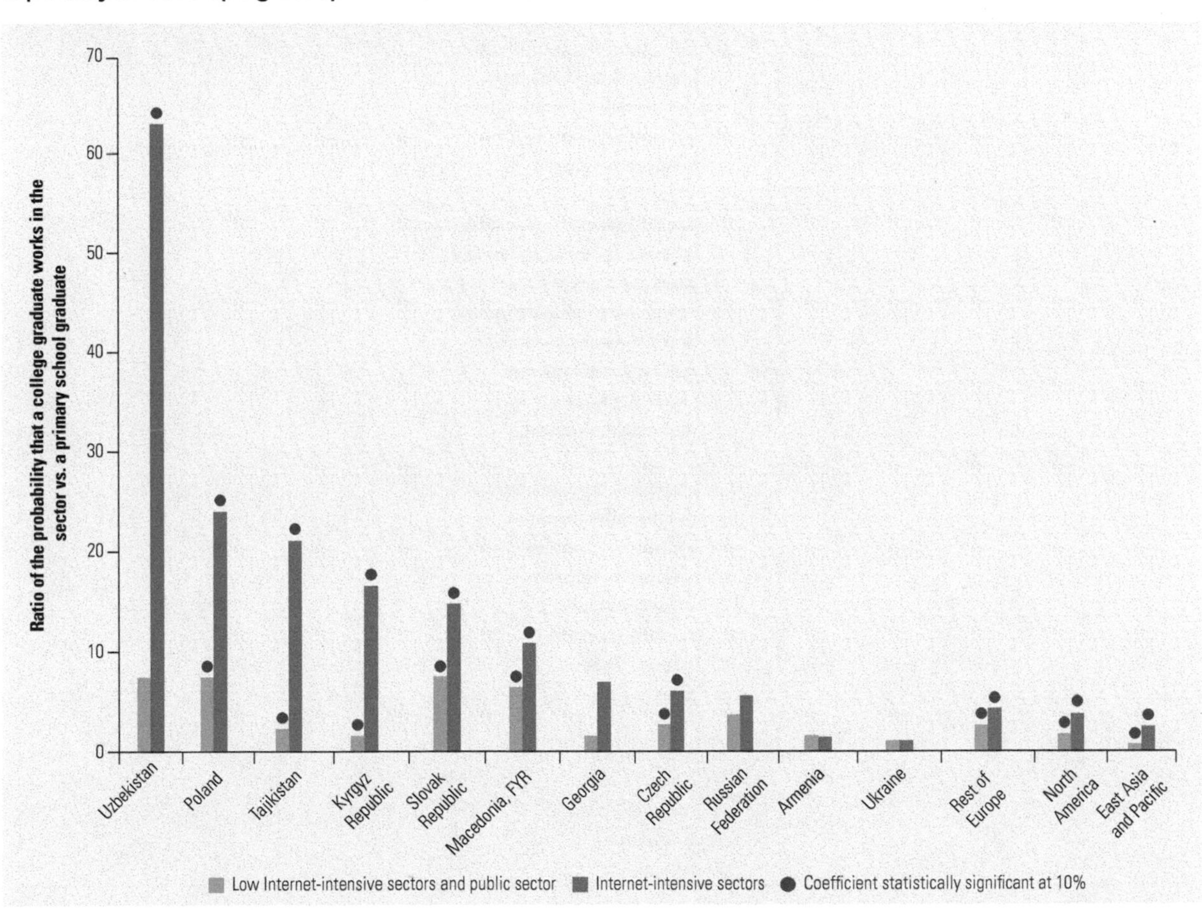

Note: Data for Georgia, Armenia, FYR Macedonia, and Ukraine are representative of urban areas only. Each bar represents the odds ratio, that is, the probability for college graduates of working in the sector over the probability of primary school graduates of working in the sector. The coefficients were estimated using a multinomial logit regression for three employment outcomes of individuals ages 25 to 64 years: not working, employed in low-Internet intensive sectors or the public sector, and employed in Internet-intensive sectors. High Internet-intensive sectors are those with a share of workers using computers higher than the median sector. The omitted category is primary education. The microdata come from PIAAC (Western Europe, East Asia, Russia, Central Europe, and Estonia), STEP (urban areas of Armenia, Georgia, Ukraine, and FYR Macedonia), and CALISS (Kyrgyz Republic, Tajikistan, and Uzbekistan).

The Shortage of Advanced Computer Skills

The limited supply of advanced computer skills also constrains Internet adoption. While most adults in ECA can carry out basic computer tasks, few are advanced users (figure 5.11). For instance, while 80 percent of adults in Serbia have basic computer skills, only 10 percent of them have carried out more advanced

FIGURE 5.11
Most people in Europe and Central Asia have some computer skills, but few are advanced users

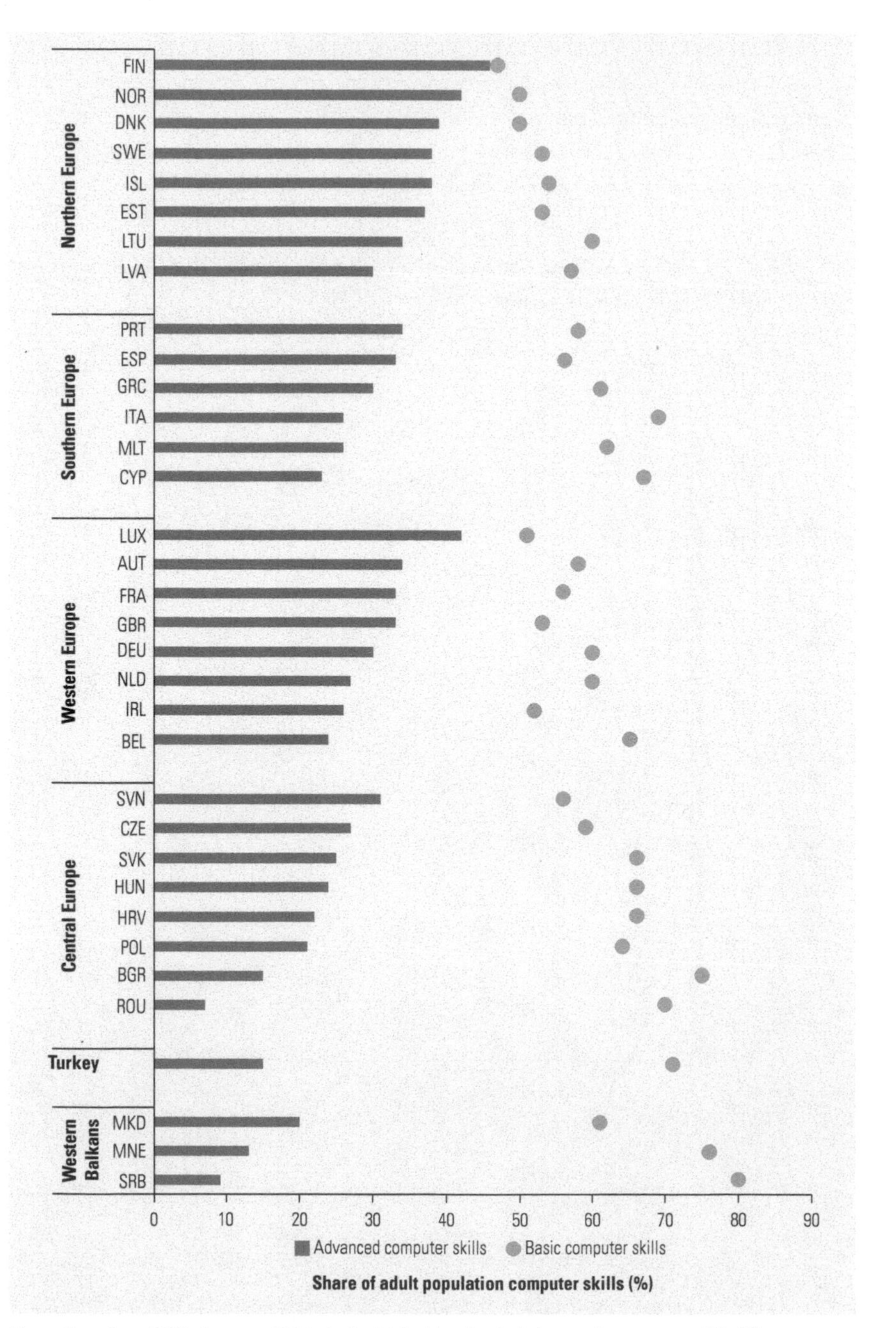

Note: Data from CSIS, Eurostat 2014. An individual has basic (advanced) computer skills if he performed 1 or 2 (5 or 6) computer activities of the list. The list of activities includes the following: launching a program, copying and pasting, performing basic arithmetic using spreadsheets, compressing files, programming, and connecting and installing devices.

computer tasks. At the other extreme, almost all adults in Finland have computer skills, with 50 percent of them being advanced users.

The limited supply of computer skills largely reflects levels of education and income. Figure 5.12 shows the estimated gap in computer skills, using a regression of a binary variable equal to 1 if the individual is an advanced user. In Central Europe, college graduates are 44 and 28 percent more likely than primary school and high school graduates, respectively, to have advanced computer skills. In Central, Northern, and Southern Europe, individuals in the top 25 percent of the income distribution are about 20 percent more likely than the poorest individuals to be advanced users. In contrast, the gaps by education and income are considerably smaller in Western Europe. The urban-rural and gender divides in computer skills are small across the region when controlling for educational attainment and income. For example, women are 10 percent less likely than men to be

FIGURE 5.12 Computer skills' gaps are largely driven by education

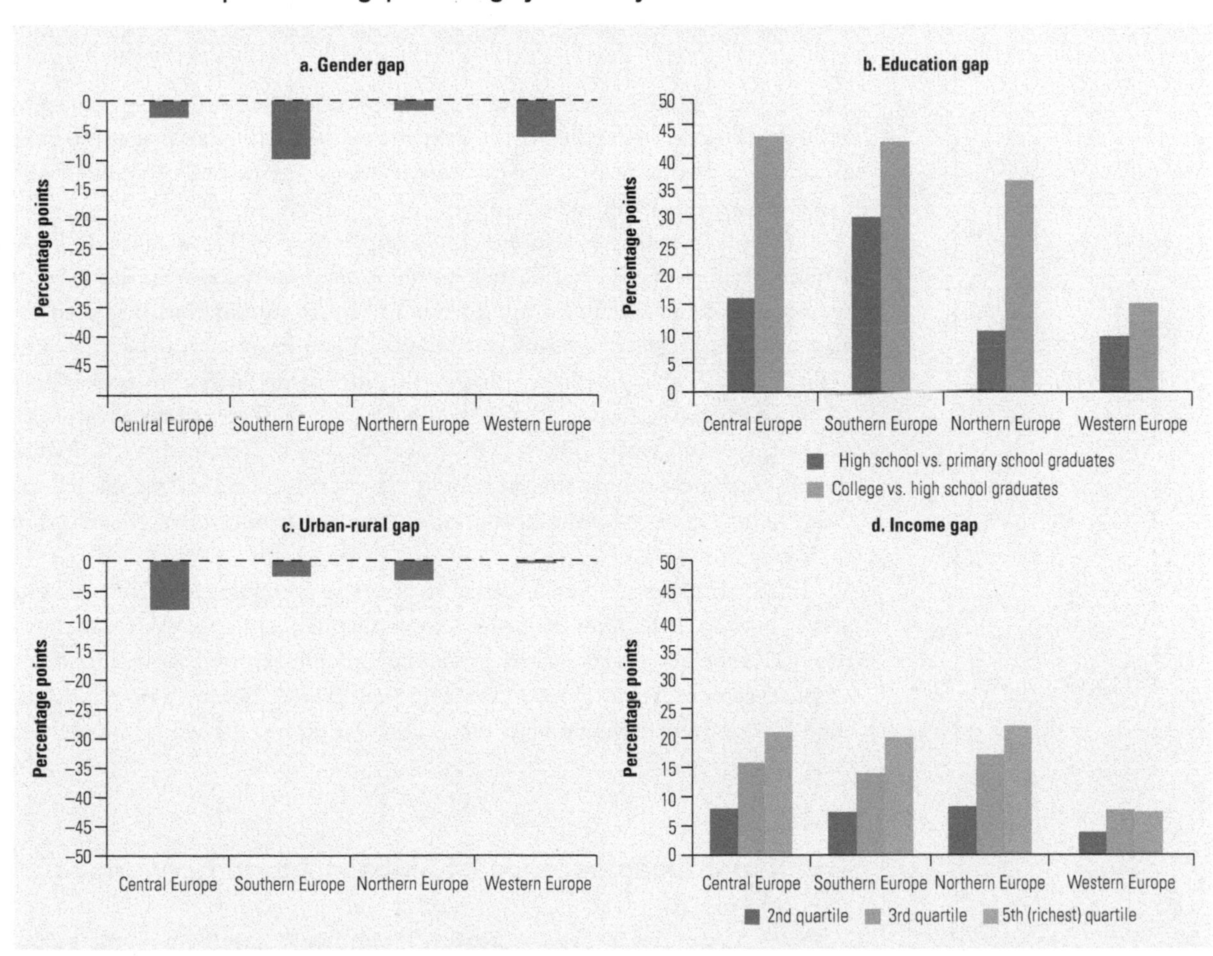

Note: Based on microdata from CSIS 2012, Eurostat. Each bar represents the estimated coefficients from an ordinary least squares regression of a dummy variable equal to one if the individual has advanced computer skills, on a set of explanatory variables including education, gender, income quartile, rural area, age, and country fixed effects. An individual has advanced computer skills if he or she performed five or six computer activities in the list. The list of activities includes the following: launching a program, copying and pasting, performing basic arithmetic using spreadsheets, compressing files, programming, and connecting and installing devices. All the coefficients are statistically significant at the 1 percent level. For panel d, the 1st quartile is the omitted category.

FIGURE 5.13
The share of ICT specialists in total employment has increased almost everywhere during the past 10 years

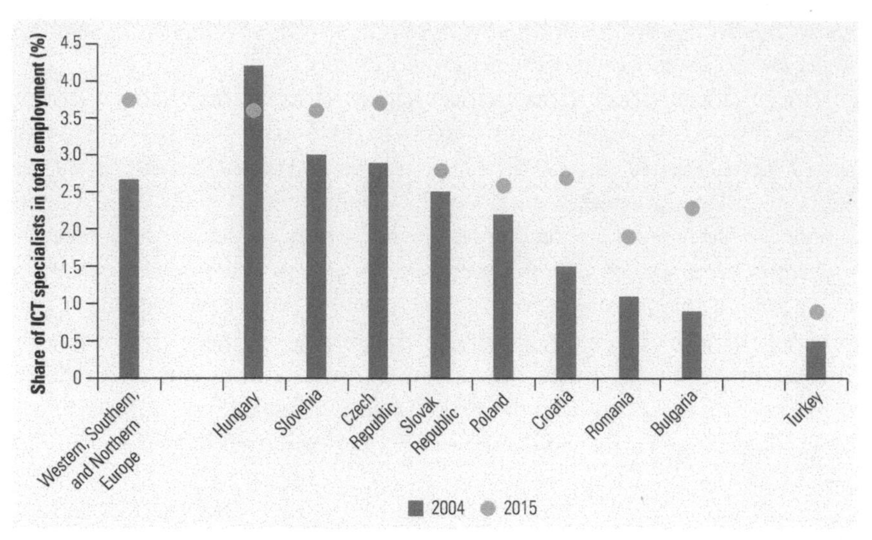

Note: Based on data from CSIS Eurostat.

advanced computer users in Western Europe. And individuals living in rural areas in Central Europe are 8 percent less likely than individuals in urban areas to have advanced computer skills. These gaps are relatively narrow when compared with those related to income and education.

Labor market surveys also indicate a shortage of ICT specialists in ECA. While the share of ICT specialists in total employment has increased in most countries with adequate data (figure 5.13), most studies find that a shortage of ICT specialists creates bottlenecks even in developed economies (Falk 2002). A study by the European Commission (Attström and others 2014) finds that ICT specialists are one of the top three occupations with the largest skills bottlenecks in Europe.[2] Within ICT specialists, software developers and systems analysts are in the shortest supply. The study finds that the main reason for the bottleneck is the lack of applicants meeting the skills requirements.

These shortages are manifested in large wage premiums for ICT specialists across Europe. Even after controlling for individual, job, and firm characteristics, ICT specialists earn wages at least 5 percent higher than comparable employees in most countries in the sample (figure 5.14). In some countries, such as Hungary, Romania, and the United Kingdom, the wage premium is close to 20 percent.

The Rapid Obsolescence of Skills in Aging Economies

Even when countries manage to provide a good skills base through the school and university system, rapid technological change could make those skills

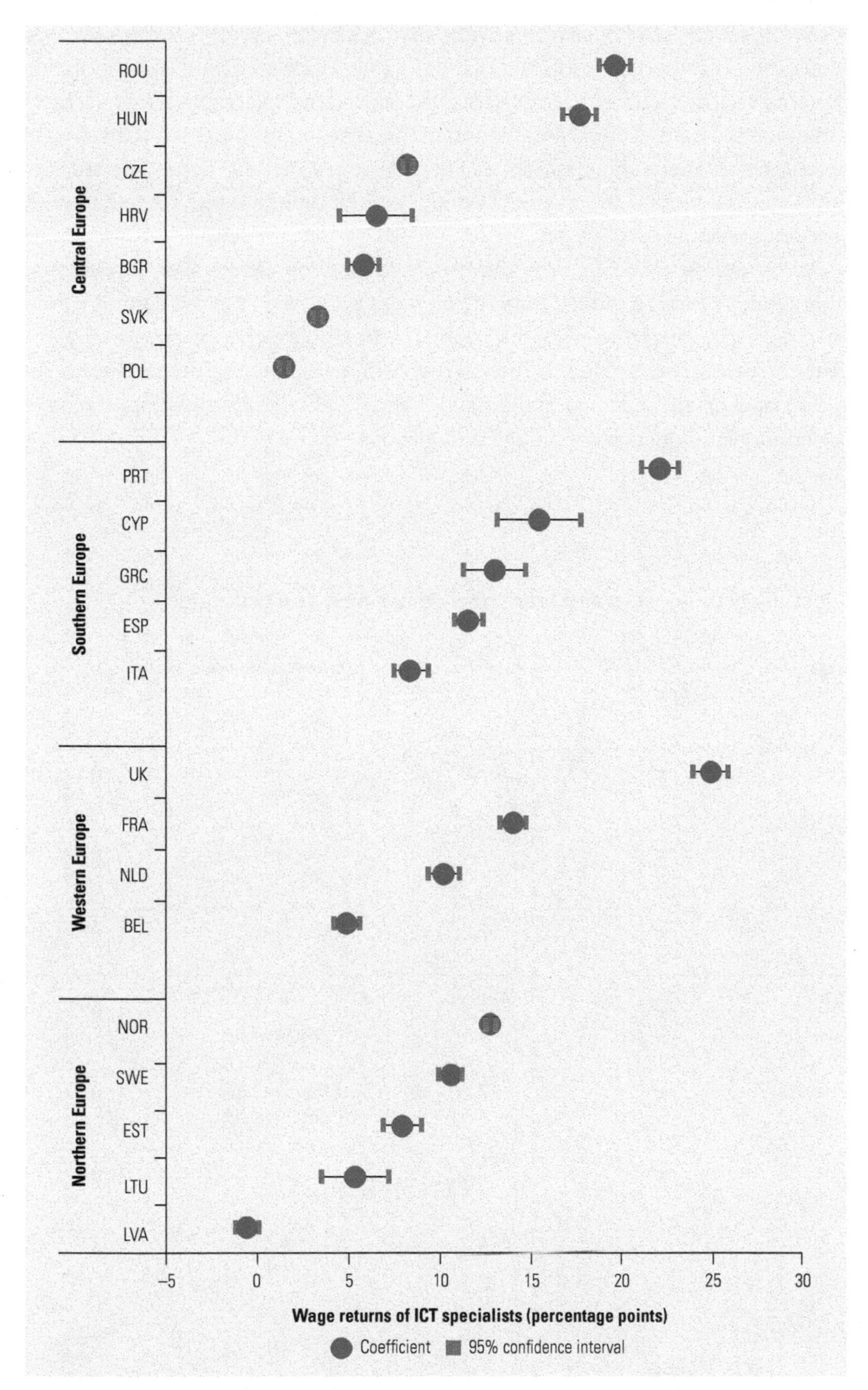

FIGURE 5.14
Even after controlling for education, ICT skills pay in the labor market

Note: Based on microdata from SES 2010 Eurostat. Estimations using an ordinary least squares regression of log hourly wages on several individual characteristics (education, age, gender, firm size, and sector) and a dummy variable equal to 1 for ICT specialists.

obsolete for adults who have completed their education cycle. This obsolescence could be exacerbated by the fact that old age is associated with a deterioration in some cognitive and socioemotional skills, which tend to complement new technologies (Bussolo, Koettl, and Sinnott 2015). These are important concerns in economies that are aging rapidly, as they may imply that the aggregate stock of skills may deteriorate rapidly as individuals' skills decline with age and older workers' skills become less relevant.

In Central Asia and Europe, the rise in educational levels over time implies that older individuals in the future will be more educated than the current cohort of old people. Nevertheless, if the pace of technological change accelerates, the skills of older cohorts could fail to keep up with the changing demand for skills. Using data on advanced computer skills, figure 5.15 illustrates the importance of distinguishing between cohort and age effects.

FIGURE 5.15 Older cohorts are less likely to possess and learn how to use new technologies

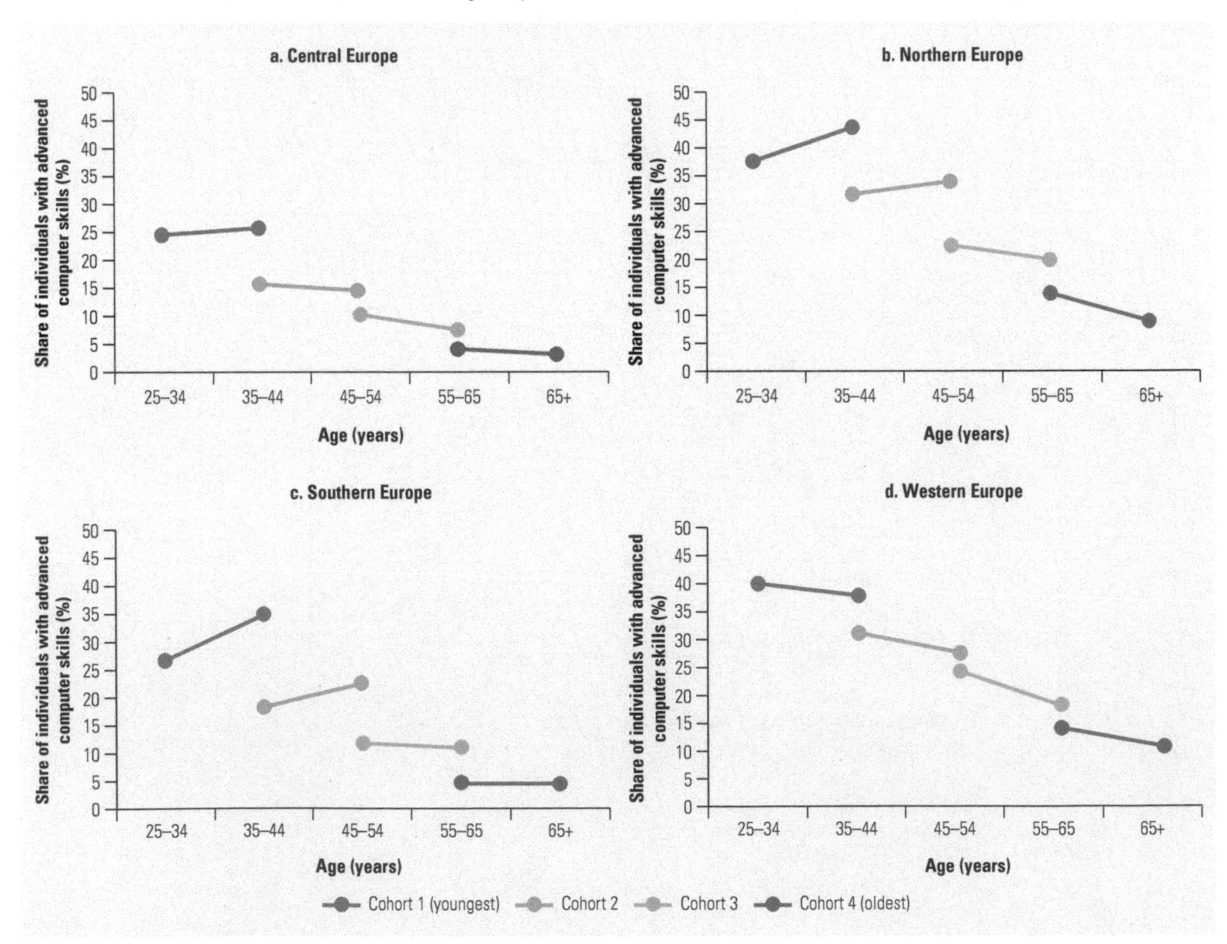

Note: Based on microdata from CSIS 2005 and 2014, Eurostat. An individual has advanced computer skills if he performed 5 or 6 computer activities of the list. The list of activities includes launching a program, copy and paste, basic arithmetic using spreadsheets, compressing files, programming, and connecting and installing devices. The initial (final) observation for each cohort corresponds to the year 2005 (2014).

In Central and Western Europe, most of the age gap in computer skills is driven by aging, that is, the fact that individuals are less likely to use computer skills as they become older. In Northern and Southern Europe, the age gap is driven by a mix of generational change and age effects, as younger cohorts are more likely to have and to learn advanced computer skills than older cohorts (their profiles have a positive slope). Given the current state of lifelong learning institutions in the region, older cohorts in the future may have a hard time adapting to new technologies, as their skills-age profile looks rather flat. At the same time, the experience of workers age 35–44 in Northern and Southern Europe shows that it is possible to adapt to new technologies even after completing a formal education.

The gap between old and young workers also exists when analyzing cognitive and socioemotional skills. Younger generations in ECA have better cognitive abilities—literacy, numeracy, and problem solving—than older generations (Bussolo, Koettl, and Sinnott 2015). At the same time, aging also affects personality traits and noncognitive skills, some of which are important complements to new technologies. The most common model for personality traits is the Big Five model, a taxonomy of broad families of personality traits along the following dimensions: openness to experience, conscientiousness, extraversion, agreeableness, and emotional stability (John and Srivastava 1999). There is some agreement in the literature that openness and extroversion are socioemotional skills that tend to decline with age. Being open to new ideas is a crucial component of being adaptable to new technologies and learning new skills. Data for Central Asia are consistent with this hypothesis, as older individuals are less likely to score high on measures of openness than their younger peers (figure 5.16).

FIGURE 5.16
Internet-complementary skills decline rapidly with age in Central Asia

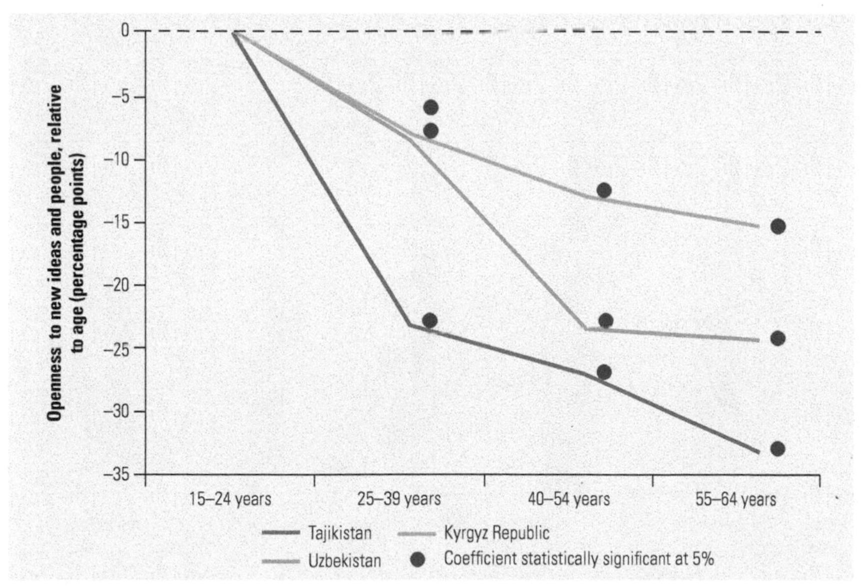

Note: Based on microdata from CALISS, circa 2013. Estimated using a linear probability model, where the dependent variable is equal to one if the individual scored in the top 25 percent in openness. Explanatory variables include education, age, and gender. Each line shows the coefficients associated with the age dummy variables.

The Changing Demand for Skills May Affect Shared Prosperity and Inequality in ECA

The Internet may be associated with increased inequality. The Internet increases the demand for workers performing nonroutine cognitive tasks, who are more likely to be in the top 60 percent of the income distribution, and reduces the demand for workers performing routine tasks, who are typically in the middle of the income distribution (figure 5.17). Moreover, these data reflect only persons who are employed; individuals with routine skills who are excluded from the labor market are more likely to be in the bottom 40 percent. Nonroutine manual workers, who are more likely to be low income, may also benefit from an increase in demand. However, this increased demand may not materialize in higher wages, as their supply may expand quickly as routine workers move into manual tasks (WDR16). In addition, the rise of ICTs is associated with a decline in the share of national income going to workers across the world (Eden and Gaggl 2015; Karabarbounis and Neiman 2013), and most countries in ECA have witnessed a dramatic decline in the share of income going to labor since 1995 (figure 5.18). As capital income is distributed more unequally than labor income, a decline in the labor share of income could contribute to a rise in inequality. Simulations using an economic model calibrated to match features of European economies indicate that the Internet could have a significant impact on inequality (annex 5A).

FIGURE 5.17 Workers in the bottom 40 of the income distribution are less likely to have occupations intensive in advanced tasks

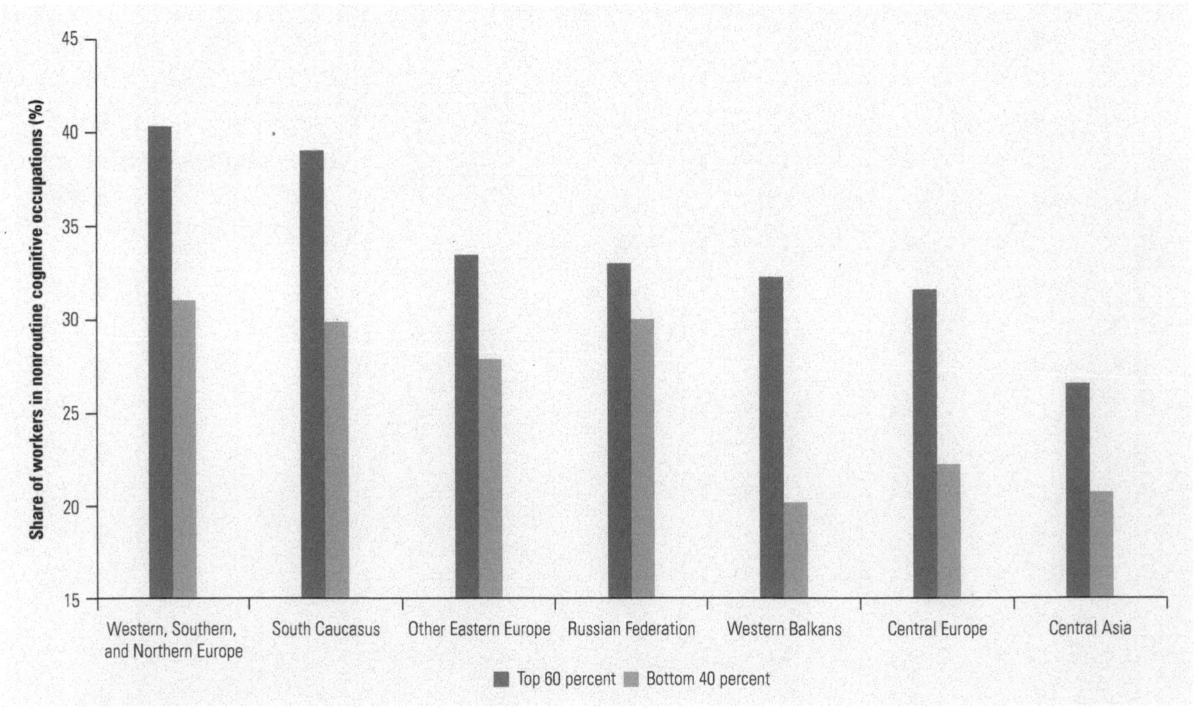

Source: LITS II 2010.
Note: Sample excludes agricultural workers. The blue bars show the difference between the bottom 40 and the top 60 in the share of workers in occupations intensive in nonroutine cognitive tasks. For example, in Hungary, workers in the bottom 40 are about 8 percent less likely to be in these occupations than are workers in the top 60.

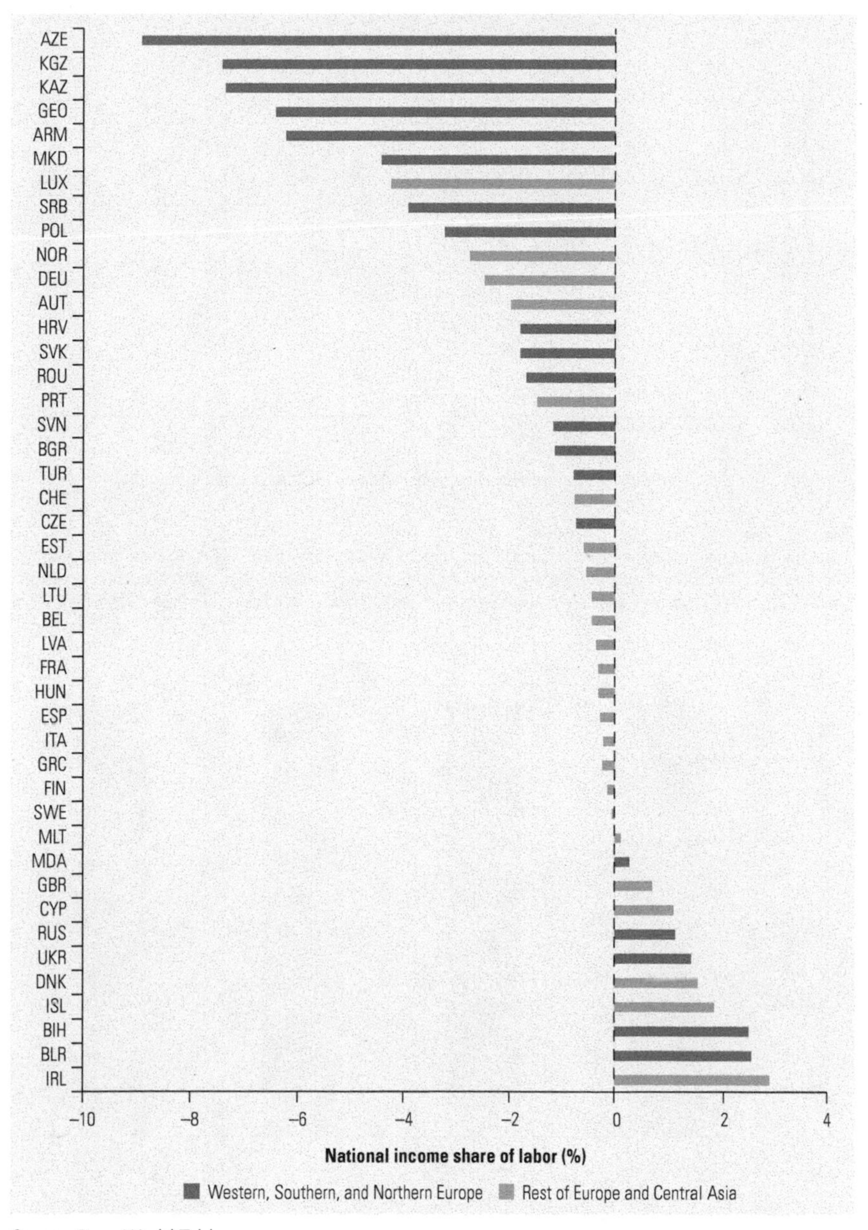

FIGURE 5.18
Labor shares of income are declining in most ECA countries

Source: Penn World Tables.
Note: Each bar represents the average 10-year trend between 1995 and 2011.

Further evidence of the impact of the Internet on inequality is that ICT-intensive firms pay wages about 14–60 percent higher than other firms (figure 5.19).[3] This difference is important, as the dispersion of wages across firms, rather than individual or job characteristics, is the main contributor to wage inequality in the region (figure 5.20). If the use of ICT improves labor productivity, unequal access to these technologies could contribute to productivity and wage inequality in Europe, especially in Central and Northern Europe. In fact, the fraction of total wage inequality explained by wage gaps across firms (as opposed to within firms) is higher in countries where the wage premium associated with working in ICT-intensive firms is also higher.

FIGURE 5.19
The technological divide between firms is correlated with wage inequality across firms

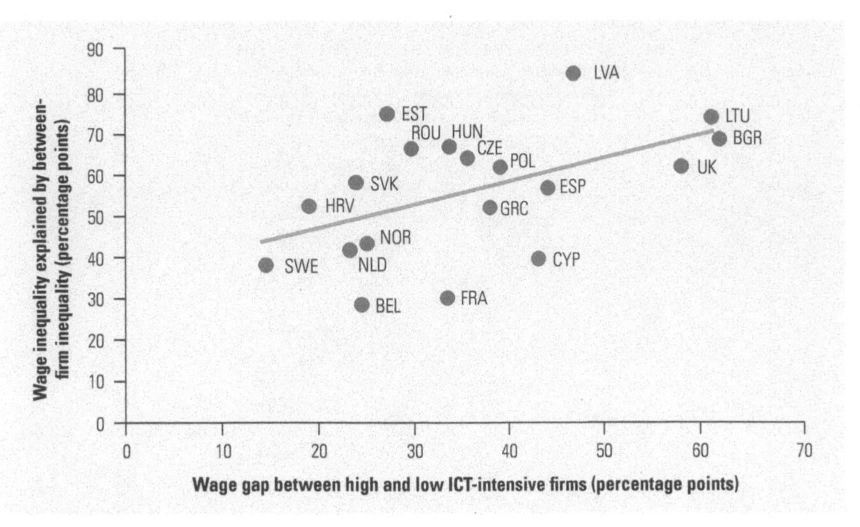

Source: Based on SES 2010.
Note: Firms are divided into high and low ICT-intensive by whether they have ICT specialists employed. See figure 5.20 for a description of the wage inequality measure. The wage gap is the difference between the average wage paid by high ICT-intensive firms and low ICT-intensive firms, as a percentage of the latter.

FIGURE 5.20
Firms' characteristics account for a large share of wage inequality, especially in Central Europe and Northern Europe

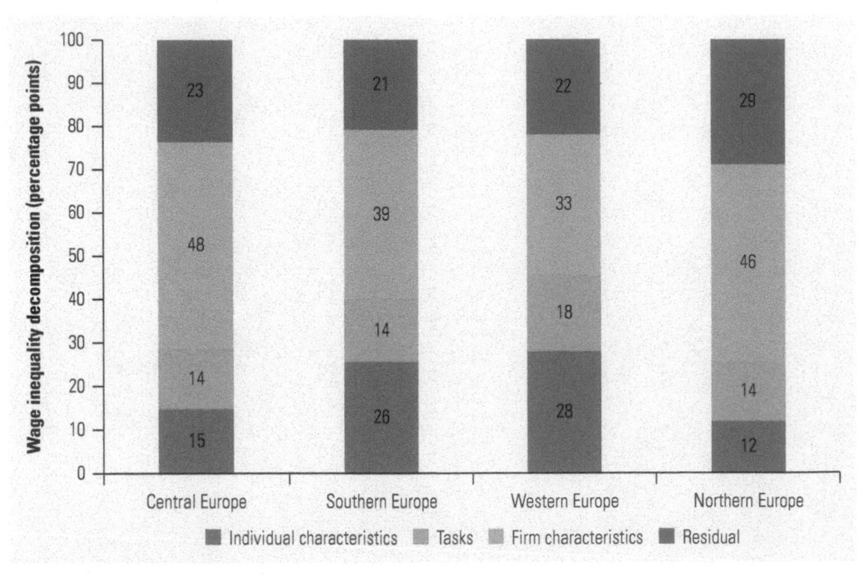

Source: Based on SES 2010.
Note: Based on a simple Fields' decomposition of the coefficient of variation of the logarithm of hourly wages. Individual characteristics include age, gender, education, full-time job, temporary job, age, age squared, tenure, and tenure squared. Firm characteristics are simply the firm fixed effects (the SES is conducted at the firm level). Tasks are the dummy variables for nonroutine cognitive and routine tasks. Some values do not add up to 100 because of rounding.

Policies for Developing the Skills of the Digital Economy

While increasing the supply of new economy skills could alleviate some of the skills bottlenecks affecting ECA countries, the ranking of policy priorities that would generate the largest benefits at the least cost is unclear. The WDR16 provides a useful framework for this discussion. In emerging digital countries, such as those in the South Caucasus, Central Asia, and Turkey, there is a shortage of basic foundational skills as well as basic digital literacy. Transitioning countries, such as Russia, Ukraine, and some Central European economies, have done fairly well in building basic skills, but they have a shortage of higher-order skills that complement new technologies. Finally, transforming countries, such as those in Northern and Western Europe, the Czech Republic, and Slovenia, which are aging rapidly, may put more emphasis on developing advanced technical skills (that is, ICT and science, technology, engineering, and mathematics [STEM] specialties) and lifelong learning systems.

Classifying countries into emerging, transitioning, and transforming stages is far from straightforward. Figure 5.21 provides a framework by grouping countries according to their adaptability to new technologies and the expected labor market disruption. Quality-adjusted years of education are used to measure adaptability, that is, the ease by which workers will benefit from rapid Internet adoption. Expected disruption is approximated by Internet penetration, as the biggest changes in the demand for skills are yet to come in countries that have not yet fully embraced the Internet. Table 5.1 provides the country groupings and policies suggested by this framework. However, some caution is needed when analyzing particular countries, as there are no clear jumps or discontinuities in these indicators across groups.

FIGURE 5.21
Emerging, transitioning, and transforming countries for skills policies

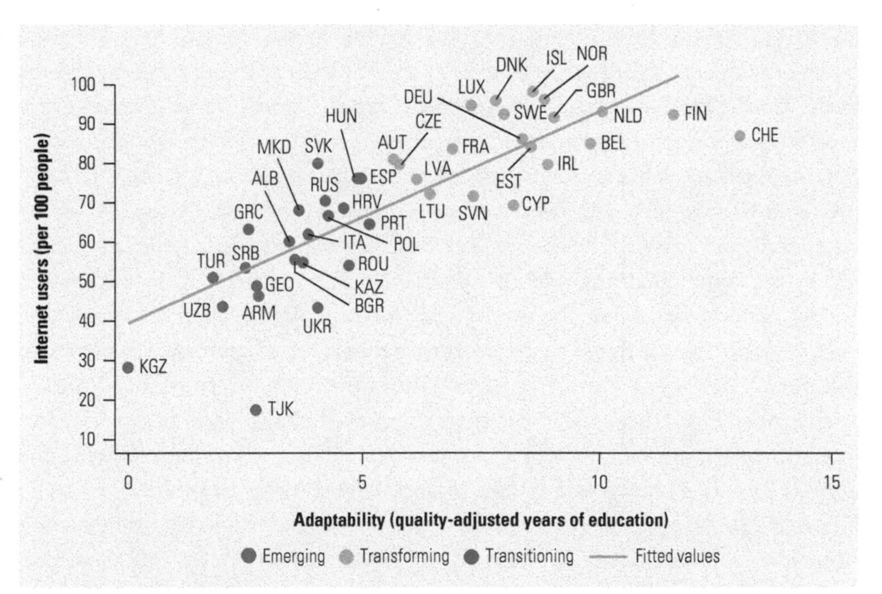

Sources: Calculations are based on using data on Internet users from World Development Indicators 2014 and quality-adjusted years of education from World Development Report, World Bank 2016.
Note: The quality-adjusted years of education are constructed by adjusting average years of education for each country with the World Economic Forum's quality-of-education indicator. For example, if a country has, on average, 10 years of education and scores 3.5 on the indicator (which ranges from 0 to 7), its quality-adjusted years of education are 5. See Monroy-Taborda, Moreno, and Santos, forthcoming, for the WDR 16.

TABLE 5.1 Policy Priorities for Emerging, Transitioning, and Transforming Countries

Groups	Countries	Policy priorities
Emerging	Armenia, Georgia, Kyrgyz Republic, Serbia, Tajikistan, Turkey, Uzbekistan	• Foster early childhood development • Improve quality of primary and secondary education
Transitioning	Albania, Bulgaria, Croatia, Greece, Hungary, Italy, Kazakhstan, FYR Macedonia, Poland, Portugal, Romania, Russian Federation, Slovak Republic, Spain, Ukraine	• Achieve better synchronization between skills demanded and skills taught at upper-secondary and tertiary levels • Use curricular reforms to introduce more innovative and creative learning • Make labor regulations more flexible to foster employer-provided training
Transforming	Cyprus, Czech Republic, Northern Europe, Slovenia, Western Europe	• Use ICT for more creative learning • Address gender gaps in STEM education • Foster lifelong learning • Reduce barriers to immigration of workers

Note: ICT = information and communication technology; STEM = science, technology, engineering, and mathematics. Countries are grouped based on their quality-adjusted years of education and Internet use, shown in figure 5.21. Data on years of education are not available for Azerbaijan, Bosnia and Herzegovina, Moldova, and Montenegro.

Building Foundational Skills in Emerging Countries

Most countries in the ECA region have achieved high levels of educational attainment (Sondergaard and others 2012), and many countries in the region deliver relatively good-quality education in the early grades. For example, most Central European economies and Russia performed very well on the Progress in International Reading Literacy Study (PIRLS) at the fourth-grade level in both 2006 and 2011.[4] Nevertheless, emerging digital economies are failing to deliver a good quality of education for their young children. For example, fourth graders in the South Caucasus, FYR Macedonia, and Moldova score well below their peers in Western Europe. And enrollment in preprimary education, which is important for the development of the socioemotional and cognitive skills that are essential complements for new technologies, is particularly low in the South Caucasus, Central Asia, and the Western Balkans, as well as in Croatia and Turkey (figure 5.22).[5] Moreover, some countries, such as Azerbaijan, FYR Macedonia, Tajikistan, and Turkey, perform worse than the average Sub-Saharan African country in this area.[6]

Similar patterns emerge when analyzing measures of educational quality at the secondary level. Data from the Programme for International Student Assessment (PISA) show that more than 50 percent of adolescents age 15 years in Central Asia are functionally illiterate—that is, they know how to read and write, but cannot make inferences or understand forms of indirect meaning (figure 5.23). Teenagers in the Western Balkans are also failing to achieve levels of functional literacy close to those of high-income economies. The same patterns hold when examining performance in mathematics and science. In contrast, Russia and many economies in Central Europe—except Bulgaria and Romania—are performing close to the level of developed economies.

Emerging countries also face the challenge of digital illiteracy. As seen in this chapter, while a large share of individuals in ECA know how to carry out very basic computer tasks, only a minority have the computer skills to perform more advanced tasks such as using spreadsheets or installing devices. At least in

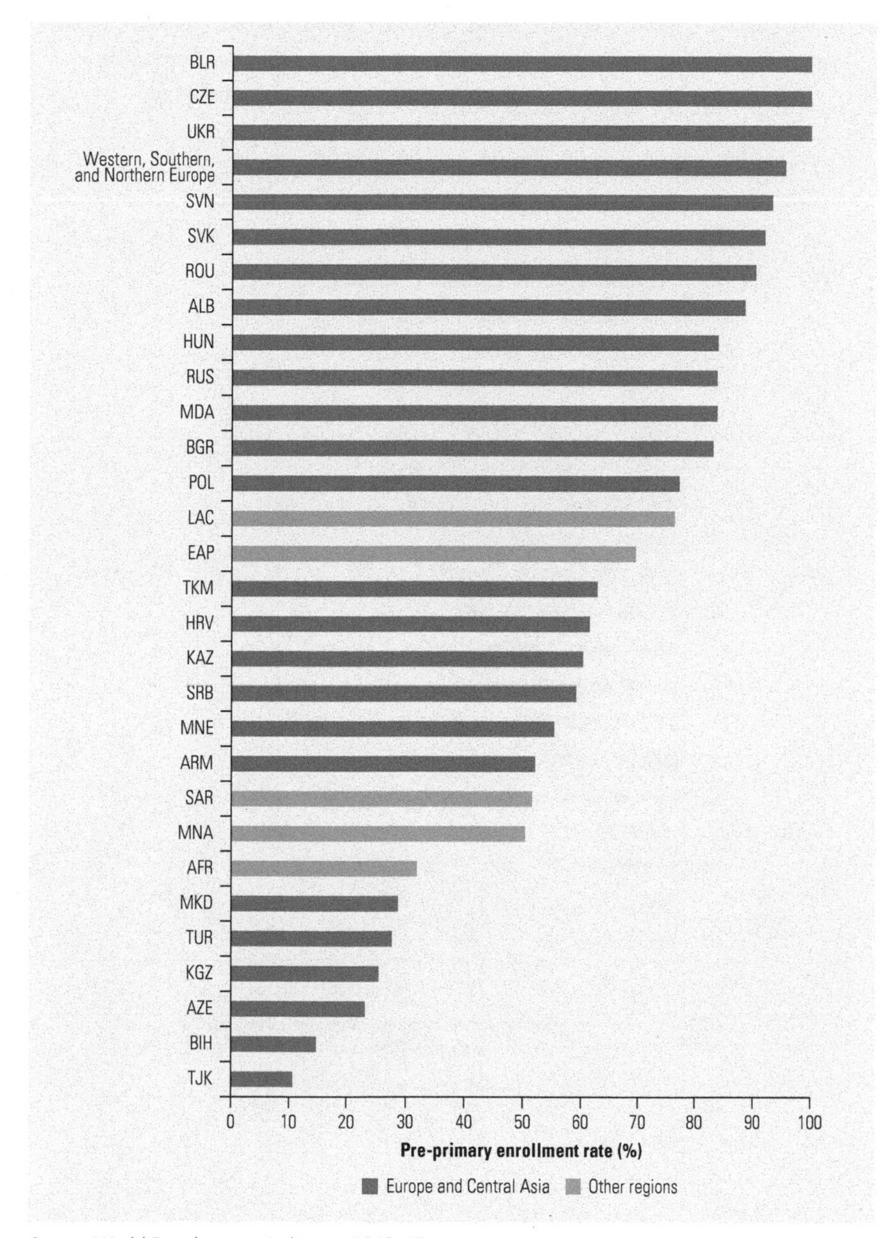

FIGURE 5.22
Emerging digital economies are falling behind in early childhood education

Source: World Development Indicators 2012–15.

Europe, gaps in computer skills are driven largely by differences in educational attainment. Thus, emerging digital economies may be able to catch up with the level of ICT skills of transforming digital economies as they invest in building up their foundational skills. At the same time, while in transitioning and transforming economies a significant share of prime-age advanced computer users learned their skills through formal education, an even larger share acquired their skills through learning by doing or informal assistance from colleagues, relatives, and friends (figure 5.24). Higher educational attainment can improve digital literacy directly by including computer classes in the curriculum, but also indirectly by creating the skills necessary to learn independently.

FIGURE 5.23
A large share of 15-year-olds is functionally illiterate in emerging digital economies

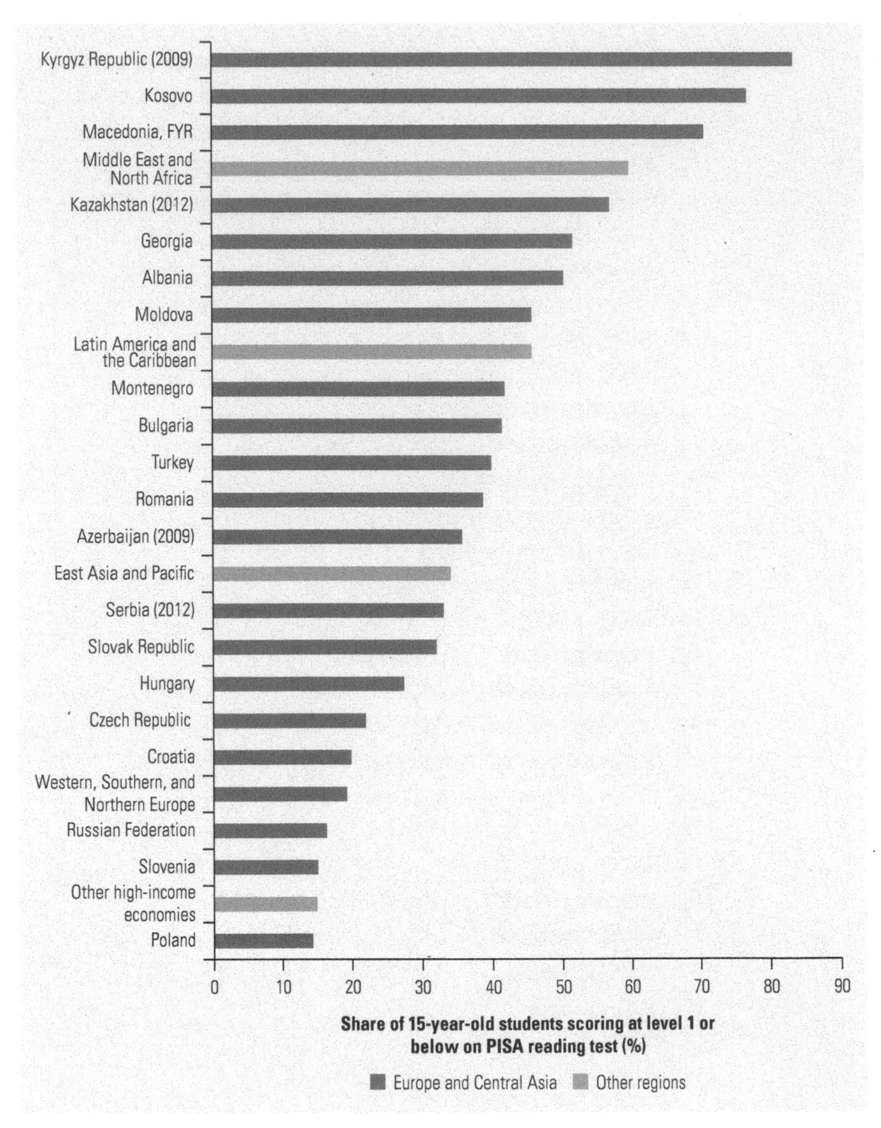

Source: PISA, latest available data (2015, 2012, and 2009).

More specific programs to develop computer skills have shown mixed success. The One Laptop per Child program, whose main objective is to empower children by providing them with digital devices at an affordable price, has had very small positive effects on academic performance. Moreover, the effect was basically zero when the program was not accompanied by some level of guidance. As the WDR16 argues, digital technologies are more likely to have an impact when the focus is not on hardware and software, but on how they contribute to learning. Technology-assisted learning, together with teacher training programs, have among the largest positive effects on learning outcomes. A recent series of World Bank pilot programs in Tajikistan also illustrates the importance of complementary factors when trying to use new technologies to build skills (box 5.2).

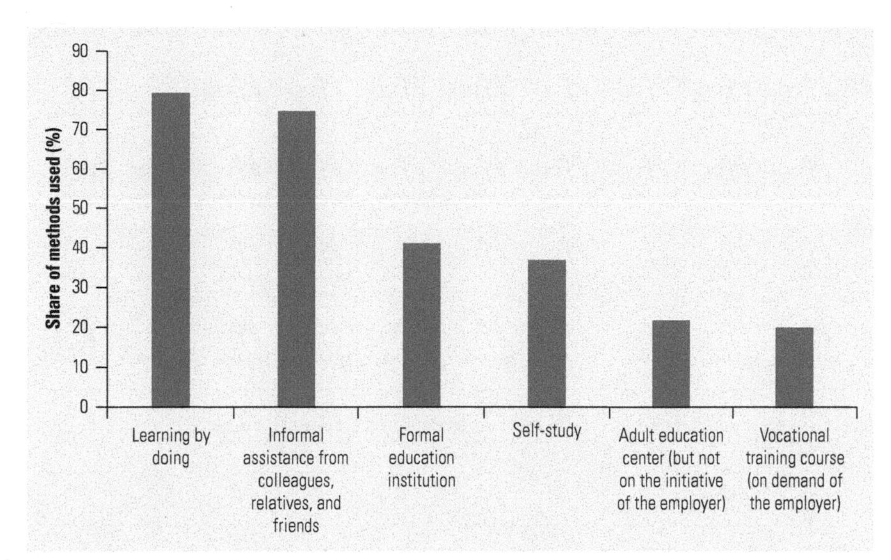

FIGURE 5.24
Both formal and informal methods matter to learn advanced computer skills in ECA

Note: Data from CSIS 2011 Eurostat. Sample includes individuals 25 to 54 years old. An individual has advanced computer skills if he or she performed five or six computer activities of the list. The list of activities includes the following: launching a program, copying and pasting, performing basic arithmetic using spreadsheets, compressing files, programming, and connecting and installing devices.

BOX 5.2 How Can the Internet Be Used to Build Skills?

The rise of massive online open courses (MOOCs) offers a unique opportunity to build skills across the world. MOOCs are characterized by open enroll ment, online assessment, and an interactive forum. For instance, in April 2015, eight of Russia's leading higher education institutions created a noncommercial organization—the Russian National Platform for Open Education—to develop and deliver MOOCs. The platform will provide any student in Russia with access to the best MOOCs. Online education could help to overcome some of the barriers faced by certain demographic groups, such as persons living in remote areas, those with disabilities, and those unable to afford a full-time schedule in a formal educational institution.

However, as a recent series of pilot studies in Tajikistan illustrates, online education platforms are far from being the silver bullet unless they are accompanied by other factors. A recent World Bank study supported a small project to pilot four ICT-based solution models with five partners (higher education institutions and nongovernmental organizations) to tackle challenges in higher education. These pilots included interventions such as creating online education modules for teacher training, lifelong learning, university courses, and firm training as well as building awareness about existing online courses. The results from these interventions highlighted the factors that may limit the positive effects of online education on developing countries:

- Poorer areas that may benefit the most from distant learning often lack not only proper Internet access but also electricity.
- More capacity needs to be developed for teachers and information technology (IT) departments in universities to provide support for online course delivery. In particular, since teachers tend to be reluctant to learn new technologies, it is important to provide them with enough support to acquire new pedagogy

(Continued)

BOX 5.2 How Can the Internet Be Used to Build Skills? *(continued)*

using technologies as well as an opportunity to develop online courses.

- Lack of diploma accreditation weakens the incentives to enroll and complete the courses. Intuit, a Russian online course platform, provides an interesting example. This platform issues certificates after course completion that are recognized in Russia, where most of the graduates from Tajikistan go for work. Most students take courses in Russian or Tajik, as their English language skills are weak. Yet, too little online content is available in Tajik.
- A quality assurance mechanism for online content is a key to ensuring that distance education provides a high quality of education. Copyright issues of the online education resources also need to be addressed.

Source: World Bank 2014, with input from Kirill Vasiliev and Dmitry Chugunov.

Transitioning Countries: Build Skills, Not Just Diplomas

A World Bank assessment of skills shortages in the region finds that the quality of upper-secondary and tertiary education in many countries is not keeping up with the changing demand for skills (Sondergaard and others 2012). OECD (2013) claims that this is the case in Russia, where the highly specialized and compartmentalized nature of the tertiary education system, often closely affiliated with specific sectors of the economy, leaves graduates with skills that have limited transferability across occupations or sectors. Moreover, the lack of engagement of employers with the education system in Russia and Ukraine contributes to the mismatch between the changing demand for skills and supply of new graduates. This could be improved, for example, by having students spend a significant part of their program in the workplace (as in Denmark and Switzerland).

At the same time, curricular reforms to shift the focus from routine learning to applying knowledge more creatively and innovatively will be needed to foster the development of socioemotional and cognitive skills, which tend to complement new technologies. For example, Finland has been streamlining new economy skills into the traditional curricula by promoting student autonomy as well as collaborative classroom practices by using multidisciplinary and project-based studies where several teachers work with any given number of students simultaneously (World Bank 2016).

Labor regulation reforms can also help to increase the incentives for firms to provide training to their workers in order to keep up with technological change. For example, high labor taxes and inflexible contractual arrangements in Ukraine could be limiting the flexibility of employers to adapt to changing needs and to broaden employment opportunities (Del Carpio and others 2017).

Transforming Countries: Building Advanced Technical Skills and Lifelong Learning

Northern and Western European countries as well as most of the Central European economies also need to focus on developing higher-order cognitive and socio-emotional skills. However, as most of them have achieved higher levels of educational attainment and quality of education, they also may need to focus more on the next-mile challenges—that is, developing more advanced technical skills.

As noted, ICT specialists represent one of the groups with the most severe skills bottlenecks in Central and Western Europe. In order to address the increasing demand for ICT skills, some advanced economies have already introduced coding as a subject in the general education curriculum (WDR16). While not everyone needs to become an ICT specialist, these policies could strengthen critical thinking if training in coding is used as a vehicle for solving problems. ICT skills can also be learned informally when the skill base is solid, as in transforming countries. As seen in this chapter, most advanced computer users in Central and Western Europe have learned their skills through learning by doing or with their colleagues' assistance. Transforming countries may also satisfy their demand for ICT specialists by facilitating the immigration of high-skill foreign workers. For instance, immigrant workers are overrepresented among IT occupations in Canada, when compared with their share of total employment (OECD 2004). Moreover, transforming countries could also explore innovative ways of using the Internet to build skills (box 5.2).

Addressing gender gaps in STEM education could help to boost the supply of ICT specialists in ECA. In most European countries, only about a third or less of ICT specialists are women, and the gender gap has risen in almost every country during the last 10 years.[7] These gender gaps start early in life, as girls perform worse than boys in mathematics (OECD 2015). Encouraging parents and teachers to become more aware of these gender gaps could increase the participation of women in the Internet economy and reduce bottlenecks in ICT skills.

The other challenge faced by transforming countries in ECA is the rapidly aging labor force. The share of people older than age 64 in the total population increased from 6 percent in 1950 to 12 percent in 2015 (Bussolo, Koettl, and Sinnott 2015). According to simple extrapolations based on the United Nations medium-fertility demographic scenario, by 2050 the share of older people could reach 21 percent. This could affect the way economies embrace new technologies, since a large fraction of young workers in Europe have learned advanced computer skills in the formal education system. Unless firms are able to provide training to their workers throughout their careers or institutions provide sufficient opportunities for lifelong learning, workers who have already completed their formal education may not be able to develop skills demanded by a rapidly changing labor market.

Formal education seems to matter not only for learning how to use digital technologies, but also for improving the capacity to keep up with the changing demand for skills throughout the life cycle. As seen in figure 5.25, while older individuals are less likely to have advanced computer skills, the age gap is much larger for those with less education. For instance, among persons age 55–65 years

FIGURE 5.25 Advanced computer skills are more stable throughout the life cycle for skilled workers, especially in Western Europe

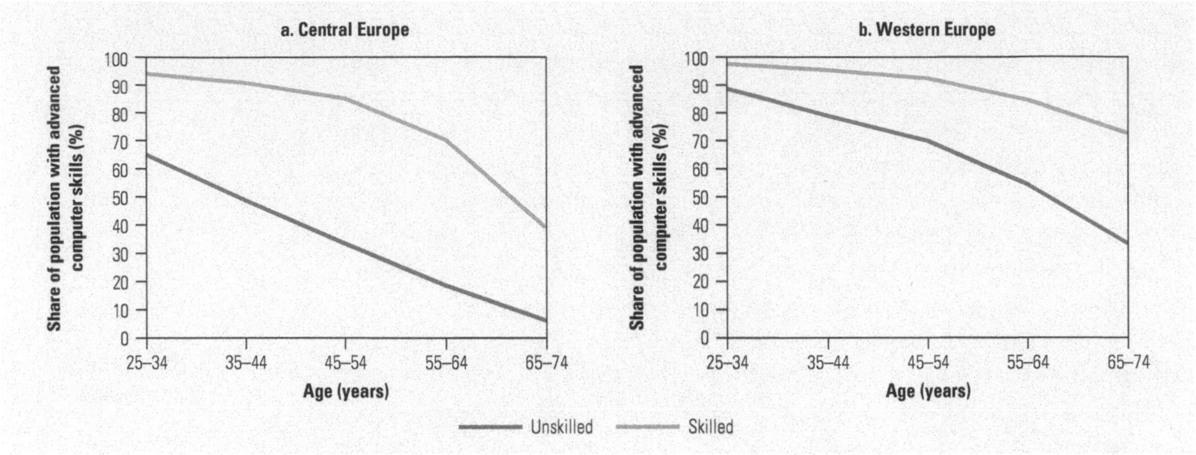

Source: CSIS 2011 Eurostat. The vertical axis measures the proportion of people in each age bracket with advanced computer skills. Skilled workers have college degrees; the rest are unskilled workers.

in Central Europe, about 70 percent of college graduates have advanced computer skills, compared with only 19 percent of noncollege graduates. The case of Western European economies is particularly interesting, as computer skills barely decline with age among skilled workers. Moreover, even unskilled workers across age groups in Western Europe perform better than their peers in Central Europe. Differences in the stage of development of lifelong learning institutions may explain this gap, as these institutions are far more developed in the former group of countries than in the latter. While less than 5 percent of individuals age 54–65 in most European countries participate in education and training, many Western European economies have participation rates above 10 percent.[8] Improving the quantity and quality of secondary and tertiary education, as well as developing lifelong learning institutions, could help workers to keep up with the changing demand for skills.

The Skills Challenge of the Internet Economy in ECA

The diversity of the ECA region provides a microcosm of the variety of challenges facing countries at very different stages of development during times of rapid technological progress. While the lack of skills is preventing the adoption of new technologies in the poorer countries of the region, Internet adoption in transitioning and transforming countries is benefiting many workers, but leaving behind those without the right skill set. While building basic foundational skills in the South Caucasus and Central Asia is the priority for closing the gap with more advanced economies, building high-order cognitive and socioemotional skills is the main challenge facing transitioning economies if they want to avoid increasing income inequality and polarization. Finally, transforming economies face the challenges of scarce ICT specialists and a rapidly aging workforce. Releasing these skills bottlenecks will not only increase the economic benefits of the Internet, but also facilitate their more equal distribution across socioeconomic groups.

Annex 5A. Impact Studies

The Impact of Telecommunications Reform on the Share of Nonroutine Workers

To measure the labor market effects of telecommunications reforms, we use disaggregated labor force data at the sector of activity level for 28 countries between 1995 and 2013.[9] Using these data, it is possible to control for other countrywide factors that may affect both the timing of the reforms and the demand for skills. This framework allows us to test a prediction from economic theory, that is, if reforms aimed at making the Internet more accessible and affordable change the demand for skills, more dramatic changes would be expected in sectors that are more Internet-intensive than in other sectors.

In addition to liberalization of the sector, three other reforms are considered: (1) privatization of the sector, measured by the year in which the government first sells a majority stake in the state-owned telecommunications provider; (2) regulatory independence, measured by the year in which the regulatory authority was separated from direct political oversight; and (3) regulatory depoliticization, measured by when a regulatory authority is judged to be fully autonomous from the executive branch.[10]

Telecommunications reforms affect the educational structure of employees differently across sectors. As seen in figure 5A.1, all four indicators of

FIGURE 5A.1
Telecommunications reforms are associated with higher demand for skilled workers in ECA

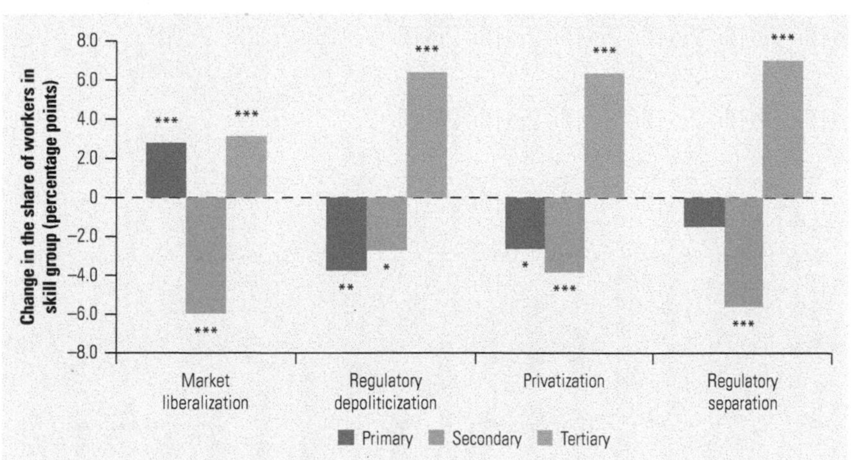

Note: Based on a country-sector level regression for European countries for the period between 1994 and 2012, using the share of workers with each level of education in each sector-country-year cell as a dependent variable. Each bar shows the coefficient associated with the interaction between the reform dummy variable and the ICT intensity of the sector. The value of each bar shows the increase in the share of workers with that level of education after the reform, in sectors with an ICT-intensity 50 percentage points larger than the rest. The asterisks denote that the results are statistically significant at (at least) the 10 percent level. See background paper for this report (Vazquez and Winkler 2017).

telecommunications reforms are associated with a larger share of employees with a university degree and a lower share of employees with a high school diploma in ICT-intensive sectors. For instance, regulatory separation is associated with a 6 percentage point increase in the share of college-educated employees in ICT-intensive sectors than in the rest of the economy. The share of employees with primary education also decreased relatively more in ICT-intensive sectors as countries introduced reforms related to the privatization and depoliticization of the telecommunications sector. These results are consistent with existing empirical evidence showing that changes in the task content of jobs in Central Europe were facilitated by an increase in educational attainment during this period (Lewandowski, Hardy, and Keister 2016).

These results are robust to a series of alternative specifications and robustness checks. See Vazquez and Winkler (2017) for more details.

The Impact of the Internet on Inequality

Simulations using an economic model indicate that the Internet could have a significant impact on inequality (figure 5A.2).[11] As economies accumulate ICT capital, the gains from reallocating labor toward nonroutine occupations increase rapidly, as nonroutine occupations become relatively scarcer. In Norway and Sweden, workers carrying out nonroutine tasks would experience a wage increase 3 percentage points higher than routine workers when ICT capital accumulation rises 10 percent. Bulgaria and Romania would experience the smallest changes. Figure 5A.3 sheds light on the drivers of these results across countries. Panel a suggests that the impact of ICT on nonroutine wages is higher in countries that already have higher levels of ICT capital, such as Sweden and the United Kingdom. In contrast, the impact would be smaller in countries that have not accumulated as much ICT capital, such as those in Central and Southern Europe. The scatter plot in panel b suggests that the

FIGURE 5A.2
The wage premium for nonroutine tasks increases with ICT capital accumulation

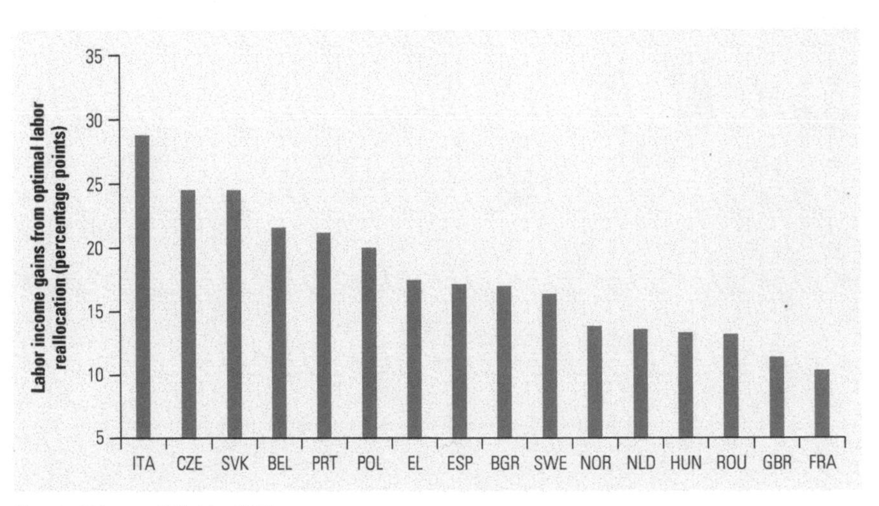

Source: Eden and Winkler 2016.
Note: Each bar represents the change in the returns to nonroutine tasks (percentage points) when ICT capital increases by 10 percent, using a calibrated production function.

FIGURE 5A.3 The effect of ICT on nonroutine wages is higher in countries with already high rates of ICT capital accumulation and where nonroutine workers are already more scarce

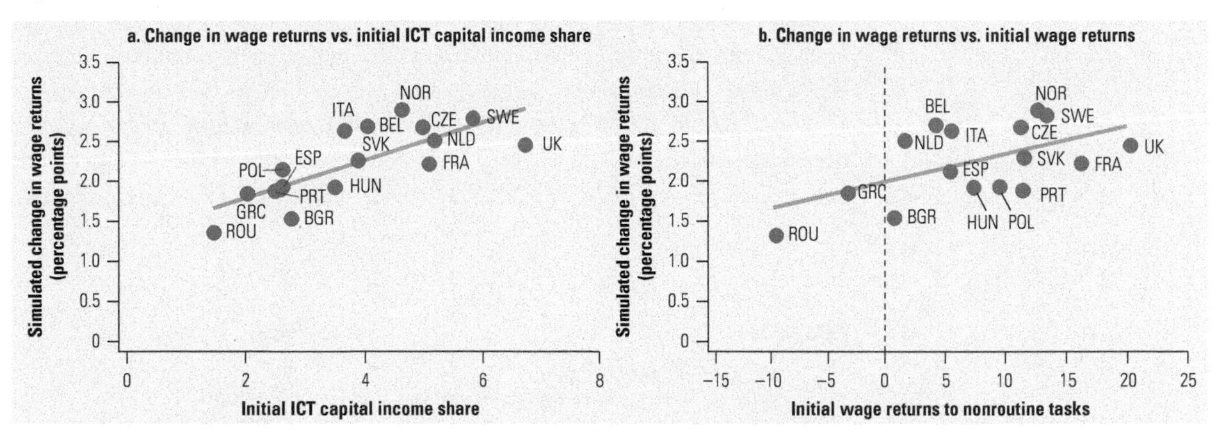

Source: Eden and Winkler 2016.
Note: The *y* axis represents the change in the returns to nonroutine tasks (percentage points) when ICT capital increases by 10 percent, using a calibrated production function. The *x* axis of panel a is the initial ICT capital income share, whereas that of panel b is the actual values of the wage returns to nonroutine tasks.

FIGURE 5A.4
The labor income gains from increasing the supply of nonroutine workers could be high, especially in Central and Southern Europe

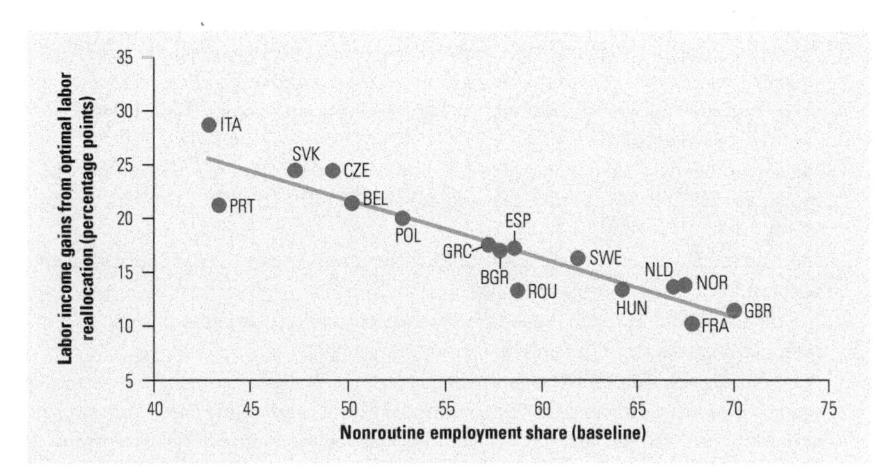

Source: Eden and Winkler 2016.
Note: The labor income gains are static gains obtained from equalizing the marginal product of labor across routine and nonroutine occupations, holding ICT and non-ICT capital stocks fixed.

impact of ICT on nonroutine wages tends to be larger in countries where nonroutine workers are already scarcer (because the returns to nonroutine occupations are already higher). In other words, while these results suggest that ICT capital accumulation may affect inequality in most countries (as nonroutine workers tend to earn more than routine workers), the effects will be larger in more advanced economies.

These results also suggest that the gains from increasing the supply of nonroutine workers will rise as countries invest in ICT. To illustrate this point, Eden and Winker (2016) estimate the welfare implications from increasing the supply of nonroutine workers until their marginal product equals that of routine workers.[12] As seen in figure 5A.4, the efficiency gains could be large, especially in Central and Southern Europe. Countries such as the Czech Republic, Italy, Poland, Portugal, and the Slovak Republic could experience an increase in total labor income of about 20 percent or more. Gains in total output would also be large, equivalent to about 10 percent or more of GDP in Central and Southern Europe. While these estimates should be interpreted with caution, as they do not take into account possible adjustments in ICT and non-ICT capital, they do suggest that the gains from increasing the supply of nonroutine workers would be larger in the east of ECA.

Notes

1. Based on Eurostat, results available from the author upon request.
2. The bottleneck index is based on the duration of vacancy filling, employer statements regarding the difficulty of filling vacancies, and employer estimates of the difficulty of filling vacancies in the near future.
3. The measurement of the ICT intensity of firms is based on whether the firms employ ICT. Thus, firms that outsource these activities may be classified as being low ICT-intensive, even though they use ICT.
4. Based on PIRLS test score data from the EdStats database.
5. A growing body of literature shows empirical evidence that improving early childhood education has significant positive effects later in life through better educational attainment, better employment, and higher wages (Currie and Almond 2011; Heckman, Pinto, and Savelyev 2012).
6. For a comprehensive skills diagnostics of Central Asian countries, see Ajwad, Abdulloev, and others (2014); Ajwad, de Laat, and others (2014); and Ajwad, Hut, and others (2014).
7. The data are from Community Statistics for Information Society (CSIS) Eurostat, results available from the author upon request.
8. The data are from Eurostat, based on labor force surveys, results available from the author upon request.
9. This section is based on Winkler and Vazquez (2016). More specifically, the sectors of economic activity are first classified in terms of the degree of Internet intensity using data from a developed economy. If sectors that are technologically more prone to adopt the Internet are more likely to do so as reforms are introduced, their demand for skills and tasks may be more likely to be affected as well. The estimation includes time-invariant sector characteristics (sector fixed effects) and fluctuations at the country level (country-year pair fixed effects) as control variables. This last set of fixed effects would absorb any other broad policy changes at the country level during this period.
10. The data on telecommunications reforms were obtained from Howard and Mazaheri (2009). While the labor market data cover the period from 1994 to 2012, the data on reforms cover the period from 1990 to 2007. Almost all of the countries in our sample implemented these reforms before 2007.
11. This section is based on Eden and Winkler (2016).
12. This static computation ignores dynamic adjustments in the ICT and non-ICT capital stock. In practice, an optimal allocation of labor is likely to lead to more accumulation of non-ICT capital and perhaps also to more accumulation of ICT capital. Thus, the static gains from labor reallocation should be viewed as a lower bound to the long-run gains.

References

Acemoglu, D., and D. Autor. 2011. "Skills, Tasks, and Technologies: Implications for Employment and Earnings." In *Handbook of Labor Economics*, vol. 4, part B, edited by D. Card and O. Ashenfelter, 1043–71. Amsterdam: North-Holland Elsevier.

Ajwad, M. I., Abdulloev, I., Audy, R., Hut, S., de Laat, J., Kheyfets, I., and Torracchi, F. 2014. "The Skills Road: Skills for Employability in Uzbekistan." https://openknowledge.worldbank.org/handle/10986/20389.

Ajwad, M. I., de Laat, J., Hut, S., Larrison, J., Abdulloev, I., Audy, R., and Torracchi, F. 2014. "The Skills Road: Skills for Employability in the Kyrgyz Republic." http://documents.worldbank.org/curated/en/311261468016850877/The-skills-road-skills-for-employability-in-the-Kyrgyz-Republic.

Ajwad, M. I., Hut, S., Abdulloev, I., Audy, R., de Laat, J., Kataoka, S., and Torracchi, F. 2014. "The Skills Road: Skills for Employability in Tajikistan." https://openknowledge.worldbank.org/handle/10986/20388.

Attström, K., S. Niedlich, K. Sandvliet, H.-M. Kuhn, and E. Beavor. 2014. "Mapping and Analysing Bottleneck Vacancies on the EU Labour Markets." European Commission, Brussels.

Autor, D. 2014. "Polanyi's Paradox and the Shape of Employment Growth." NBER Working Paper 20485, National Bureau of Economic Research, Cambridge, MA.

Autor, D., and David Dorn. 2013. "The Growth of Low-Skill Service Jobs and the Polarization of the U.S. Labor Market." *The American Economic Review* 103 (5): 1553–1597.

Autor, D., L. F. Katz, and M. S. Kearney. 2008. "Trends in U.S. Wage Inequality: Revising the Revisionists." *Review of Economics and Statistics* 90 (2): 300–23.

Bussolo, M., J. Koettl, and E. Sinnott. 2015. *Golden Aging: Prospects for Healthy, Active, and Prosperous Aging in Europe and Central Asia.* Washington, DC: World Bank.

CALISS. 2013. Jobs, Skills, Migration, Consumption Survey. Collected in the Kyrgyz Republic, Tajikistan, and Uzbekistan. World Bank-GIZ Survey.

CSIS 2011, 2012. Eurostat. http://ec.europa.eu/eurostat/web/microdata/community-statistics-on-information-society.

Currie, J., and D. Almond. 2011. "Human Capital Development before Age Five." *Handbook of Labor Economics* 4 (2011): 1315–486.

Del Carpio, X., O. Kupets, N. Muller, and A. Olefir. 2017. *Skills for a Modern Ukraine.* Directions in Development—Human Development. Washington, DC: World Bank.

Eden, M., and P. Gaggl. 2015. "On the Welfare Implications of Automation." Policy Research Working Paper 7487, World Bank, Washington, DC.

Eden, M., and H. Winkler. 2016. "The Implications of ICT for the Demand of Tasks in Western and Central Europe." World Bank, Washington, DC.

Eurostat. (various years). Community Statistics on Information Society. http://ec.europa.eu/eurostat/web/microdata/community-statistics-on-information-society.

Eurostat. (various years). Structure of Earnings Survey. http://ec.europa.eu/eurostat/web/microdata/structure-of-earnings-survey.

Falk, M. 2002. "What Drives the Vacancy Rate for Information Technology Workers?" *Jahrbucher fur Nationalokonomie und Statistik* 222 (4): 401–20.

Heckman, J. J., R. Pinto, and P. A. Savelyev. 2012. "Understanding the Mechanisms through Which an Influential Early Childhood Program Boosted Adult Outcomes." NBER Working Paper w18581, National Bureau of Economic Research, Cambridge, MA.

Howard, P. N., and N. Mazaheri. 2009. "Telecommunications Reform, Internet Use, and Mobile Phone Adoption in the Developing World." *World Development* 37 (7): 1159–69.

ILO (International Labour Organization). Various years. Key Indicators of the Labour Market (KILM). ILO, Geneva, http://www.ilo.org/empelm/what/WCMS_114240/lang--en/index.htm.

John, O. P., and S. Srivastava. 1999. "The Big Five Trait Taxonomy: History, Measurement, and Theoretical Perspectives." In *Handbook of Personality: Theory and Research*, edited by L. A. Pervin and O. P. John. New York: Guilford Press.

Karabarbounis, L., and B. Neiman. 2013. "The Global Decline of the Labor Share." *Quarterly Journal of Economics* 129 (1): 61–103.

Lewandowski, P., W. Hardy, and R. Keister. 2016. "Technology or Upskilling? Trends in the Task Composition of Jobs in Central and Eastern Europe." IBS Working Paper 01/2016, Instytut Badań Strukturalnych, Warsaw.

LITS II. Life in Transition Survey, 2010. EBRD-World Bank.

Monroy-Taborda, Sebastian, Martin Moreno, and Indhira Santos. Forthcoming. "Technology Use and Changing Skills Demands: New Evidence from Developing Countries." Background paper for the World Development Report 2016, World Bank, Washington, DC.

OECD (Organisation for Economic Co-operation and Development). 2004. *Information Technology Outlook, 2004*. Paris: OECD.

———. 2013. "Russia: Modernising the Economy." Better Policies Series, OECD, Paris, April.

———. 2015. "The ABC of Gender Equality in Education: Aptitude, Behavior, and Confidence." OECD, Paris.

Penn World Tables. http://cid.econ.ucdavis.edu/pwt.html.

PIAAC (Programme for the International Assessment of Adult Competencies). Various years. OECD, Paris, http://www.oecd.org/site/piaac/.

PISA (Programme for International Student Assessment). Various years. OECD, Paris.

SES 2010. Structural Earnings Survey 2010, Eurostat. http://ec.europa.eu/eurostat/web/microdata/structure-of-earnings-survey/.

Sondergaard, L., M. Murthi, D. Abu-Ghaida, C. Bodewig, and J. Rutkowski. 2012. *Skills, Not Just Diplomas: Managing Education for Results in Eastern Europe and Central Asia*. Washington, DC: World Bank.

STEP (Skills Towards Employability and Productivity). Various years. Skills Towards Employability and Productivity (STEP) household surveys (database), World Bank, Washington, DC, http://microdata.worldbank.org/index.php/catalog/step/about.

Telegeography. Regulatory Status Dataset. www.telegeography.com.

Vazquez, Emmanuel Jose, and Hernan Winkler. 2017. "How Is the Internet Changing Labor Market Arrangements? Evidence from Telecommunications Reforms in Europe." Policy Research Working Paper No. 7976, World Bank, Washington, DC.

Winkler, H. 2016. "How Does the Internet Affect Migration Decisions?" *Applied Economics Letters* 1–5.

World Bank. 2014. "Transforming Higher Education: E-Learning, Distance Education, and Technologies in Tajikistan." Background paper for *Tajikistan: Higher Education Sector Study*, World Bank, Washington, DC.

———. 2016. *World Development Report 2016: Digital Dividends*. Washington, DC: World Bank.

———. Various years. EdStats: Education Statistics. Washington, DC: World Bank. http://datatopics.worldbank.org/education/.

World Development Indicators. 2012-2015. World Bank, Washington, DC. http://data.worldbank.org/data-catalog/world-development-indicators.

World Development Indicators. 2014. World Bank, Washington, DC. http://data.worldbank.org/data-catalog/world-development-indicators.

6

The Internet Is Changing the Labor Market Arrangements

The Internet has dramatically facilitated the rise of new forms of work. According to some estimates, more than 24 million independent workers in Europe and the United States provide services on online platforms. This is an important development, especially considering the backlash that some sharing-economy platforms have experienced. For example, Uber has faced violent protests and legal battles in almost every country it has entered. However, similar platforms for online freelance work have not generated these types of reactions. Why?

One of the main differences between Uber and online freelance platforms is the reaction of the incumbents. While the taxi industry disrupted by Uber is a textbook example of a highly regulated industry with strong trade unions, freelancers tend to be more scattered and lack such market power. These contrasting experiences highlight the trade-offs of new technologies. While Uber has reduced income volatility for many workers, it has hurt the labor market outcomes of taxi drivers, who in many cases have had to use their savings to pay for their licenses, endure days of training, and comply with strict regulations. Since the tides of technological change are difficult to stop, this chapter argues that countries should instead embrace change, while minimizing its disruptive effects, for example, by facilitating the labor market transition of workers from the incumbent industry and by bringing sharing-economy workers into the formal economy. As this chapter argues, postponing these adjustments could have dramatic consequences.

Introduction and Main Messages

The Internet is not only affecting the demand for and supply of skills, but also generating revolutionary new forms of work. Today, even someone living in a remote rural area of Romania could be working in the comfort of her living room for a company in Bucharest or even in a foreign country, as long as she has the right skills and a reliable and affordable Internet connection.

The increased ability to transfer information across vast distances in real time has been accompanied by an increase in more flexible forms of work. For example, the incidence of part-time work increased between 2 and 4 percent more in high Internet-dependent sectors—compared with low Internet-dependent sectors—after opening the telecommunications sector to competition. The share of employees working from home, a proxy for telecommuting, almost doubled between 2002 and 2015 in Northern European economies, although the share of employees using this work arrangement has increased little in the other Europe and Central Asia (ECA) subregions.

The Internet improves labor market efficiency by lowering the information costs involved in job search and increasing the quality of job matching. The rise of online labor markets has provided greater flexibility to workers, while increasing employers' access to, and information on, lower-wage employees from abroad. Online work can offer cost savings by reducing congestion and increasing flexibility, increasing the opportunities for interactions among skilled workers by generating innovation spillovers, and lowering the barriers to entrepreneurship by increasing the market for discrete tasks. The rapid rise in online labor markets largely reflects firms in high-income countries that are hiring workers from developing countries.

Participation in online markets is limited in many ECA countries, but is increasing rapidly. The quality of the enabling environment, including the cost and reliability of broadband access and the quality of education, is important for participation in online work. A limited ability to communicate in English may limit participation in online markets in many ECA countries.

Technology platforms that facilitate the exchange of goods, assets, and services between individuals, referred to as the sharing economy, are becoming increasingly important. While online platforms represent a small fraction of overall incomes, the share of individuals participating in these platforms is large in many European countries. The sharing economy can improve efficiency by enabling individuals to use assets that would otherwise be idle (for example, renting out a room through Airbnb), improve the environment by increasing the sharing, rather than the production, of assets (ride sharing), and improve information and incentives by creating a system of ratings and reviews.

The increase in online work is associated with conflicting labor market trends. On the one hand, the increase in part-time work facilitated by the Internet may be associated with reduced job satisfaction, as it has been accompanied by a rise in the share of part-time workers (particularly among young workers) who would prefer, but are unable to find, full-time positions. On the other hand, older workers and women, who often face significant barriers to labor market participation, may benefit from the more flexible work arrangements afforded by online markets. This could play an important role in increasing labor force participation rates in ECA, which are among the lowest in the world. Telecommuting also may benefit workers who need more flexible work arrangements, such as those with children too young to attend school. Finally, online work may be subject to more limited legal protection and encourage greater informality.

Online work could significantly affect income distribution, both between and within countries. Online freelancing could play a role similar to that of international migration in reducing income disparities between countries, by increasing opportunities for workers and financial inflows in poor countries. However, the availability of online work could increase income disparities if poorer countries face severe limits on participation due to inadequate Internet infrastructure or low levels of education. Online freelancing is likely to widen income disparities within countries by increasing the return to skills.

The policies that ECA countries should adopt to benefit from online work vary, depending on their ability to use the Internet and to adapt to online freelancing. Emerging countries lack adequate access to broadband Internet and new economy skills; to exploit the full opportunities afforded by online work, these countries should focus on removing these barriers. Transitioning countries, where online freelancing and the sharing economy are starting to emerge, could see a rise in informality. These countries should focus on reducing labor market rigidities that encourage informal work arrangements, improving the legal system to protect online workers, and strengthening public administration to improve tax enforcement and the provision of government services to businesses. Transforming countries need to address the challenges that the Internet economy creates for labor regulations and social protection. The effectiveness of current arrangements, where workers benefit from regulations governing working conditions, as well as employer-based pension and health insurance, will erode as more workers are viewed as independent contractors lacking such protections. Shifting to a new model of social protection that links contributions and benefits to the individual would increase the portability of benefits, reducing distortions related to job mobility, entry, and exit, while at the same time providing some insurance against risks. Transforming countries should also consider changing licensing requirements to reap the efficiencies generated by Internet-enabled services.

Overall, this chapter argues that, while the Internet can create opportunities for many individuals who are out of work in ECA, the very same factors that hold back the labor markets of the region could challenge the emergence of more flexible work arrangements enabled by the Internet. For example, the rise of telecommuting will fail to bring older individuals back to work if workers retire at an early age. Moreover, new technologies will be unable to create new and more productive jobs if vested interest groups keep blocking the spread of technological change across occupations. At the same time, this chapter argues that the Internet poses new challenges, such as the need to adjust regulations or data collection methods for better evidence-based policy making in the new economy. Postponing such adjustments could increase the income gaps with developed economies and foster rising inequality and informality. Policy makers will have to evaluate whether inaction is worth such consequences.

The Internet Is Changing the Labor Market Structure in ECA

The emergence of the Internet at the workplace is transforming labor markets in the ECA region. Telecommunications reforms that helped to raise Internet penetration (Howard and Mazaheri 2009; Winkler 2016) have also been associated with a larger expansion in employment levels in more Internet-intensive sectors in Europe. Econometric analysis, which controls for other changes at the economy-wide level and sector characteristics that may be correlated with these reforms and labor market outcomes, confirms these results. Within a period of 10 years after the reform was introduced, employment grew 35 percent more in sectors that are inherently more Internet-intensive than in the rest of the economy.[1] The Internet may be driving a rise in alternative work arrangements in ECA—in particular, greater freelancing.[2] Since the early 2000s, the share of employees in temporary contracts has risen across most of Europe, while the share of workers in part-time employment has increased, particularly in Southern and Western Europe as well as Turkey (figure 6.1).

At the same time, the share of workers in self-employment has remained stable or even declined in some economies. However, current labor force surveys (LFSs) may not be the right statistical tool for capturing all of the new labor market developments.[3] Alternative measures indicate that the Internet is associated with a higher incidence of alternative work arrangements in Europe. For example, the incidence of part-time work increased between 2 and four percentage points more in high Internet-dependent sectors than in low Internet-dependent sectors after opening the telecommunications sector to competition. The results for temporary work, in contrast, are not very clear (Vazquez and Winkler 2017).

As one would expect, the Internet is increasing the share of employees who are telecommuting or working away from their employer's premises through the use of computer networks and telecommunications. According to some estimates, the share of employees involved in telework in the European Union (EU) member states increased from about 5 percent in 2000 to 7 percent in 2005, with Northern European countries leading the ranking (Welz and Wolf 2010).

Data on trends in the incidence of teleworking in ECA are not available. However, the share of workers teleworking is correlated with the share working from home (figure 6.2), and this share almost doubled from 2002 to 2015 in Northern European economies (figure 6.3, panel a). Still, the rise was more moderate for workers who *usually* work at home (figure 6.3, panel b), and the share of employees using this work arrangement increased little in the other ECA subregions. Regions with a higher rate of broadband penetration have a higher share of employees working from home (figure 6.4), and econometric results suggest that the share of employees working from home is 1.5 to 3 percentage points higher in highly Internet-intensive sectors than in the rest of the economy after the introduction of telecommunications reforms (Vazquez and Winkler 2017).

FIGURE 6.1 Official figures show a mixed picture regarding the rise of alternative work arrangements in ECA

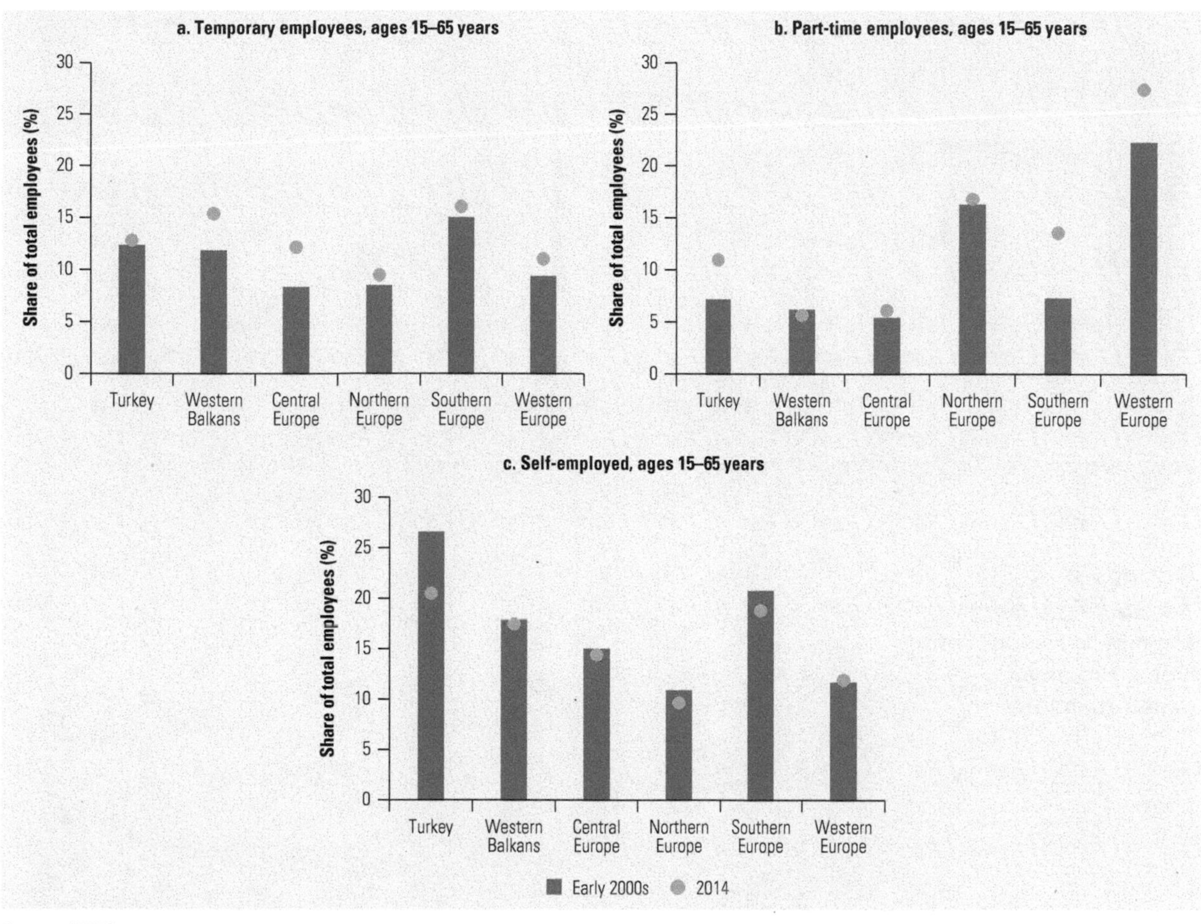

Source: LFS Eurostat.

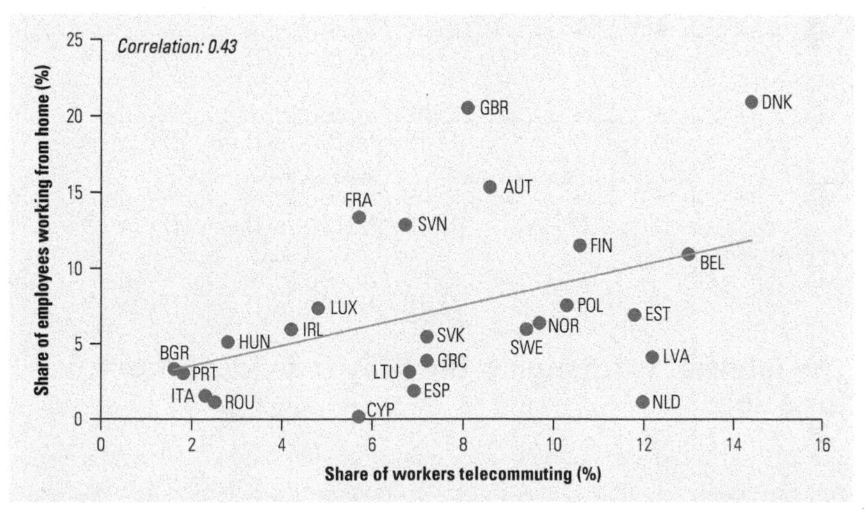

FIGURE 6.2
The share of teleworkers is correlated with the share of employees working from home, 2005

Sources: The *y* axis measures the share of employees working from home (usually or sometimes) in 2005 and 2006, estimated using microdata from Labor Force Surveys. The estimates of teleworkers in the *x* axis come from Welz and Wolf 2010.

FIGURE 6.3 Teleworking is rapidly increasing in Northern European economies

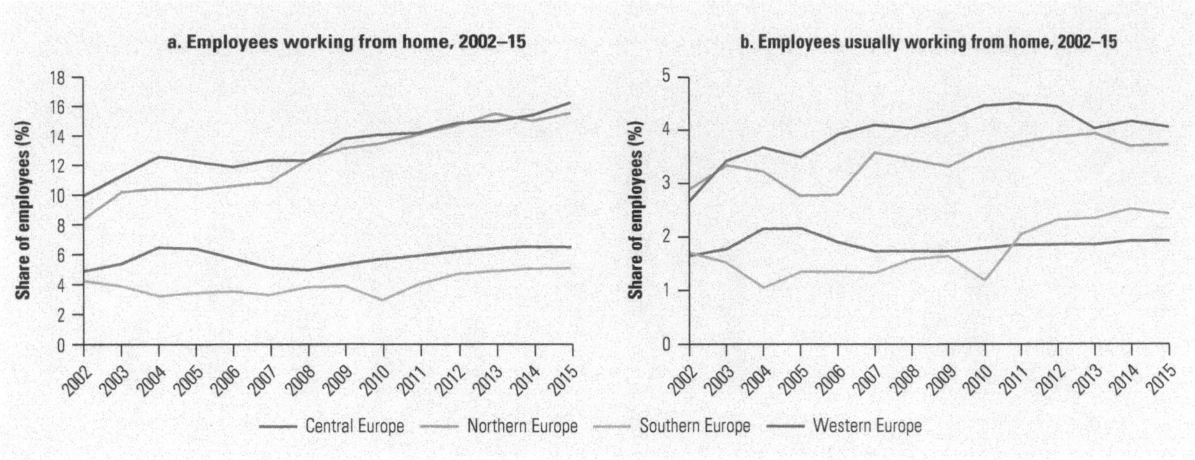

a. Employees working from home, 2002–15

b. Employees usually working from home, 2002–15

Central Europe — Northern Europe — Southern Europe — Western Europe

Notes: Based on LFS Eurostat. The percentage shown is a simple average of employees working at home in each country within the region.

FIGURE 6.4 Teleworking is more common in regions with higher broadband penetration rates

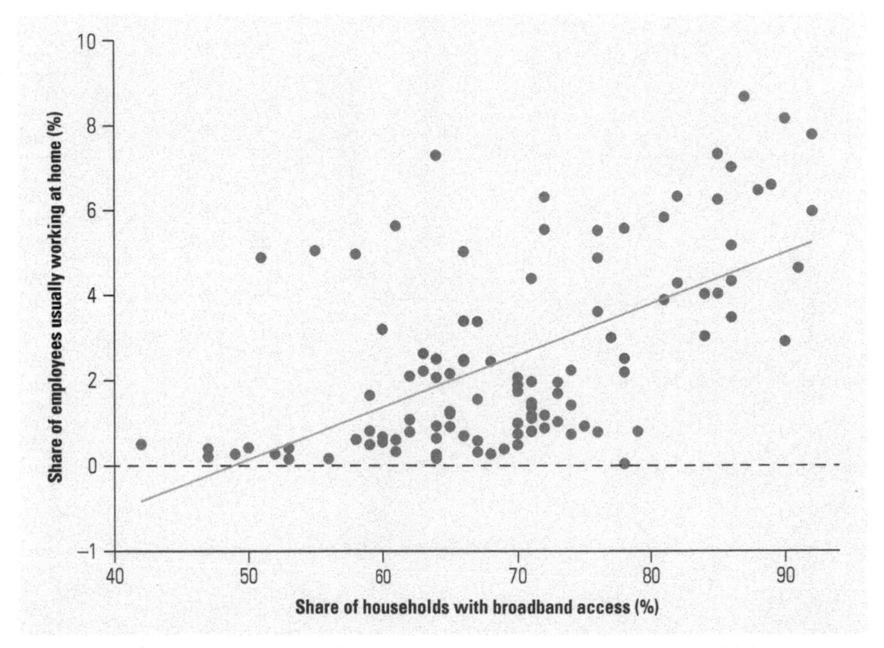

Sources: EU Labour Force Survey Database and Eurostat. Each point represents a NUTS 2 region within Europe.

The Internet Is Changing the Way Individuals Look for a Job

The Internet could also improve labor market efficiency by lowering information costs for potential employees and by improving the quality of job matching through online screening of potential employees for firms (Autor 2001). However, since electronic communication makes it easy for workers to apply for many jobs,

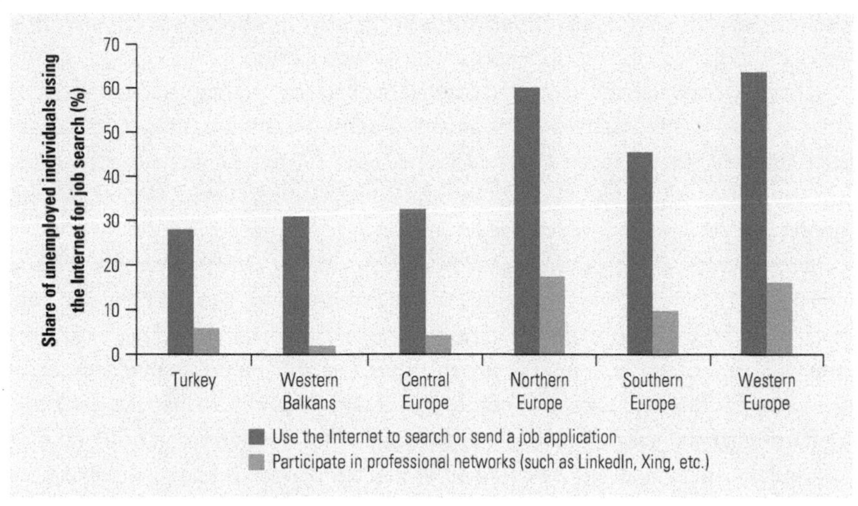

FIGURE 6.5
The unemployed are using the Internet to find a job, especially in Western ECA

Source: CSIS Eurostat 2015.

recruiting on the Internet may increase the screening costs for firms (Oyer and Schaefer 2011). Unemployed persons in the United States who searched for a job online found a job 25 percent faster than workers who did not search online, after controlling for a rich set of covariates (Kuhn and Mansour 2014). In Italy, the implementation of a job board by universities reduced unemployment by 1.6 percentage points and increased wages by 3 percentage points (Bagues and Labini 2009). In any event, the Internet saves time and lowers information barriers for workers looking for a job, while those without Internet access or skills may have to rely on more time-consuming job search methods. While more than half of unemployed individuals in Northern, Southern, and Western Europe are using the Internet to search for a job, less than a third of them are doing so in Central Europe, Western Balkans, and Turkey (figure 6.5).

The Internet Is Not Only Changing Existing Jobs, but Also Creating New Forms of Work

The Internet has become a significant source of new forms of work arrangements. For workers, online labor markets facilitate flexible work arrangements that purportedly reduce stress, improve familial relations, and, more generally, help them to manage the competing demands of paid labor, unpaid labor, and leisure. Evidence on the benefits of online work is limited, likely due to the lack of data, the relative novelty of this work structure, and vast differences in types of online service providers. However, based on the considerable literature on work arrangements, online outsourcing to freelancers may benefit employees and freelancers in developing countries by opening up work opportunities to suit workers' lifestyle preferences. Workers who prefer to work from different locations, from home, with more flexibility, and on more differentiated types of work are now part of the workforce (Greene and Mamic 2015). Online labor markets enable employers to replace high-wage workers with like-quality, low-wage workers abroad (Roach 2004) and provide companies with cheaper, broader search options. Business owners can browse detailed work samples and customer ratings for thousands of vendors in service categories ranging from accounting to web design.

Online work offers the cost-saving benefits of telecommuting, such as reduced congestion and increased flexibility, as well as some advantages unique to the way such markets appear to be structured. The rapid mixture of workers across and between firms might speed up innovation spillovers, creating a kind of pseudo geographic co-location, which has been shown to increase productivity in other contexts (Horton 2010). Further, by allowing firms to buy "small" amounts of labor, online labor markets lower the barriers to entrepreneurship. The flexibility and freedom afforded by Internet communications can also induce individuals who were previously out of the labor market to participate through the online marketplace. In the United States, an increase in home high-speed Internet use leads to a 4.1 percentage point increase in labor force participation for married women. These markets also create incentives for people otherwise disconnected from the global labor market to invest in their human capital (for example, as seen in call centers in India).

While self-employment has often been linked with exclusion from salaried work, Shevchuk and Strebkov (2012b) emphasize that the attractiveness of these work arrangements explains a large share of why individuals choose this form of work in Eastern Europe. Moreover, online freelancers seem to be different from the typical worker in the region in many respects. For example, they are more entrepreneurial, with 42 percent of them creating their own businesses and hiring employees within five years. In the EU, most freelancers reported consciously choosing this work, expressing preferences for having a certain lifestyle and for not being an employee (Leighton and Brown 2013).

The online work market has risen sharply in recent years, mostly involving employers from high-income countries and workers from developing countries. From 2009 to 2012, the number of employers in online markets from high-income countries increased almost 900 percent, and the number of workers in online markets from developing countries increased almost 1,000 percent (Agrawal and others 2013).

Online freelancing provides jobs where they are needed the most. Countries with higher unemployment rates tend to have higher rates of online freelancing (figure 6.6), an important consideration in ECA, where unemployment rates tend to be higher than average. Also, the incidence of online freelancing tends to be higher in countries where women may face larger barriers to employment and in countries with a low level of economic development.

The quality of the enabling environment is important for participation in online work. The share of online freelancers is higher in countries with a higher rate of Internet penetration and a larger share of college graduates in the population (figure 6.7). English literacy is crucial to communicating with potential employers, and countries with a larger share of English speakers are more likely to engage in online freelancing. The lack of English literacy may be holding back ECA countries, where residents are less likely to have English-speaking skills than residents in comparable countries. More disaggregated data by task show that English-speaking skills are particularly correlated with the share of online freelancers offering writing, administrative, and marketing services. In contrast, college education seems to be more correlated with the share of online freelancers in design, multimedia, and information and communication technology (ICT) services.

FIGURE 6.6 Countries with higher unemployment, higher gender gaps, and lower incomes are more likely to engage in online freelancing

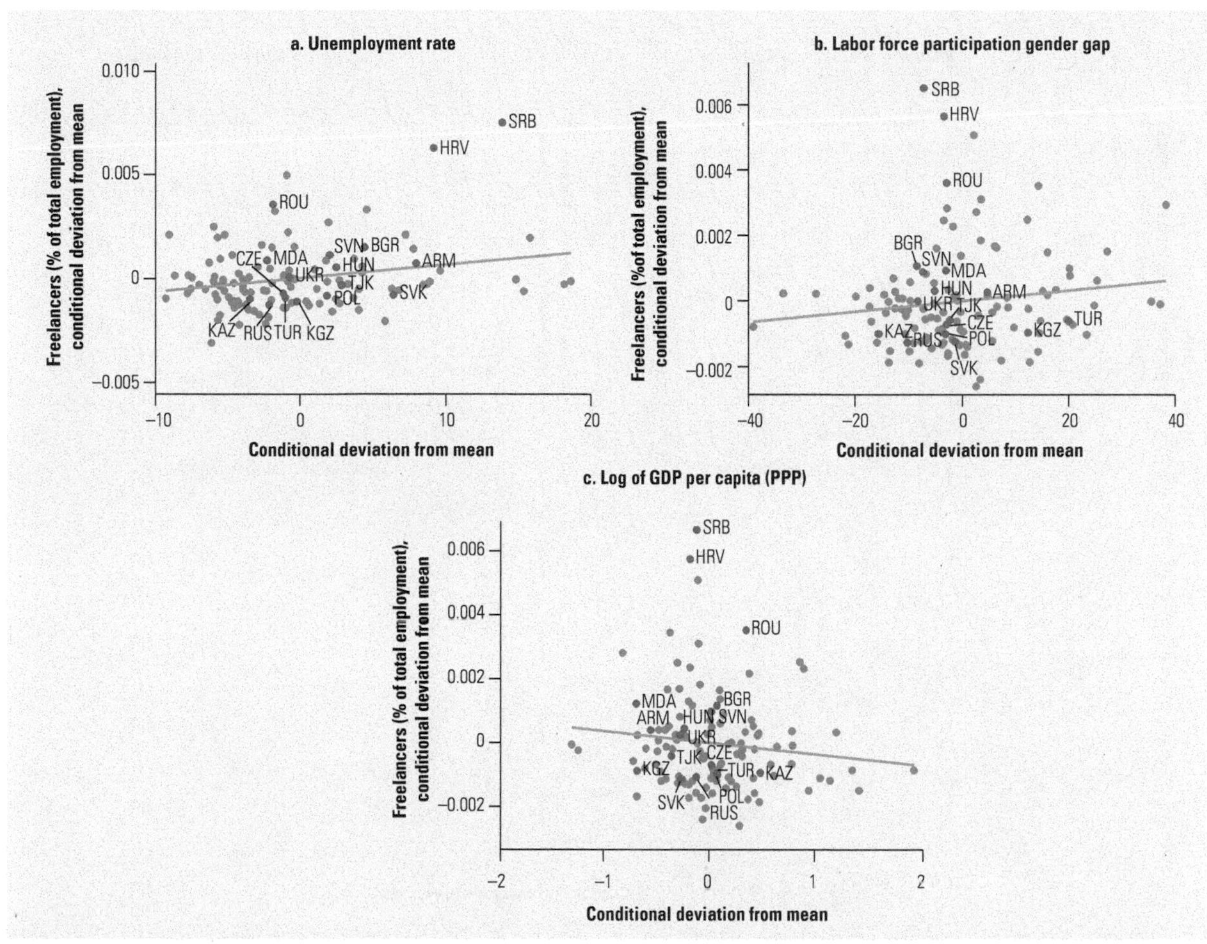

Note: GDP = gross domestic product; PPP = purchasing power parity. The vertical axis shows the share of online freelancers in total employment, while the horizontal axis in each graph (clockwise) measures the unemployment rate, gender gap in labor force participation (males' minus females'), and the logarithm of GDP per capita PPP. Freelance estimates were obtained from Elance.com (at the time the data was collected, Elance users represented 16 percent of total freelancers in platforms with international coverage (see World Development Indicators [database] 2015, http://data.worldbank.org/data-catalog/world-development-indicators), while the other variables come from the World Development Indicators. The graph shows the residuals of a cross-country regression of each variable in the horizontal and vertical axis, on a set of covariates including the unemployment rate, gender gap in labor force participation (males' minus females'), the logarithm of GDP per capita PPP, ease of doing business indicator, share of college graduates, Internet penetration, and English-speaking population. The share of English-speaking population for all countries comes from https://en.wikipedia.org/wiki/List_of_countries_by_English-speaking_population. Because the cross-country comparability of these measures is not clear, its quality was double-checked by regressing it on a dummy variable indicating if English is an official language, a dummy variable indicating if the country is included in the English Proficiency Index 2013 (EPI 2013, which is available for a smaller subset of countries, see http://www.ef.edu/epi/), and the values of the EPI 2013. All the coefficients were positive and statistically significant, showing that the relative country rankings suggested by this variable are supported by harmonized data.

Some ECA countries are important suppliers of global online talent. Online freelancers registered on Elance.com represent 0.7 percent of employed individuals in the Western Balkans and 0.3 percent of employed individuals in Central Europe (figure 6.8). While these numbers are small, these types of jobs did not even exist a few years ago, and recent trends point to a dramatic increase over a short period of time. In addition to ICT, frequent online tasks include writing and translation, design and multimedia, and administrative support. Online freelancing is limited in Central Asia, the Russian Federation, and Turkey, perhaps due to

FIGURE 6.7 Internet provision, education, and English-speaking skills are enabling factors of online freelancing

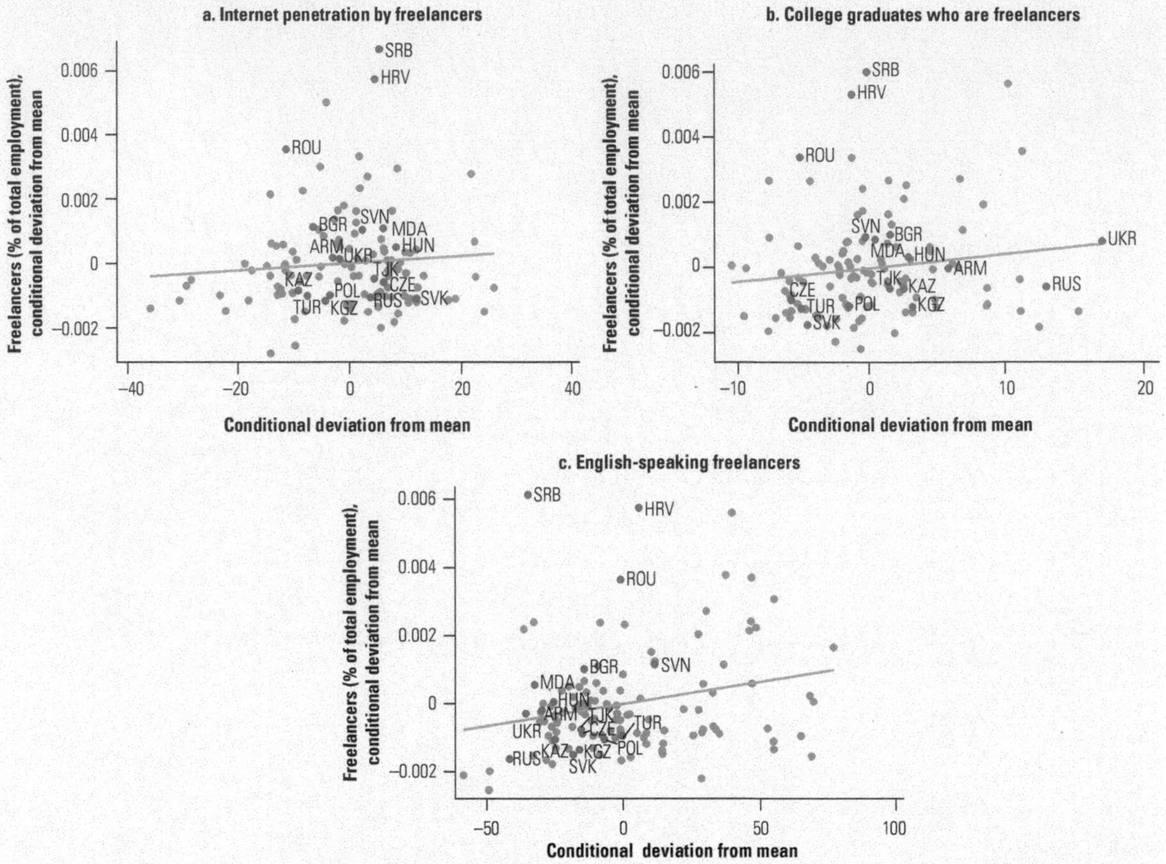

Note: GDP = gross domestic product; PPP = purchasing power parity. The vertical axis shows the share of online freelancers in total employment, while the horizontal axis in each graph (clockwise) measures the unemployment rate, gender gap in labor force participation (males' minus females'), and the logarithm of GDP per capita PPP. Freelance estimates were obtained from Elance.com (at the time the data was collected, Elance users represented 16 percent of total freelancers in platforms with international coverage (see World Development Indicators [database] 2015, http://data.worldbank.org/data-catalog/world-development-indicators), while the other variables come from the World Development Indicators. The graph shows the residuals of a cross-country regression of each variable in the horizontal and vertical axis, on a set of covariates including the unemployment rate, gender gap in labor force participation (males' minus females'), the logarithm of GDP per capita PPP, ease of doing business indicator, share of college graduates, Internet penetration, and English-speaking population. The share of English-speaking population for all countries comes from https://en.wikipedia.org/wiki/List_of_countries_by_English-speaking_population. Because the cross-country comparability of these measures is not clear, its quality was double-checked by regressing it on a dummy variable indicating if English is an official language, a dummy variable indicating if the country is included in the English Proficiency Index 2013 (EPI 2013, which is available for a smaller subset of countries, see http://www.ef.edu/epi/), and the values of the EPI 2013. All the coefficients were positive and statistically significant, showing that the relative country rankings suggested by this variable are supported by harmonized data.

language barriers in all countries and poor Internet access in Central Asia. However, these figures may understate participation in Central Asia and Russia, where workers may rely more on Russian-speaking platforms.[4]

Online freelancing is just one part of a broader trend, the sharing economy. Broadly, the sharing economy is organized around a technology platform that facilitates the exchange of goods, assets, and services between individuals across a varied and dynamic collection of sectors. Examples of companies that facilitate

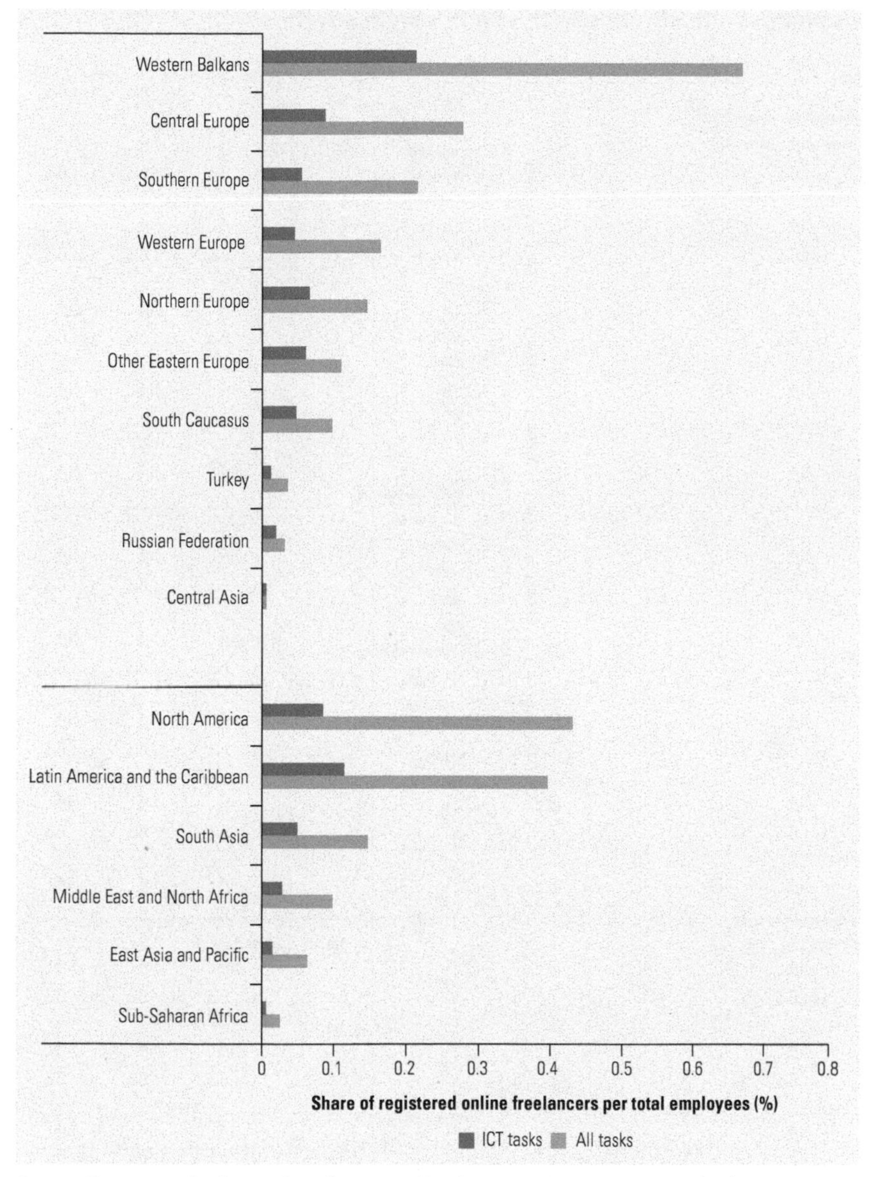

FIGURE 6.8
Some ECA countries are big suppliers of global online talent

Source: Elance.com, for the number of registered freelancers; ILOSTAT, International Labour Organization (http://www.ilo.org/ilostat), for the number of employed individuals.
Note: ICT = information and communication technology.

exchange of property or space include Airbnb, RelayRides, Getaround, and Liquid; examples of companies that facilitate exchanges of labor include Uber, Lyft, Taskrabbit, Handy, and Instacart. Data on the sharing economy are scarce, but a recent study estimates that the revenue of these platforms has grown dramatically. In the EU, their total revenue increased from around €1 billion in 2013 to €3.6 billion in 2015. Even though this estimate of sharing-economy revenues equals only 0.2 percent of EU gross domestic product (GDP), recent trends indicate that they are expanding rapidly (Vaughan and Daverio 2016).

While online platforms represent a small fraction of overall incomes, the share of individuals participating in these platforms is large in many European

FIGURE 6.9 A large share of Europeans is already using sharing economy platforms

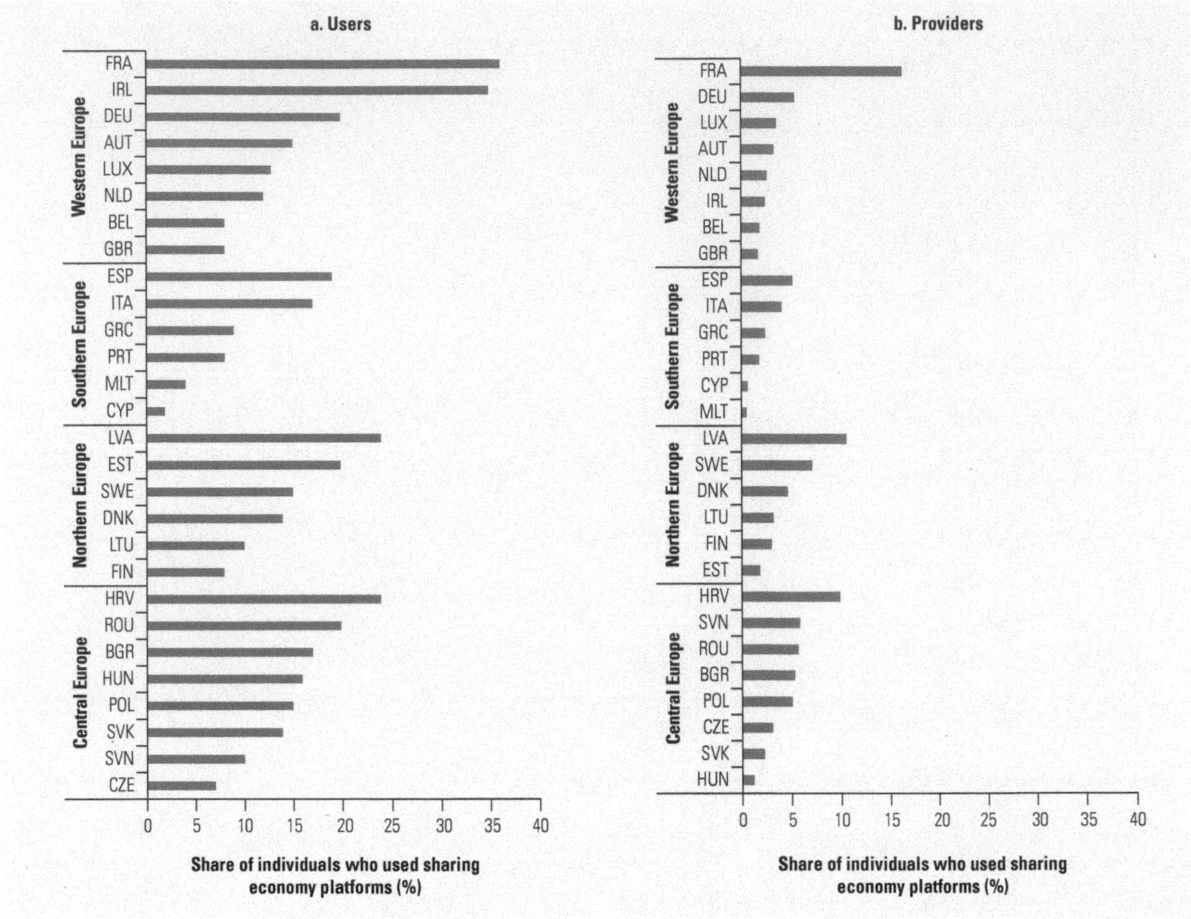

Source: Based on data from European Commission 2016.

countries—for example, about a third in France and Ireland and at least 10 percent in Central and Northern Europe (figure 6.9). The share of the population that has used these platforms to offer services and earn an income is much lower in all ECA countries, but is 10 percent or more in Croatia, France, and Latvia. However, most of these individuals have used the platforms only sporadically.

The sharing economy can bring efficiency gains by enabling individuals to use assets that would otherwise be idle. It can also bring environmental benefits, since assets can be shared by multiple users and fewer resources are needed to produce them.[5] Moreover, the system of ratings and reviews helps to lower information asymmetries by creating a mechanism to penalize bad performers and reward good ones. Evidence on the impact of the sharing economy on economic outcomes is still scarce. However, Cramer and Krueger (2016) find that Uber drivers spend a much larger share of their time driving and drive more miles with a passenger in the car than do taxi drivers. They argue that Uber is more efficient because (1) the use of Internet and smartphones allows for a much better matching of drivers and passengers than the outdated technology used by taxis;

(2) inefficient taxi regulations in some cities allow taxi drivers to drop off passengers outside their license area, but do not allow them to pick up another passenger there; and (3) Uber's flexible labor supply and prices allow for a better matching of supply and demand during high- and low-demand periods. Evidence for an accommodation-sharing platform, Airbnb, also shows important benefits for consumers, as the platform's additional competition helps to bring down hotel rates (Zervas, Proserpio, and Byers 2016).

Are Labor Market Developments Benefiting Everyone Equally in ECA?

The increased labor market flexibility that accompanied Internet adoption in the region was not necessarily accompanied by higher job satisfaction, especially among young workers. At least a third of part-time workers in Central Europe, Southern Europe, and the Western Balkans are involuntary part-time workers—that is, they are unable to find full-time positions (figure 6.10). The share of involuntary part-time workers is even higher for workers age 20–29 and has increased in almost every country of Central, Southern, and Western Europe during the last 15 years. Similarly, most prime-age, temporary workers in Europe are in those positions because they could not find a permanent job (figure 6.11), and the share of temporary workers who would prefer a permanent job increased in almost every country between 2000 and 2014. An important exception is Northern Europe, where involuntary temporary work declined.

Older workers and women, who often face higher barriers to labor market participation, may benefit from the more flexible work arrangements afforded by online markets. In ECA, employees older than age 64 are more likely than their younger peers to work from home (figure 6.12). Moreover, in countries where older

FIGURE 6.10 A large share of part-time workers would like permanent full-time jobs

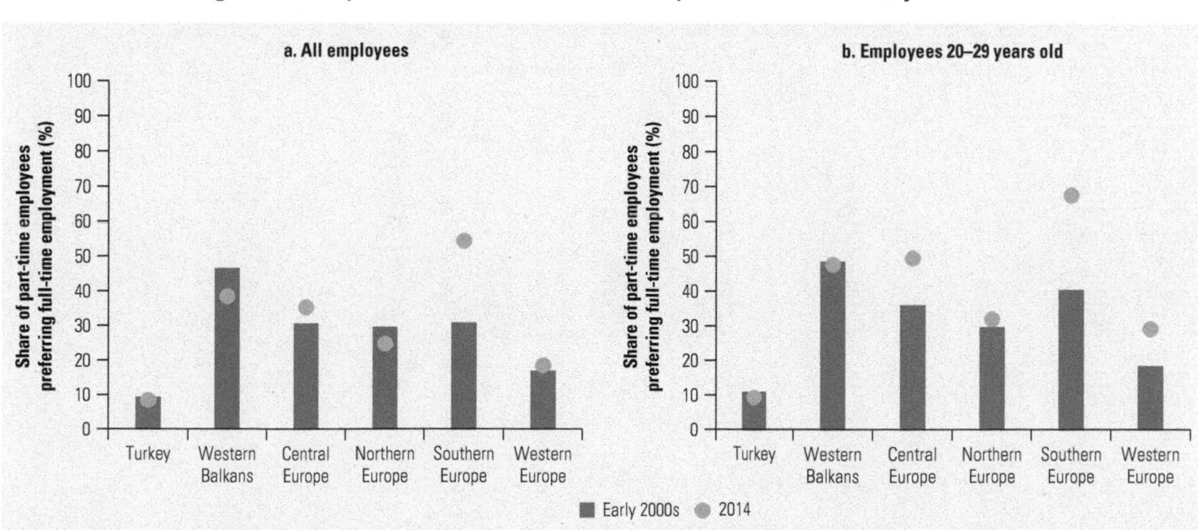

Source: LFS Eurostat.
Note: An individual is involuntarily in part-time employment when he or she could not find a full-time job.

FIGURE 6.11 Involuntary temporary work is on the rise across most of Europe

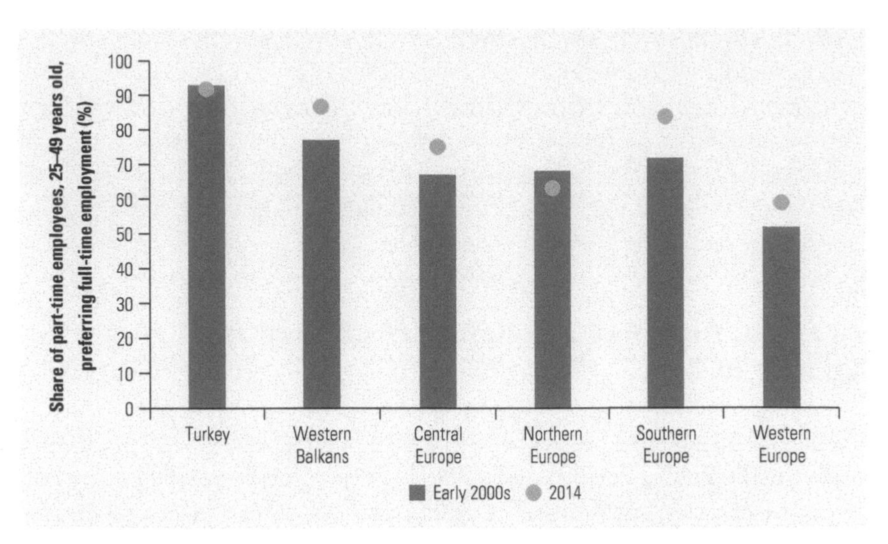

Source: LFS Eurostat.
Note: An individual is involuntarily in temporary employment when he or she could not find a permanent job.

FIGURE 6.12 Older and female employees are more likely to work from home

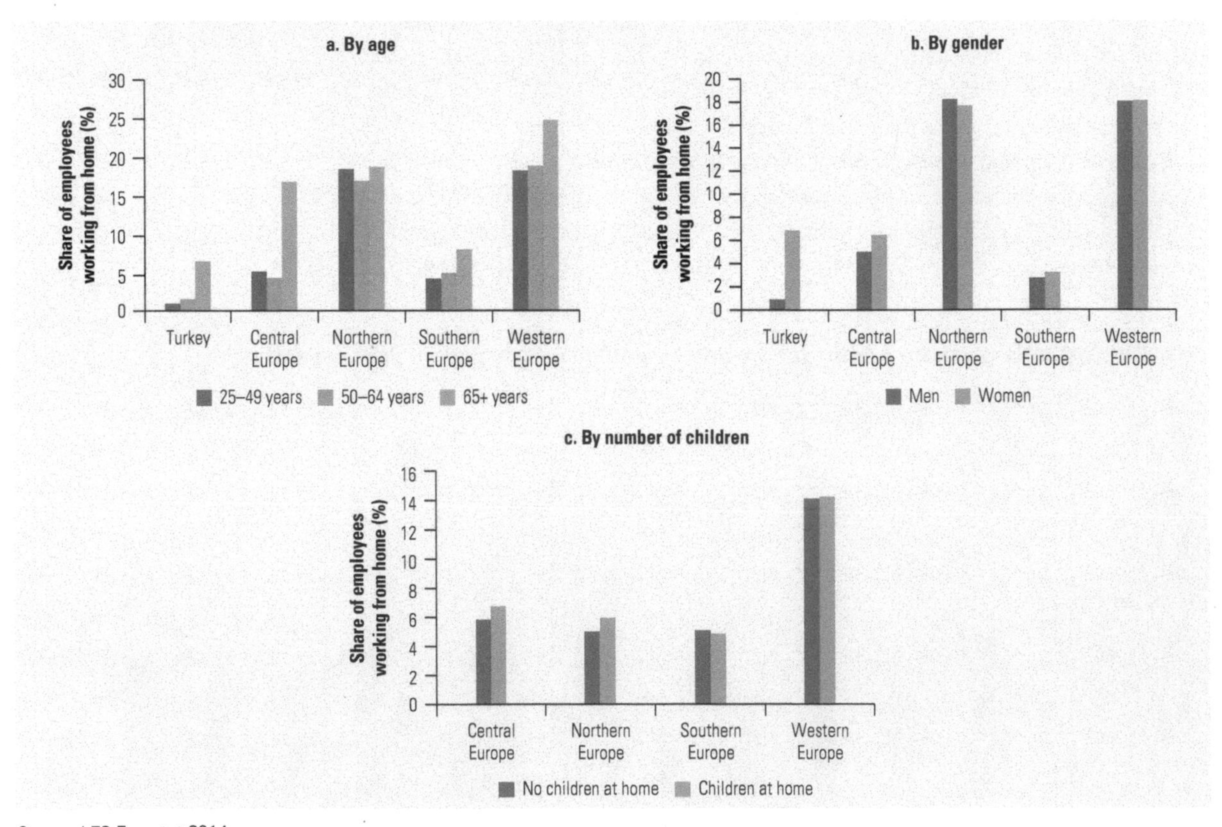

Source: LFS Eurostat 2014.
Note: In panel c, we consider children ages 4 years or younger.

FIGURE 6.13 Older and female employees are more likely to take advantage of telecommuting in countries where they face lower participation barriers

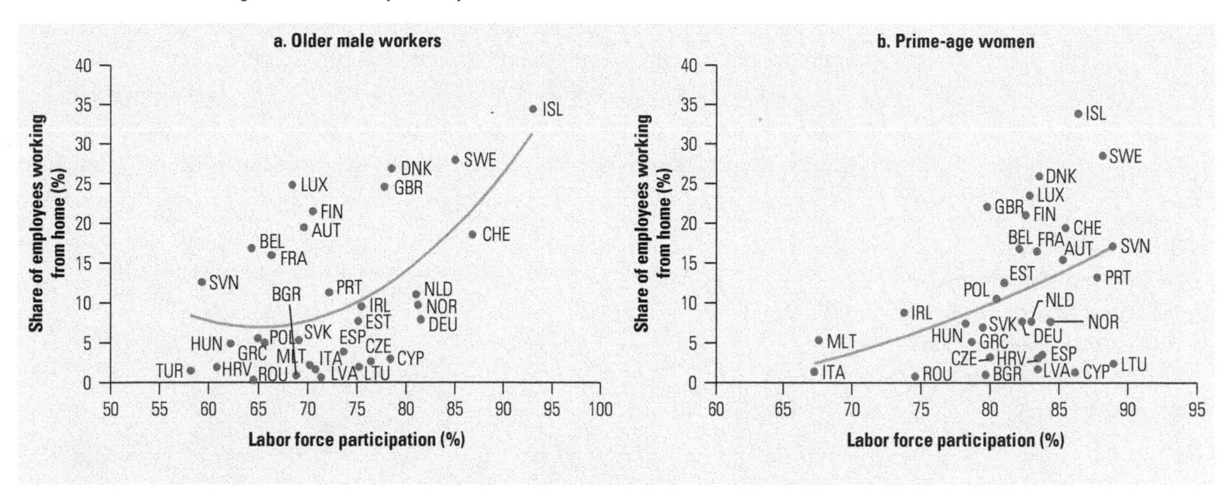

Source: LFS Eurostat 2014.
Note: Turkey was removed from panel b because it is an outlier. Older male workers are those individuals between 50 and 64 years old. Prime-age women are those individuals 25 to 49 years old.

FIGURE 6.14
Older workers are more likely to use flexible work arrangements in countries with older retirement ages

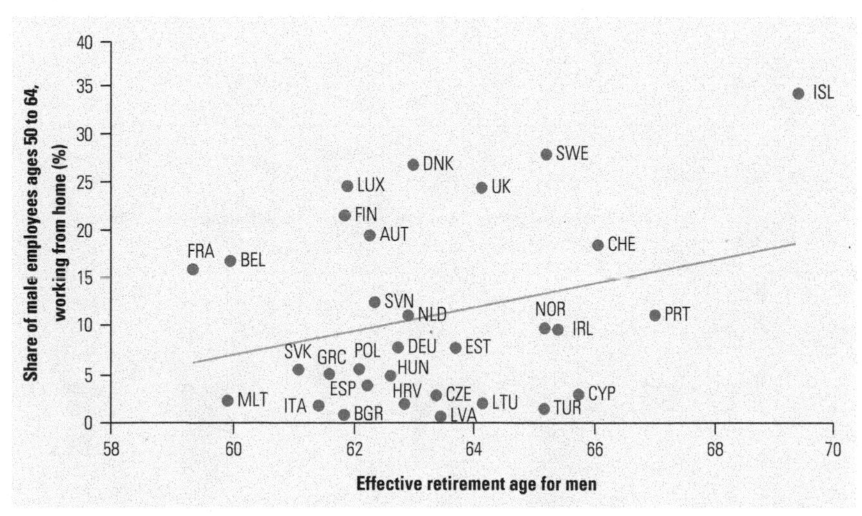

Sources: LFS Eurostat. The effective retirement ages are from OECD, estimated on the basis of the results of national labor force surveys, the European Union Labor Force Survey and, for earlier years in some countries, national censuses.

individuals are less likely to be in the labor market and where the retirement age is lower, older employees are less likely to work from home (figures 6.13 and 6.14). Also, in many countries female employees are more likely than men to work from home. In Turkey, where female labor force participation is low, 7.5 percent of female employees work from home, as opposed to only 1 percent of male employees. Countries where women are more likely to be working or looking for a job also display higher rates of telecommuting among females (figure 6.13). In Northern Europe, the rates of individuals working from home are not very different by age or gender, as the level of telecommuting is generally high across demographic groups. Telecommuting may benefit workers who need more flexible work arrangements, such as those with children too young to attend school. For example, workers in Central and Northern Europe with children below age 5 are more likely to work from home than those without children.

Education and location may also be important barriers to freelancing. Unskilled workers are less likely than skilled ones to use the Internet to find a job or to participate in professional networks (figure 6.15). And individuals living in rural areas

FIGURE 6.15
Unskilled individuals, older individuals, and those living in rural areas are less likely to use the Internet to find job opportunities

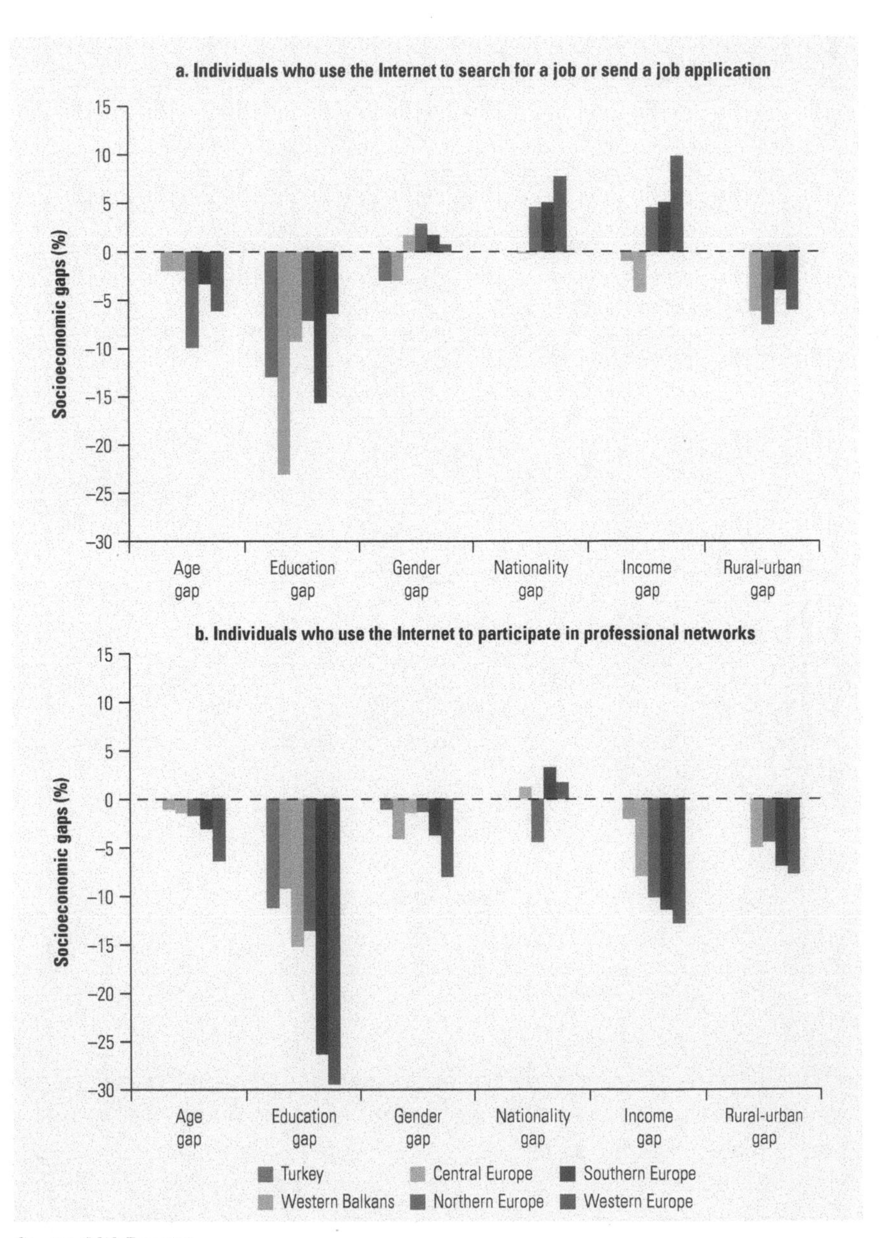

Source: CSIS Eurostat.
Note: Age gap: average for old individuals (older than 65 years) minus average for young individuals (ages 25 to 64); Education gap: average for low-educated individuals minus average for college graduates; Gender gap: average for females minus that of males; Nationality gap: average for non-EU born minus average for EU-born; Income gap: average for first income quartile minus average for fourth income quartile; Rural-urban gap: average for rural minus average for urban.

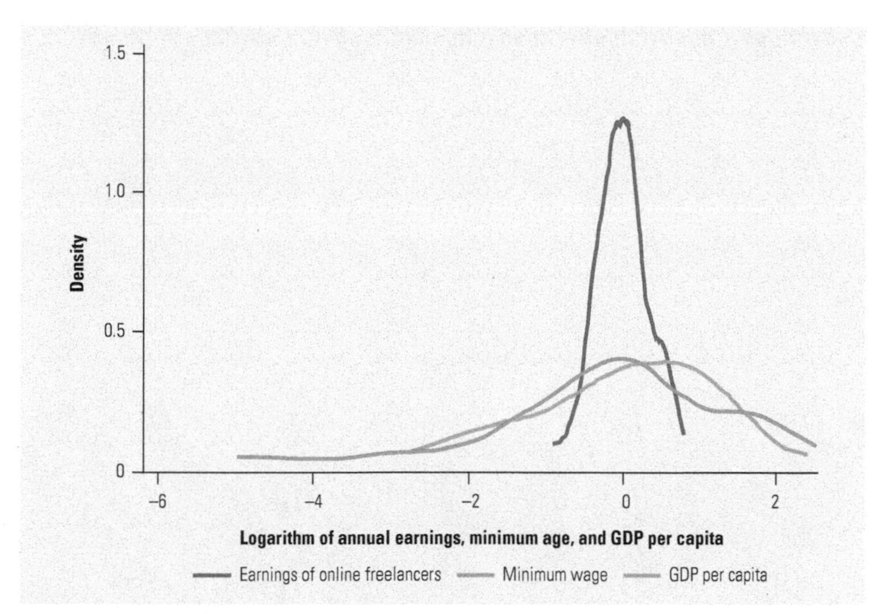

FIGURE 6.16
The distribution of online earnings across countries is less unequal than that of national incomes or wages

Source: Elance.com, for the hourly earnings of online freelancers; World Development Indicators (database), for GDP per capita PPP; Doing Business Indicators, for the minimum wage data.
Note: Each line shows the kernel density estimation of the distribution of each variable in 2013. All variables are annualized and expressed in logarithms. Their means have been subtracted so that they are centered on zero.

of ECA are also less likely than those living in urban areas to use the Internet to learn about job opportunities.

As with most technological innovations, online freelancing could have a significant impact on income disparity, both between and within countries. On the one hand, online freelancing could have an impact similar to that of migration in reducing income disparity between countries. The distribution of average country earnings of online freelancers is much more concentrated than that of GDP per capita or minimum wages (figure 6.16). For example, while the minimum wage in Serbia is equivalent to only 16 percent of the minimum wage in the United States, an online freelancer in Serbia earns an hourly rate equivalent to about half that of an online freelancer in the United States. This is probably because the skills of online freelancers are more similar across countries than the skills of the average resident. However, unlike migrant workers, online freelancers earn (and likely spend) their income in their country of residence, which may reduce the income gap across economies more rapidly than migration, which also involves high costs of mobility and remittances. Online work may increase income differences across countries if workers in poorer countries are less able to take advantage of these opportunities, perhaps because of limited Internet access or inadequate education.

Online freelancing is likely to widen income disparities within countries. Online freelancers' earnings are equivalent to at least seven times the minimum wage in many ECA countries (figure 6.17). Moreover, this gap between online earnings and minimum wages is higher in poorer countries with lower levels of human capital. In the South Caucasus and Central Asia, an online freelancer earns at least 40 times the minimum wage, while in Western Europe online freelancers make, on average, 3 times the minimum wage. In other words, if only individuals with good technical

FIGURE 6.17 Online freelancers' earnings are much higher than minimum wages, especially in poorer countries

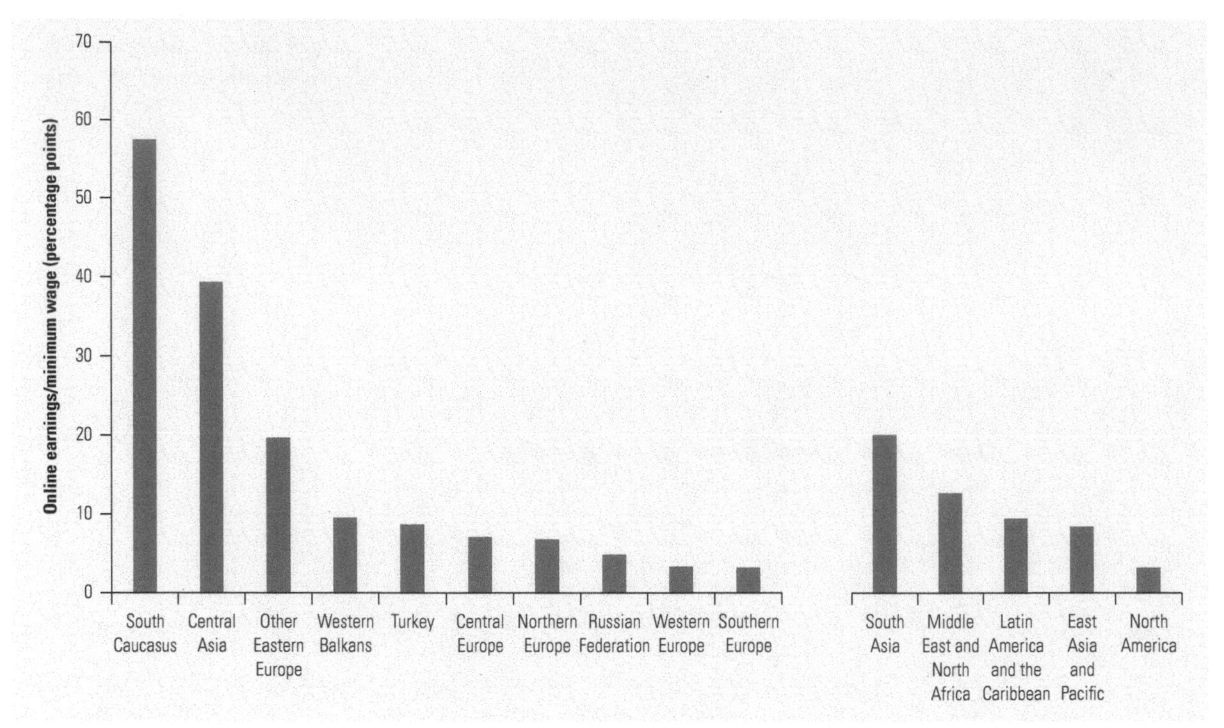

Source: Elance.com, for the hourly earnings of online freelancers; Doing Business Indicators, for the minimum wage data.
Note: Each bar shows the ratio between online freelancers' earnings (annualized) to minimum wages (annualized).

and English language skills and Internet access will benefit from the new forms of work brought about by the Internet, online work could increase income inequality within a country, especially in countries with large differences in access to education or new technologies.

Online freelancing also poses important challenges to existing labor market regulations. Online freelancing is rarely governed by legal contracts. For example, in a sample composed mostly of Russians and Ukrainians, only about 12 percent of online freelancers had a full legal contract with their clients (figure 6.18), while 34 percent relied on fully informal and 52 percent relied on mixed agreements. According to the same survey, about 70 percent of online freelancers reported that their clients broke a contract during the last year, largely because contract enforcement is not feasible in this area (only 1 percent of affected individuals who pursued legal action were successful). The system of reviews and ratings, as well as escrow payment systems, which are often embedded in online platforms, aim to minimize such abuses, but these institutions are just emerging in Russia (Shevchuk and Strebkov 2012b). Moreover, this information may indicate that the informality of online freelancers extends to other areas such as tax evasion and lack of social security and pension contributions.

Finally, while the sharing economy may bring economic inclusion, it may also contribute to economic disparities. A recent report from the Pew Research Center (2016) shows that ride-hailing platforms may help to bring transportation services to low-income or less centrally located neighborhoods where taxis are often scarce

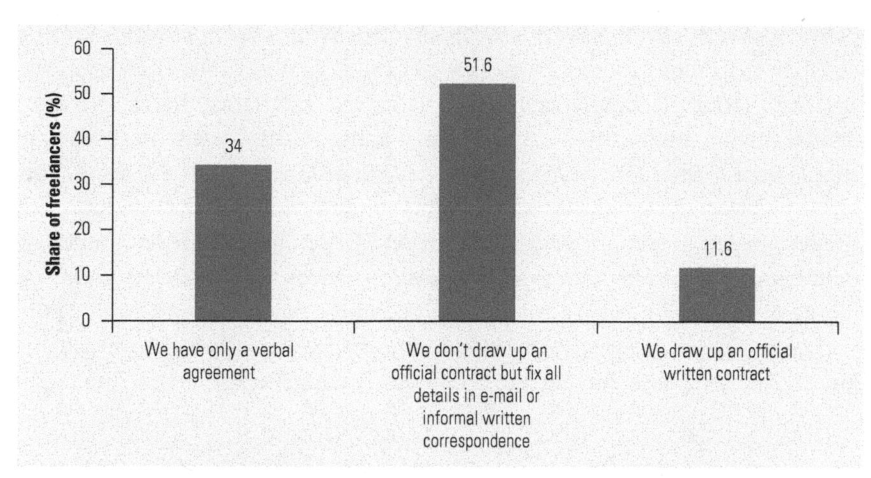

FIGURE 6.18 Most online freelancers in Eastern Europe are in the informal sector

Source: Shevchuk and Strebkov 2012a, based on a sample of online freelancers from Russia (70 percent), Ukraine (11 percent), Belarus (3 percent), and other countries (16 percent).

FIGURE 6.19 Large disparities exist in the use of the sharing economy in the EU

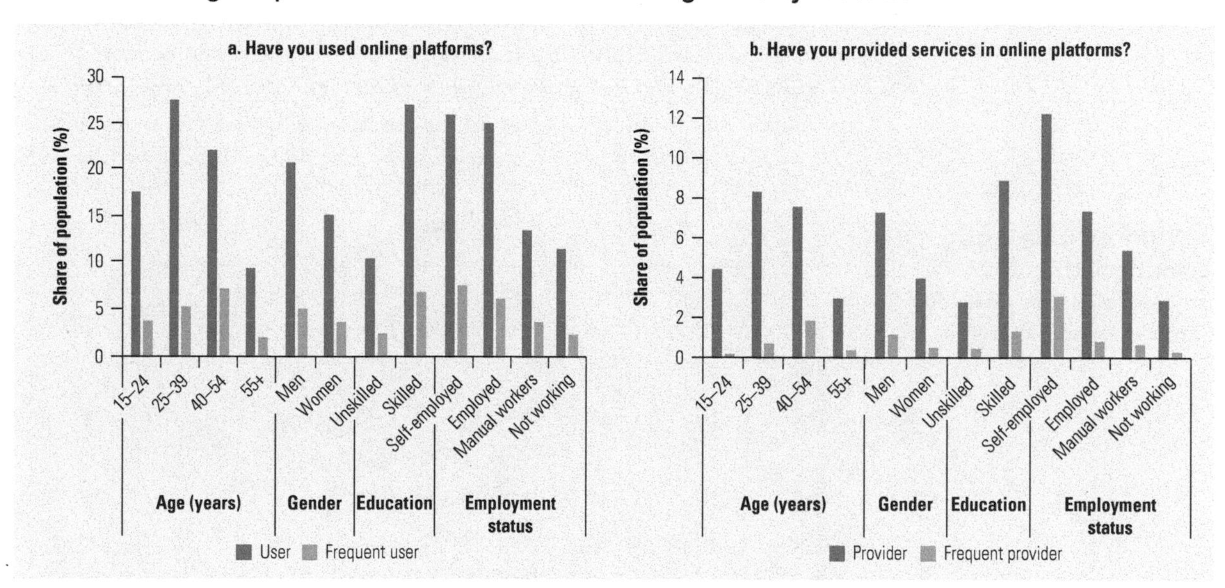

Source: Estimates are based on data from Flash Eurobarometer 438, March 2016 (http://www.gesis.org/eurobarometer-data-service/survey-series/flash-eb).
Note: Data for EU economies. A frequent user or provider is someone who uses or provides services every month.

and may improve service for individuals who have difficulty getting a taxi because of their race or appearance. Nevertheless, the same report finds important disparities across socioeconomic groups regarding access to these platforms. Similar gaps are observed in the European Union, where about 10 percent of unskilled individuals have used sharing-economy platforms, compared with 27 percent of skilled individuals (figure 6.19, panel a). Also, younger and male individuals are more likely to use these platforms than their older and female peers.

A study from the JP Morgan Chase Institute finds interesting trade-offs when analyzing a sample of individuals who have used the platforms to work and earn an income (Farrell and Greig 2016). Individuals who worked in labor-sharing platforms—such as ride-hailing or online freelance ones—were more likely to do so when their other sources of income declined. That is, the sharing economy seems to reduce income volatility for this group. In contrast, those who used capital-sharing platforms—such as home sharing or eBay—were more likely to supplement their other income sources in a more stable way. At the same time, those involved in capital-sharing platforms tended to be richer and older than those involved in labor-sharing platforms. In ECA, prime-age, skilled, and self-employed workers are more likely to provide services through online platforms (figure 6.19, panel b). While the sharing economy may help certain workers to cope with income volatility, it may also contribute to inequality by allowing richer individuals to increase the return to their existing capital stock.

Labor Market Policies for the Digital Economy

Various policies would help ECA countries to benefit from online work, depending on their ability to use the Internet and to adapt to online freelancing. Internet use is measured by the number of broadband subscriptions per capita, while adaptability is measured by the degree of formalization of the online economy. Since data are lacking, the formalization of the online economy is approximated by a combination of the degree of labor market informality and the importance of online work (figure 6.20).[6] Countries with moderate rates of online work

FIGURE 6.20 Emerging, transitioning, and transforming countries for labor market policies

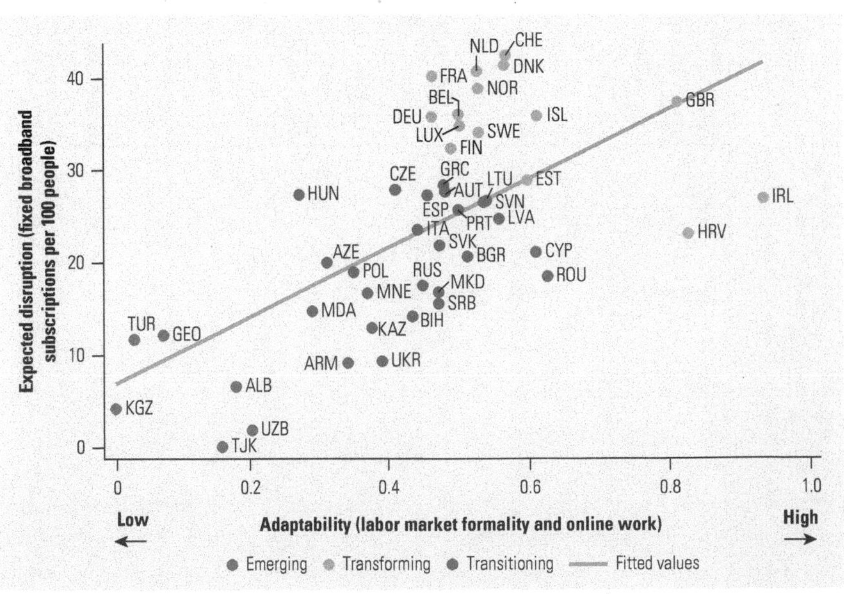

Source: World Development Indicators 2014, for broadband subscription numbers; Life in Transition Survey 2010, for the index of labor market formality; Elance.com, for the number of online freelancers and jobs posted, per capita.
Note: The formality index is the share of workers that are employees with a written job contract or self-employed persons in a professional occupation or large enterprise. The online work index is the the average number of online jobs and online freelancers, per capita. Both indexes were rescaled to have zero mean and a standard deviation of 1. The index of labor market formality and online work is the average of both indexes. Data on informality are not available for Western European countries except Sweden, Germany, Italy, and France. The rest of North European economies are imputed a formality rate equal to the average of Sweden, Germany, and France.

TABLE 6.1 Policy Priorities for Emerging, Transitioning, and Transforming Countries

Group	Countries	Policy priorities
Emerging	Albania, Georgia, Kyrgyz Republic, Tajikistan, Turkey, Uzbekistan	• Build foundational skills and improve Internet access
Transitioning	Armenia, Azerbaijan, Bosnia and Herzegovina, Bulgaria, Cyprus, Czech Republic, Greece, Hungary, Italy, Kazakhstan, Latvia, Lithuania, FYR Macedonia, Moldova, Montenegro, Poland, Portugal, Romania, Russian Federation, Serbia, Slovak Republic, Slovenia, Spain, Ukraine	• Reduce barriers to formal work • Increase the efficiency of courts and legal resolution systems • Promote active aging
Transforming	Croatia, Estonia, Western Europe, rest of Northern Europe	• Reduce the number of inefficient occupational regulations • Update labor market regulations to include online freelancing and the sharing economy • Improve data collection efforts to quantify the magnitude of online freelancing and sharing economy

Note: Countries are classified using the index of labor market efficiency and formality and the number of broadband subscriptions per 100 people.

and informality are at risk of witnessing an increase in the informalization of their economies as broadband penetration increases. Table 6.1 displays the final classification of countries and the policy priorities across groups. The policy priorities should be viewed as changing gradually across country groups, rather than as radical differences between priorities in, for example, emerging versus transitioning economies.

Emerging digital economies are not taking full advantage of the new forms of work brought about by the Internet. Even though there are no comparable data on telecommuting or online job search for these countries, data from an online freelance platform suggest that workers in these economies are not part of this new trend, perhaps due to barriers to Internet adoption and limited availability of new economy skills (see chapters 2 and 5). For example, in countries such as Azerbaijan and the Kyrgyz Republic, limited broadband access and a lack of digital and Internet-complementary skills are preventing individuals from realizing the gains from increasing the use of digital technologies in the labor market. Removing these barriers should be prioritized before focusing on the next-generation challenges faced by the labor markets of more advanced economies (box 6.1).

Transitioning economies, where online freelancing and the sharing economy are starting to emerge, could face the risk of increasing informality, as even traditional jobs are more likely to be in the informal sector when compared with their Northern European peers. Labor market rigidities, such as excessively high taxes or social security contributions, may increase the incentives for ICT specialists to work as online freelancers informally. For example, most employers in Ukraine complain about the lack of ICT skills, even though the country is among the top suppliers of information technology online workers for firms in developed economies. At the same time, Ukraine has high taxes on labor, high social security contributions, and a high bias toward taxation of employers (Del Carpio and others 2017). Packard, Koettl, and Montenegro (2012) argue that removing minor taxes and minimizing exemptions and loopholes would reduce compliance costs and facilitate the formalization of independent contractors. Integrating the

BOX 6.1 Connecting Kosovar Women to the Global Digital Economy and Work Opportunities

Rinora Tasholli is a 23-year-old university graduate with an English degree. She now works online, offering her web development and coding skills through a range of global online work portals that connect her with employers from around the world. Doing so allows her to work and earn without leaving her home in the village of Gadime, located an hour from the capital, Pristina.

Rinora is one of approximately 100 participants in a pilot program called Women in Online Work (WoW) set up by the Ministry of Economic Development of Kosovo with funding from the World Bank and the Korea Green Growth Trust Fund. The pilot currently operates in two municipalities and aims to train underemployed young women in skills that are in demand by the ever-growing online freelancing market. Five months into the six-month pilot, 55 trainees have collectively earned more than US$10,000. They compete globally for jobs in coding, web research, and data entry. The government is now planning to build on these initial successes to expand the program under its Kosovo Digital Economy initiative.

Online work could show women across the ECA region a way to connect to more, higher-paying, and inclusive work opportunities. Across the region, women tend to lag men in labor force participation, while facing persistent wage discrepancies and sector and occupational segregation.[a] Online work could address these gaps in countries where women have access to education, technology, and financial services and where complementary infrastructures such as electricity are functional. Even where cultural and religious barriers constrain traditional employment for women, online work could present a "safe harbor" by allowing flexible work schedules even while permitting engagement with a global marketplace.

Challenges remain in ensuring success. The pilot in Kosovo has shown that trainees need a basic set of skills before they can move onto complex—and hence higher-paying—digital tasks. One option is for pilot projects to focus on skilled but unemployed workers, but a more inclusive option is to invest in remedial training. This option might require more time and resources. Also, trainees new to online work benefit from coaching to build their communication skills, especially in English, and to overcome confidence barriers that arise when they start competing for jobs online. A focus on soft skills and intensive "hand holding" can help participants, but means that such programs might take longer to scale. Other challenges relate to the limits to global demand for online work, and it might be necessary to manage expectations and, simultaneously, to promote the capacity for self-employment and entrepreneurship, building on the skills imparted in the training program.

The WoW pilot shows that even developing, fragile countries such as Kosovo could use innovative approaches to connect more women to work. The digital foundations and analog complements need to be in place, while programs such as WoW (and others in countries such as Bhutan, Lebanon, and Pakistan) can help to build skills and expand access to technology. Rinora's efforts paid off, even with only some of these enablers in place. In just two months, she earned more than US$400 online. She is on her way to economic independence, competing in the global digital economy. Building a vibrant digital economy could include more people and expand such opportunities across the ECA region.

Source: Contributed by Zhenia Viatchaninova, Christine Shepherd, Siddhartha Raja, and Natalija Gelvanovska.

a. The EU and some EU-acceding countries are an exception to this statement (http://www.worldbank.org/en/news/feature/2014/05/20/empowering-women---where-does-europe-and-central-asia-stand).

collection and auditing of taxes and social insurance contributions would create administrative synergies and enable cross-checking and verification. Increasing the availability of e-government services would facilitate the payment of taxes and contributions, while also increasing transparency. Moreover, some online freelancing and sharing-economy platforms already offer the necessary online services to facilitate tax payments and even some insurance options for their workers.[7] Estonia offers a good example of cooperation between the government and online platforms. Transactions between customers and drivers in the online platform are registered, and the tax-relevant information is transmitted to authorities, who use it to prefill tax forms in order to facilitate future tax payments. Cooperation efforts between governments and online platforms to facilitate tax and social security contribution payments would help to bring workers out of the shadow economy and, at the same time, provide them with some employment protection. Improving the efficiency of the legal system also could improve confidence and increase entrepreneurship in the digital economy. Court systems in some ECA countries, especially in Central Europe and the Western Balkans, are an important barrier to doing business when compared with other non-ECA countries (figure 6.21).

Finally, the fact that older individuals and women are less likely to take advantage of telecommuting in countries where their overall labor force participation is low suggests that these groups face barriers even to accessing traditional jobs. These barriers should be addressed in order for them to take advantage of the new opportunities brought about by the Internet.

In transforming economies, the emergence of online freelancing and the sharing economy poses new challenges to existing labor laws, which were designed to regulate traditional forms of employment. Given the revolutionary nature of these changes, economic analyses and policy recommendations in this area are scarce. Moreover, there is an increasingly heated debate about how to define some of the jobs created in the sharing economy.

For example, are Uber drivers independent contractors or employees of the online platform? They are currently considered to be independent contractors, so the company is not obligated to contribute to their health insurance or pensions or to assume any other obligations typically involved in an employer-employee relationship. This could potentially add pressures to social protection systems if, for example, these new workers do not contribute to health insurance on their own but nevertheless use public health services. Moreover, as most European economies have contribution-based social security systems, if these workers do not contribute to a pension account, they may fall into poverty in old age and be eligible for social assistance, placing fiscal pressure on the social security system.

All of these challenges highlight the fact that existing social protection systems have become inadequate under the new rules of the game. Unlike the traditional employee-employer relationship, the sharing economy is not based on a centralized structure, but on a multitude of contractors providing services, sometimes through different platforms simultaneously. This suggests that the new social protection model could be one where contributions and benefits are no longer linked to a job (as in a salaried job) but to the individual (Goudin 2016). This would increase the portability of benefits, which would help to reduce distortions related

FIGURE 6.21 Inefficient court systems may hinder online entrepreneurship

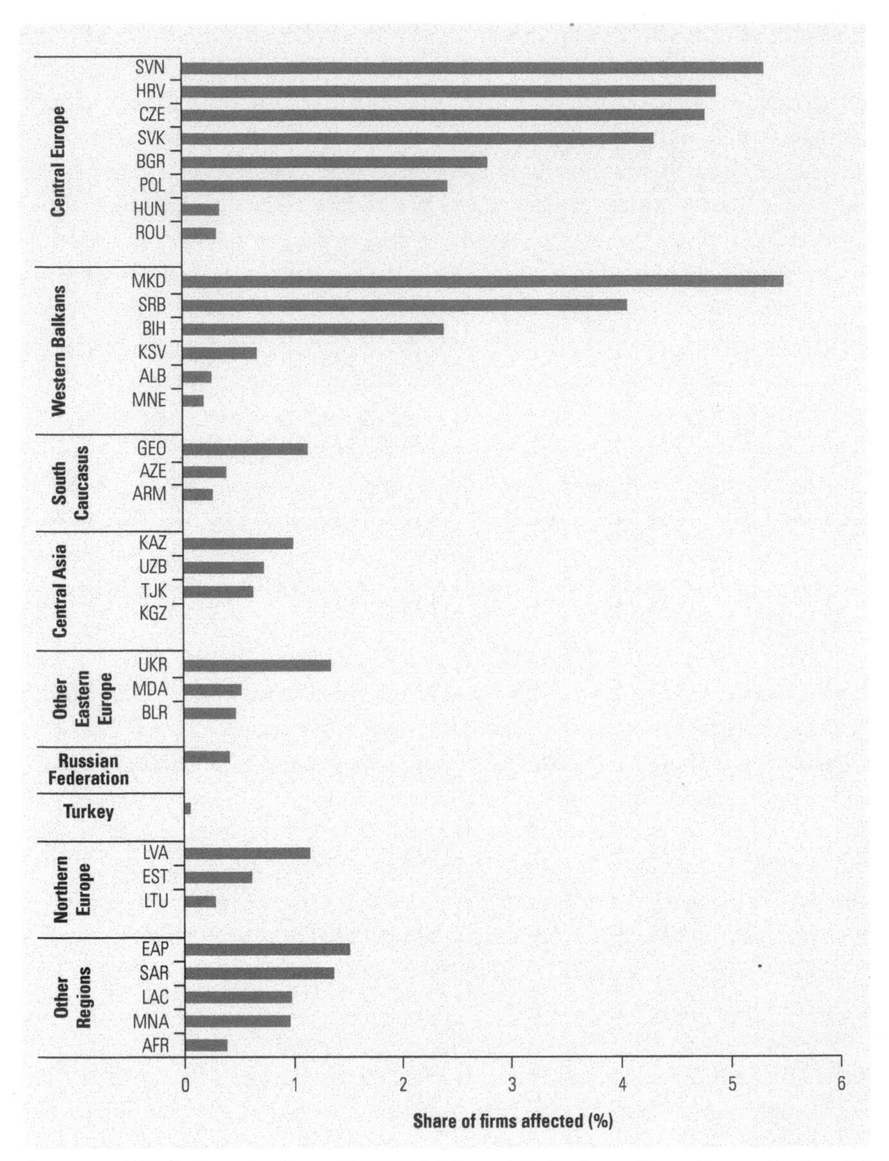

Source: World Bank Enterprise Survey 2009–2013. The *x* axis is the share of firms that mention "courts" as the most serious obstacle affecting its operations (out of 14 mutually exclusive alternatives). The sample includes 122 countries.

to job mobility, entry, and exit, while at the same time providing some insurance against risks.

The example of Uber illustrates the inefficiencies that often accompany occupational licensing. While occupational licensing is necessary in certain occupations and can bring benefits to consumers, it can also add distortions and create vested interests. Technological change could play an important role in addressing these issues (Cramer and Krueger 2016). The share of workers in occupations subject to licensing is 30 percent in the United States and between 15 and 24 percent in the European Union. Within Europe, occupational licensing is relatively high in Cyprus, the Czech Republic, Denmark, Germany, and Poland. At the other end of the spectrum, Estonia, Latvia, and Lithuania have less than 10 percent of their workers in occupations subject to licensing. Some countries in the region are introducing reforms in this area. An interesting example is Poland, where a law was passed in April 2013 to loosen access to 51 professions, including attorneys, notaries, realtors, and taxi drivers (Kleiner 2015). Reducing the entry barriers to occupations could help to reduce the existence of vested interests affecting the pace of technological change and the associated efficiency gains in certain occupations (figure 6.22).

FIGURE 6.22 The gains from the sharing economy could be large in heavily regulated occupations

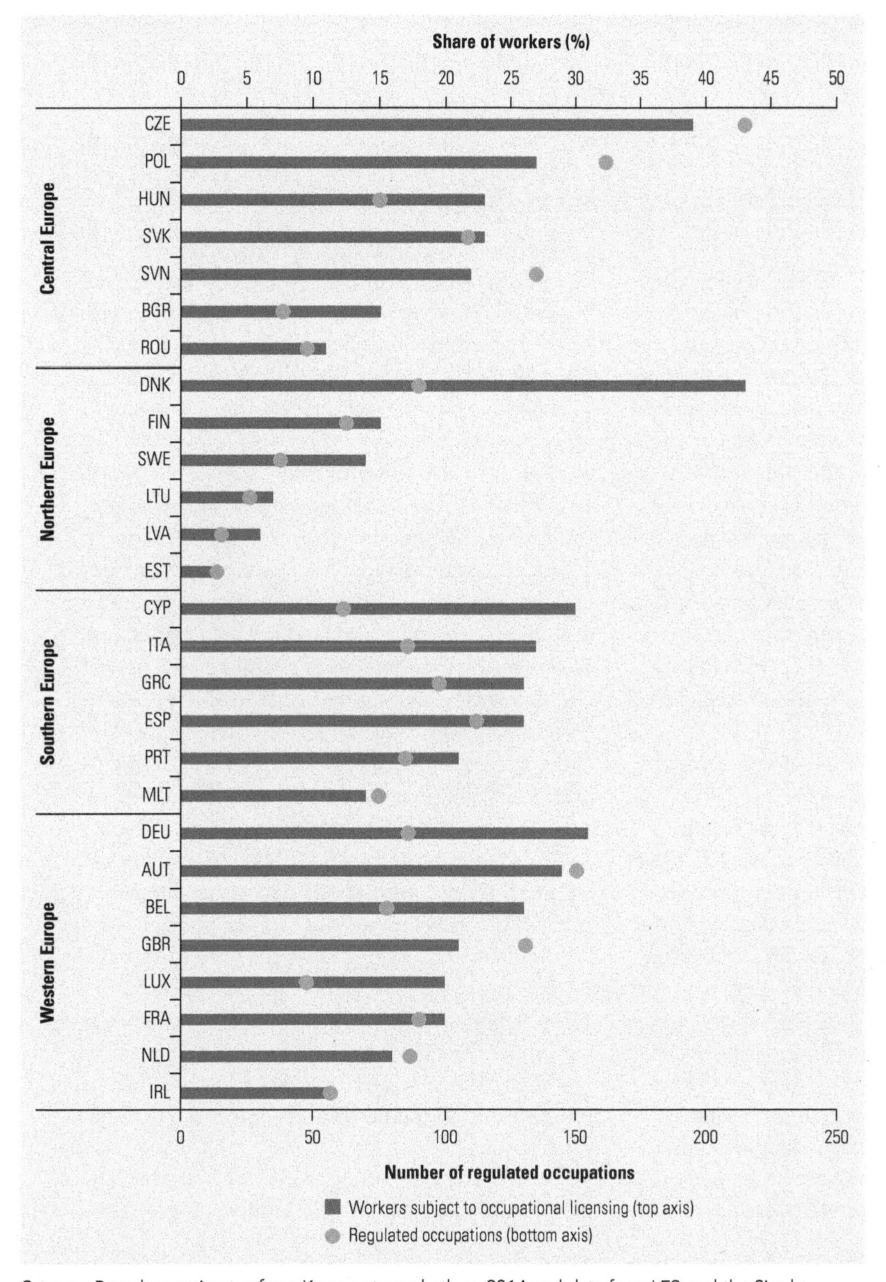

Sources: Based on estimates from Koumenta and others 2014 and data from LFS and the Single Market Regulated Professions Database from 2012.

Finally, another challenge faced by transitioning and transforming digital economies is the scarce evidence on the extent of the new labor market arrangements brought about by the Internet. As mentioned, traditional labor force surveys are poorly equipped to track these new developments. Implementing new surveys or adding new modules to existing ones (as in Katz and Krueger 2016) could help to increase our knowledge about the size of the sharing economy and alternative work arrangements and thereby allow for more evidence-based policies. The use of "big data" could provide innovative and inexpensive ways to collect massive amounts of data about the economic activities carried out through the Internet, providing useful information on economic activities that would support policy making (box 6.2).

BOX 6.2 Using Big Data for Labor Market Policies in ECA

The exponential growth of ICT in recent years, such as search engines, social media sites, online retailers, and news outlets, has rapidly expanded the amount of available data. Our Internet browsing habits are meticulously recorded and leveraged by advertising experts. Online retailers carefully collect data detailing the behavior of consumers. Websites store information on our Internet browsing habits. Data detailing our political inclinations, interests, likes, dislikes, and social connections are collected by popular social networks like Facebook, Twitter, LinkedIn, and Snapchat. These data, collected with the help of modern information technologies, are known as "big data" due to the volume of information.

Big data have some obvious advantages. Traditional methods of collecting data are extremely labor intensive, which can make the cost of procuring quality data prohibitive. This is why we often rely on governments or large international organizations to collect and disseminate economic data. However, even official data can be problematic. For example, data are often released with significant time lags, which can reduce their usefulness for policy makers who have to make decisions in a timely manner. Data can also be plagued by inaccuracies and differences in reporting standards.

However, despite advantages and availability, big data appear to be underused in policy discussions. The phrase "big data" appears in only 37 World Bank working papers and in no International Monetary Fund working papers. The top three leading economic journals have a total of 7 articles since 2000 that mention big data. The vast majority of economic papers and policy reports continue to rely on more traditional data sets. Nevertheless, studies have demonstrated the usefulness of big data in informing policy decisions. For example, Askitas and Zimmerman (2009) show that Google search data can forecast the unemployment rates in Germany using monthly data. However, unlike the official unemployment statistics, search engine data are usually available in real time and free of charge. Another study demonstrated that search engine data can forecast real economic indicators in the United States (Choi and Varian 2012). In particular, Google searches related to automobiles are a good predictor of future automotive purchases.

The increasing popularity of online retailers has significantly expanded our access to consumer price information. Price data for 22 countries are automatically collected over the Internet at the Billion Prices project at the Massachusetts Institute of Technology. These price series are available at extremely high frequencies (weekly, daily) and in real time. Furthermore, they have been shown to have very high levels of correlation with official price statistics.

The collection of labor market data, particularly in developing countries, can be notoriously costly

(Continued)

BOX 6.2 Using Big Data for Labor Market Policies in ECA *(continued)*

FIGURE B6.2.1 Online job vacancies match official indexes of economic activity in Bulgaria

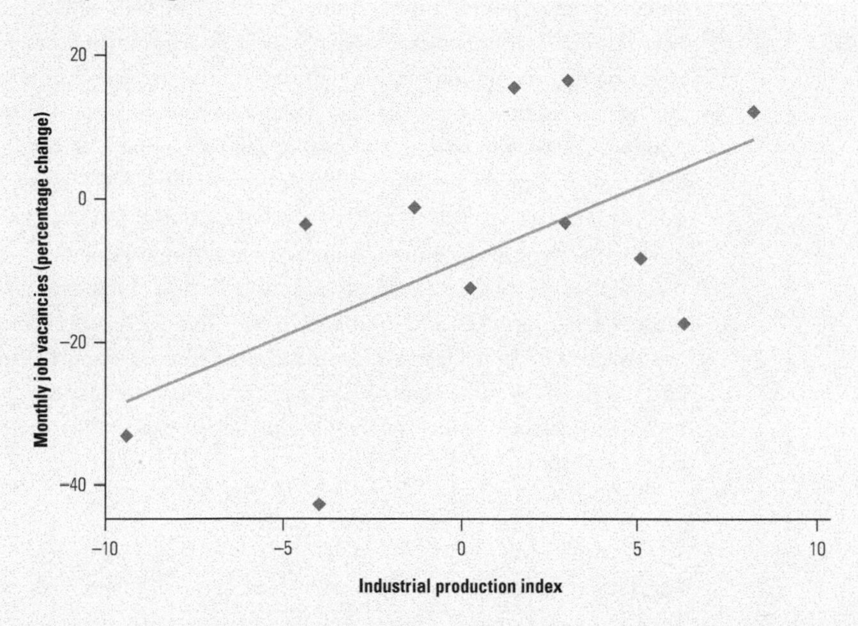

and time-consuming. However, employers increasingly make use of online job portals and advertise their job vacancies on the web. Careful analysis of the advertised job vacancy data could potentially inform policies regarding unemployment, skills mismatch, regional economic development, and other important areas, particularly in countries where reliable data are scarce. For example, job vacancy data collected from one of the leading Russian job portals closely track official statistics reported by the government. The official wage data by region have a 0.66 correlation coefficient with the average salary for all positions advertised in that region. The advertised regional wages are also negatively correlated (−0.47) with the unemployment rate in that region.

The activity on job portals could be a good proxy not only for the state of the job market but also for overall economic activity in a country. We collected monthly data for 2014 on job vacancies from one of the most popular job portals in Bulgaria. Figure B6.2.1 demonstrates the correlation (0.57) between monthly changes in the number of job posts on the website and changes in the industrial production index (not seasonally adjusted).

The use of big data in economic analysis raises some unique challenges and could pose potential difficulties for researchers. Most of what we deem as "big data" is self-reported (users voluntarily provide information about themselves). This raises the possibility of self-selection bias, and researchers should be aware about the representativeness of their data. In addition to bias, one of the greatest advantages of big data—its size—also poses unique difficulties. Spurious relationships in data sets are more likely. When analyzing thousands of variables, the probability of finding statistically significant relationships, without an underlying economic cause (spurious correlation) is large. Without properly accounting for the size of the data, economic conclusions that are based solely on statistical properties should be avoided.

Source: Contributed by Georgi Panterov.

The Labor Market Challenges of the Internet Economy in ECA

The Internet could potentially bring significant dynamism to the labor markets of ECA. The Internet can allow for job creation in Internet-intensive sectors, bring efficiency gains as sharing-economy platforms expand, and improve labor market inclusion for individuals typically excluded from the labor force. However, these benefits come with risks. On the one hand, economic polarization between and within countries may increase if not everyone has the human capital or the Internet connection speed required to take advantage of these opportunities. Moreover, the informal sector may also expand, especially for countries that have not yet managed to bring traditional workers out of the shadows. Finally, more advanced countries face the challenge of adapting their labor regulations to the new economy, while not having precise knowledge of the extent of the new labor market arrangements and who uses them. The ECA region faces the difficult challenge of embracing the Internet to inject some dynamism into its labor markets, while at the same time ensuring that these benefits reach those in the bottom 40 percent of the income distribution and avoiding the informalization of the digital economy.

Notes

1. According to data for the United States, high Internet-intensive sectors are those where workers are 50 percent more likely to use a computer at work.
2. The rise of the "gig" or freelance economy is often mentioned as one of the major changes brought about by the Internet across the world. See, for example, "Gig Economy Catches on among U.S. Workers," *Financial Times*, February 2016; "Does the Gig Economy Revolutionise the World of Work, or Is It a Storm in a Teacup?" *The Economist*, October 2015; "The Freelance Surge Is the Industrial Revolution of Our Time," *The Atlantic*, September 2011.
3. Katz and Krueger (2016) find that, while the traditional survey used to measure employment trends in the United States reported a decline in self-employment from 2005 to 2015, a survey better designed to capture alternative work arrangements and multiple job holdings, along with data on tax returns by self-employed individuals, showed a clear upward trend.
4. Such as free-lance.ru.
5. See "The Rise of the Sharing Economy," *The Economist*, March 9, 2013; Martin, Shaheen, and Lidicker (2010).
6. Online work is measured by the sum of the number of online workers in each country and the number of jobs posted by employers in each country as a ratio of the population.
7. For example, Elance.com issues the required tax forms for workers and firms based in the United States. At the same time, Elance.com offers errors and omissions insurance as well as general and professional liability protection.

References

Agrawal, A., J. J. Horton, N. Lacetera, and E. Lyons. 2013. "Digitization and the Contract Labor Market: A Research Agenda." NBER Working Paper w19525, National Bureau of Economic Research, Cambridge, MA.

Askitas, N., and K. F. Zimmermann. 2009. "Google Econometrics and Unemployment Forecasting." RatSWD Research Notes 41, German Council for Social and Economic Data, Mannheim.

Autor, D. H. 2001. "Wiring the Labor Market." *Journal of Economic Perspectives* 15 (1): 25–40.

Bagues, M. F., and M. S. Labini. 2009. "Do Online Labor Market Intermediaries Matter? The Impact of AlmaLaurea on the University-to-Work Transition." In *Studies of Labor Market Intermediation*, edited by D. Autor, 127–54. Chicago: University of Chicago Press.

Choi, H., and H. Varian. 2012. "Predicting the Present with Google Trends." *Economic Record* 88 (Suppl 1): 2–9.

Cramer, J., and A. B. Krueger. 2016. "Disruptive Change in the Taxi Business: The Case of Uber." NBER Working Paper 22083, National Bureau of Economic Research, Cambridge, MA.

CSIS (Community Statistics on Information Society), Eurostat. Various years. http://ec.europa.eu/eurostat/web/microdata/community-statistics-on-information-society.

Del Carpio, X., O. Kupets, N. Muller, and A. Olefir. 2017. *Skills for a Modern Ukraine*. Directions in Development–Human Development. Washington, DC: World Bank.

European Commission. 2016. "The Use of Collaborative Platforms." *Flash Eurobarometer* 438 (March): n.p.

Eurostat. Various years. LFS (Labor Force Surveys). http://ec.europa.eu/eurostat/web/microdata/european-union-labour-force-survey.

Farrell, D., and F. Greig. 2016. "Paychecks, Paydays, and the Online Platform Economy Big Data on Income Volatility." JP Morgan Chase Institute, San Francisco, February.

Goudin, P. 2016. "The Cost of Non-Europe in the Sharing Economy." European Parliamentary Research Service, Strasbourg.

Greene, L., and I. Mamic. 2015. *The Future of Work: Increasing Reach through Mobile Technology*. Geneva: ILO.

Horton, J. J. 2010. *Online Labor Markets*. Berlin: Springer.

Howard, P. N., and N. Mazaheri. 2009. "Telecommunications Reform, Internet Use, and Mobile Phone Adoption in the Developing World." *World Development* 37 (7): 1159–69.

ILO (International Labour Organization). Various years. ILOSTAT. ILO, Geneva, www.ilo.org/ilostat.

Katz, L., and A. Krueger. 2016. "The Rise and Nature of Alternative Work Arrangements in the United States, 1995–2015." NBER Working Paper w2267, National Bureau of Economic Research, Cambridge, MA.

Kleiner, M. M. 2015. "Reforming Occupational Licensing Policies." Brookings Institution, Washington, DC.

Koumenta, M., A. Humphris, M. Kleiner, and M. Pagliero. 2014. "Occupational Regulation in the EU and U.K.: Prevalence and Labour Market Impacts." Final Report, Department for Business, Innovation and Skills, School of Business and Management, Queen Mary University of London, London.

Kuek, S. C., Paradi-Guilford, C., Fayomi, T., Imaizumi, S., Ipeirotis, P., Pina, P., & Singh, M. (2015). The global opportunity in online outsourcing (No. 22284). The World Bank.

Kuhn, P., and H. Mansour. 2014. "Is Internet Job Search Still Ineffective?" *The Economic Journal* 124 (581): 1213–1233.

Leighton, P., and D. Brown. 2013. "Future Working: The Rise of Europe's Independent Professionals (iPros)." Report of the European Forum of Independent Professionals.

Martin, E., S. A. Shaheen, and J. Lidicker. 2010. "Impact of Carsharing on Household Vehicle Holdings Results from North American Shared-Use Vehicle Survey." *Transportation Research Record* 2143: 150–58.

Oyer, P., and S. Schaefer. 2011. "Personnel Economics: Hiring and Incentives." *Handbook of Labor Economics* 4 (2011): 1769–823.

Packard, T. G., J. Koettl, and C. Montenegro. 2012. *In from the Shadow: Integrating Europe's Informal Labor.* Washington, DC: World Bank.

Pew Research Center. 2016. "Shared, Collaborative, and On Demand: The New Digital Economy." Pew Research Center, Washington, DC.

Roach, S. 2004. "How Global Labour Arbitrage Will Shape the World Economy." Morgan Stanley, San Rafael, CA.

Shevchuk, A., and D. Strebkov. 2012a. "Freelance Contracting in the Digital Age: Informality, Virtuality, and Social Ties." Research Paper BRP 12, Higher School of Economics, National Research University, Moscow.

———. 2012b. "Freelancers in Russia: Remote Work Patterns and E-Markets." *Economic Sociology: The European Electronic Newsletter* 13 (2): 37–45.

Vazquez, Emmanuel Jose, and Winkler, Hernan. 2017. "How Is the Internet Changing Labor Market Arrangements? Evidence from Telecommunications Reforms in Europe." Policy Research Working Paper No. 7976, World Bank, Washington, DC.

Vaughan, R., and R. Daverio. 2016. "Assessing the Size and Presence of the Collaborative Economy in Europe." PriceWaterhouseCooper, Birmingham, AL.

Welz, C., and F. Wolf. 2010. "Telework in the European Union." Eurofound, Brussels.

Winkler, H. 2016. "How Does the Internet Affect Migration Decisions?" *Applied Economics Letters*: 1–5.

Zervas, G., D. Proserpio, and J. Byers. 2016. "The Rise of the Sharing Economy: Estimating the Impact of Airbnb on the Hotel Industry." Research Paper 2013-16, School of Management, Boston University, Boston, MA.

7

Toward a Digital Single Market

The first six chapters of this book identify several bottlenecks that may be preventing the emergence of the digital economy in Europe and Central Asia. While the policy priorities are different for countries at different levels of digital development, all of them point in the same direction, that is, to increase the connectivity of governments, firms, and individuals within and between countries. Improving diplomatic relationships between countries and promoting competition in the telecommunications sector in the South Caucasus and Central Asia would be crucial to lowering Internet prices and increasing the quality of access. Taking measures to reinforce cybersecurity and updating regulations to deal with the new market structures brought about by the Internet would help to increase the connectivity of firms among the rich economies of Europe. For economies in the middle, such as the Russian Federation and Ukraine, improving the stock of skills required by the new economy and introducing more incentives to work in the formal economy would increase the chances of connecting local workers with good job opportunities, at home and abroad.

Improving connectivity across Europe and Central Asia will be challenging, as the level of fragmentation regarding languages, regulations, and culture is very high in the region. A common thread across this book has been the persistence of disparities among countries, due to historical, geographical, and economic differences. The member states of the European Union have already committed to building a digital single market, with the goal of generating additional growth (European Commission 2015). Now the five countries

of the Eurasian Economic Commission (Armenia, Belarus, Kazakhstan, the Kyrgyz Republic, and the Russian Federation) have also committed to building a common digital space, Digital Eurasia, by 2025.[1] A second theme of the book has been the failure of Internet firms based in Europe to match the scale and growth rate of their peers in China and the United States. One of the reasons put forward for this failure is the lack of economies of scale in Europe that is caused by market disparities and lack of harmonization that obliges Internet companies to tackle each market separately. Thus, creating a digital single market would not only improve the customer experience, by making it more convenient to undertake digital transactions that cross the boundaries between European markets, but also facilitate the emergence of an expanded playing field for firms seeking to serve those markets.

Introduction and Main Messages

This final chapter summarizes the policy recommendations arising from the six substantive chapters of the book and assesses the challenges that lie ahead for the region as it seeks to develop a digital single market. As in earlier chapters, the analysis is based on the diagnostic framework for digital development provided in the *World Development Report 2016* (WDR16; World Bank 2016).

The first part of the book (chapters 1 and 2) focuses on the information and communication technology (ICT) sector itself and the challenge of making the Internet universally available, accessible, affordable, open, and safe. The second part (chapters 3 and 4) looks at the progress made by firms to adopt the Internet as a driver of increased productivity, more diversified markets, and more trade. The third part (chapters 5 and 6) focuses on people, especially entrepreneurs and employees, and what skills and working practices are required to adapt to the needs of the digital economy. While the first part of the book is concerned with development of the digital economy, the second and third parts look at its "analog foundations," particularly regulations, skills, and institutions.

Digital Economy: Connecting the Unconnected in ECA

Supply-Side Policies: Availability, Accessibility, and Affordability

The Europe and Central Asia (ECA) region has a relatively high level of international connectivity compared with other regions of the world: Western European countries are generally well served by submarine connectivity, and some cities are already global hubs (notably Amsterdam, Frankfurt, London, Paris, and Stockholm). But the ECA region also includes some poorly served countries in the east of the region: Central Asian economies, in particular, are landlocked and disadvantaged with regard to international connectivity. Regarding national connectivity, the ECA region lags behind in the adoption of fiber and consequently also in speed and quality, with a few notable exceptions (such as Latvia, where 60 percent of broadband subscriptions are served by fiber). ECA countries also have wide variations in mobile coverage, especially for newer technologies like fourth-generation (4G) service, which started only recently in Belarus, Kosovo, and Serbia.

But ECA has the advantage of an early start on the process of telecommunications reform. Early movers, such as Sweden and the United Kingdom, had implemented a successful formula based on market liberalization, private sector participation, and independent regulation by the mid-1980s, and this formula was generalized throughout the European Union (EU) by 1998. Countries in the east of the region soon followed, with expansion of the EU providing a model of accelerated reform for aspiring member states. Again, Central Asia is lagging, and some countries have retained oligopolistic structures and a more dominant role for the state. But even here, the direction of market liberalization tends to be one way. On the supply side, ECA countries need to implement policies that ensure improved availability, accessibility, and affordability of Internet services and help to drive down the high prices that can be a significant barrier to Internet adoption. The policy framework developed in WDR16 suggests appropriate policies for countries at different stages of development.

First, national policies need to be adapted to each country's level of development, market circumstances, and policy framework. Countries may be classified according to their level of adoption of digital technologies and the status of their analog complements (regulation, institutional accountability, skills).

- For emerging countries, the priority should be to foster telecommunication market competition by facilitating market entry in both fixed and mobile market segments, by developing a national broadband plan, and by issuing additional licenses, or authorizations, where appropriate. Some emerging countries still have oligopolistic structures in fixed-line markets or restrictions on mobile service provision. The timeline for further liberalization sometimes remains uncertain, especially where the state retains a shareholding in the historical operator or incumbent. The independence of regulatory bodies in some of these countries, though still unresolved, is surely a crucial, if elusive, factor to achieving well-functioning broadband markets. The pressure to create a digital single market could help governments to push through reforms, as has been the case among EU member states.
- For transitioning countries, further strengthening market competition, through independent light-touch regulation, and initiating national broadband plans, through stakeholder-led processes, should be priorities. This may require, for instance, public-private partnerships to build open-access networks in areas where the private sector is unwilling or unable to invest alone. Transitioning countries also frequently require steps to ensure that scarce resources—such as spectrum, rights of way, or numbering—are allocated efficiently.
- For transforming countries, as issues of availability and affordability are progressively addressed and universal access is close to being achieved, the policy attention generally shifts away from supply-side measures and toward demand-side measures, such as ensuring user privacy and cybersecurity. Most transforming countries are already well advanced on local loop unbundling, universal access, or measures to restrict market dominance. However, new policy issues are looming, in areas such as the commercial use of consumer data or open access to government data.

Beyond national policies, regional coordination and harmonization appear necessary in the ECA region, especially to open up the landlocked countries of Central Asia and Southern Caucasus or to improve the provision of broadband services in isolated and rural areas. The EU regulatory framework provides a sound basis for further coordination, notably in the deployment of infrastructure and the development of cost-effective international networks, and could be adapted for wider use throughout the region.

Demand-Side Policies: Open and Safe Internet Use

On the demand side of the Internet, the policy agenda for improving the level of Internet use in ECA includes issues of consumer privacy and data protection. Although on the supply side, there is a high degree of consensus regarding the importance of the three pillars of market liberalization, public-private participation, and independent regulation, on the demand side there is less agreement. Policy preferences vary, and countries need to find the right balance between stakeholder interests in these areas—for instance, in the trade-off between security and privacy. Ensuring that there is an enabling environment of "trust," so that Internet use can achieve its full potential, is best achieved through a multistakeholder approach (World Bank 2016, fig. O.20).

In ECA, a lack of trust among consumers is apparent, with regard to both online merchants and payment systems. As a consequence, online purchasing is not uniformly well developed, and consumer protection legislation that might contribute to creating an enabling environment is often lacking. For instance, it can be difficult to obtain refunds or to be protected against fraud when sellers are based abroad. This issue requires multistakeholder regional cooperation and can be achieved through a digital single market agenda.

Another issue of concern relates to personal privacy. Internet use and the ability to store large volumes of records in a searchable manner have raised significant concerns about the use, collection, and transfer of personal data. In ECA countries, as in the rest of the world, data are collected by private and public entities for different goals: for marketing purposes (to determine suitable advertising targets for a product or a service) as well as for government purposes (to establish a national database of taxpayers or electoral rolls). Developing policies in this area requires striking a balance between protecting the privacy of individuals and protecting them from criminals. Although the Snowden case has raised awareness about the risks of state surveillance, there is generally less awareness of the risks of commercial surveillance, which is much more widespread (Jeffreys-Jones 2016). Europe has taken the lead on the regulation of personal data use. The EU Data Protection Directive and the EU Directive on Privacy and Electronic Communications have created a regulatory framework on this subject within EU members, on top of existing national laws. Outside the EU, some national laws have also been passed, such as the Kazakhstan Law on Personal Data and Their Protection, which came into effect in 2013.

A challenge for ECA countries is to ensure the interoperability among regimes and legal systems. Interoperability will make it possible to prosecute cybercriminals across jurisdictions and to support development of a "trust" environment for citizens, firms, and governments on data protection.

Strengthening the Analog Foundations of the Digital Economy in ECA

Beyond the immediate requirements of the ICT sector itself, the ability of countries to reap digital dividends will also depend on their ability to absorb the extra capabilities offered by the Internet and to convert them into productive activities. This will require addressing the "analog complements" (regulations, skills, and institutions) that underlie digital technologies, specifically with regard to firms, individuals, and governments. Where these "analog complements" are missing, digital technologies may result in policy failures (World Bank 2016, 18), such as a higher degree of concentration among firms, an increase in inequality between different parts of the labor market, and the use of the Internet to "control" rather than empower citizens.

Firms

For firms, the Internet gives access to greater economies of scale, for instance, enabling firms to sell beyond a country's borders. But in the absence of safeguards, this ability can lead to market concentration, to the reemergence of monopolies, and thus to less future innovation. This is arguably already evident in social media and the sharing economy, where companies like Facebook or Uber have established leading brands. Thus, policies that are specific to the "digital" world need to be reinforced by "analog" policies to ensure competition, especially among "platforms" that feed on these economies of scale. In the ECA region, firms in Western Europe generally have higher labor productivity than firms in Central Asia or South Caucasus. Moreover, ICT contributed an estimated 31 percent of the productivity growth in EU countries between 2003 and 2007. Nevertheless, ECA firms do not use the Internet intensively and, when they do, use it mainly for simple tasks. More than 85 percent of firms use the Internet to send e-mails, but only 11 percent use it to provide customer service, such as through a website. Market entry by firms is also quite varied, depending on the subregion of ECA: levels of firm density, or the number of newly registered firms, are generally much higher in transforming ECA countries than in transitioning ones.

Here again, governments have an important role to play. Encouraging firms to use the Internet more intensively—to foster productivity and firm entry—will improve a country's long-term growth. However, weak management capacity and a lack of skilled labor might hinder the adoption of new technologies. But the main bottleneck that might hold back the wider application of the Internet in ECA firms is lack of competition (due to high market entry barriers, vested interest groups, or lack of incentives to innovate). Removing these bottlenecks requires an improved business environment, including lower barriers to competition and entry. In this area, governments can take the following approaches:

- In emerging countries, which generally have a relatively lower level of market competition and Internet use by firms, policy priorities should focus on setting up the fundamentals of an efficient business environment: dropping barriers to imports, to foreign direct investments, and to domestic entry as well as facilitating entrepreneurship.

- In transitioning countries, where there are generally both a moderately high level of competition and Internet use as well as the necessary elements for an enabling business environment, policies should focus on reinforcing the competition policy framework and, if not already done, establishing a competition agency to monitor and enforce these rules.
- In transforming countries, with a relatively higher level of competition, policy questions may involve adapting the regulatory environment to new issues, such as how to ensure a level playing field in multisided markets or how to promote cybersecurity.

Strategies related to improving management practices can encourage enterprises to make a more intensive use of technologies (for example, subsidized consulting services, business incubation, and mentoring). Mandating interoperability of technology can also be important. The introduction of standards and norms, for instance, in e-invoicing systems or in data exchange, would enable firms to make the most of the inherent network effects of technology and to improve their efficiency. Such harmonization measures are a mundane, but necessary, part of achieving a digital single market.

ICT can facilitate the integration of firms in the wider regional and global economy, if the regulatory framework and business environment are adequate. This can be done by increasing productivity and market competition, but also by reducing the costs of trade. Indeed, ICT can stimulate exports by reducing the transaction costs involved in trade, contribute to developing new sectors (for example, financial advice, reputation management, and data mining) in an economy, and help to make firms and their staff more mobile. This is particularly important in a digital single market.

But, despite improvements over the last five years, only a small proportion of firms in ECA countries use the Internet to sell their products (e-commerce). Moreover, online selling remains mainly local. Business-to-consumer (B2C) e-commerce sales vary: B2C sales as a percentage of gross domestic product are comparable in Belarus, the Russian Federation, Turkey, and Ukraine to the share in Australia and Canada, but less than 1 percent in Central Europe. Governments can help to promote digital trade and boost the process of structural transformation and economic growth. An enabling environment will promote digital trade by creating the right conditions for firms.

The policy priorities for ECA countries can be considered in three categories, based on the relationship between the share of firms engaged in e-commerce and the quality of infrastructure and payment systems, which captures the enabling environment for digital trade:

- In emerging countries, policy should be oriented toward ensuring that the fundamentals are in place, so that firms are able to deliver their goods and receive payment for them securely. The focus should be on improving the logistics infrastructure and process (for instance, ensuring that the postal address system is fit for purpose) and developing easier and more secure payment systems.

- In transitioning countries, where the fundamentals are generally already in place, the policy focus should be on enhancing trade facilitation measures such as reducing paperwork and simplifying procedures to facilitate exports. Additionally, greater competitive pressure would help to strengthen the incentives to engage in e-commerce.
- In transforming countries, good systems and infrastructure should generally be in place to foster digital trade, and the policy focus should be on navigating the international trading system and negotiating with trade partners to reduce barriers to trade, especially for services delivered digitally. This effort may require steps to harmonize regulations on the storage of personal consumer data.

These proposals should help to ensure an adequate regulatory environment that will leverage ICT to improve the ability of firms to compete, innovate, and trade as well as facilitate their inclusion in the world economy.

Individuals

Digital technologies automate many tasks; with the advent of artificial intelligence, many more advanced tasks, such as driving, offering investment advice, or checking passports, may also be automated. In so doing, they may eliminate many existing, routine jobs, but also create new jobs that require a higher level of creativity and digital literacy. But the job-creating and job-displacing effects of new technologies rarely take place at the same time and place. There is a danger, therefore, that more intensive use of ICT might contribute to rising inequality, rather than simply to improved efficiency, where new economy skills (that is, socioemotional and high-order cognitive skills) are lacking. Improving new economy skills is thus another "analog" complement to take into account, so that people can take full advantage of digital technologies and easily adapt to new labor market requirements, a process that may be required multiple times in the course of a career.

The net effect of changes in automation on jobs is the increasing polarization of the labor market, where routine middle-skill[2] occupations disappear, while higher-skill occupations account for a rising share of the total number of jobs, together with low-skill jobs that are hard to automate, such as cleaners, caregivers, or hairdressers. Furthermore, the sector reforms in telecommunications, which have had the effect of increasing Internet adoption, have been accompanied by a significant rise in the demand for skilled workers, at least in Central and Western Europe. Paradoxically, the demand for skills seems relatively higher in countries where Internet adoption is lower (Central Europe) and where Internet-intensive sectors absorb a larger share of skilled workers. The shortage of key skills could indeed create bottlenecks for the continued expansion of the Internet among the poorer economies of ECA.

In this context, it is significant that advanced computer skills are still scarce in many parts of ECA. While most adults have basic computer skills (at least 80 percent of adults in Serbia, for instance, compared with close to 100 percent in Finland), relatively few of them are truly advanced users (only 10 percent in Serbia and 50 percent in Finland). Categorizing ECA countries according to their adaptability to new technologies (quality-adjusted years of education) and their expected degree of labor market disruption (approximated by Internet penetration, as the biggest changes in the demand for skills are yet to come), it is possible to make the following observations on policy priorities:

- Emerging countries need to build basic foundational skills, including digital literacy. Policies should therefore focus on fostering early childhood development, improving the quality of primary and secondary education, and including the adoption of ICT in the classroom and the curriculum.
- In transitioning countries, labor regulations need to be adapted to foster vocational training, including training provided directly by employers. Curriculum reforms would help to introduce more innovative and creative learning. And a better match between skills in demand and skills taught at the secondary and tertiary levels would enable these countries to address the shortage of higher-order skills.
- Transforming countries may have a scarcity of ICT specialists, which is compounded by an aging workforce. Relevant policies would therefore focus on using ICT for more creative learning, addressing gender gaps in science, technology, engineering, and mathematics (STEM) education, fostering lifelong learning, and reducing barriers to the migration of workers. Taking a more liberal view of visas and work permits for skilled workers from elsewhere would also help, especially in the short term.

ICT is not only affecting the supply of and demand for skills but also creating new forms of work and more flexible approaches to working. Moreover, improving basic digital skills will contribute in the longer term to developing labor markets that are more responsive and better adapted to the digital revolution and to improving the scope of career opportunities and job mobility for individuals. For instance, the skills needed to participate effectively in an online meeting or to prepare an engaging visual presentation are portable between many different occupations. Public-private partnerships that engage the private sector in delivering vocational training and technical certification may also be considered in emerging and transitioning countries.

Following liberalization of the telecommunications sector in the 1980s and 1990s, most Central and Western European countries experienced net employment creation—over a period of 10 years—that was higher in Internet-intensive sectors than in non-Internet-intensive ones as well as an increase in part-time, temporary work and teleworking. In Luxembourg, Nordic countries, and the United Kingdom, for instance, at least 20 percent of employees work from home, whereas this share is almost negligible in Bulgaria, Croatia, Latvia, and Romania. The Internet has also created new forms of jobs. Online freelance jobs have developed strongly in Central Europe and Western Balkans, while Central Asia and Turkey are lagging behind. Moreover, the sharing economy continues to develop apace. Workers use sharing-economy platforms, in sectors like ride sharing, accommodation, or catering, to complement their other revenue and to combine jobs. In most ECA countries, the share of people who have used these platforms to offer services and earn an income is quite low (less than 5 percent), but it reaches 10 percent and more in Croatia, France, and Latvia.

Policy and regulatory frameworks will need to adapt, in alignment with the relative level of country development.

- In emerging countries, workers may not be able to take full advantage of the new forms of work brought about by the Internet. For that reason, policy priorities should focus on the supply side, removing barriers to Internet adoption and developing new economy skills, including both digital literacy and basic business skills.
- In transitioning economies, online freelancing and the sharing economy have started to emerge, which can increase the level of informality in the economy. Thus, government should focus on reducing the barriers to formal work, while at the same time providing safeguards, such as improving the efficiency of courts and dispute resolution systems. Providing safeguards would help to improve confidence and foster digital entrepreneurship. Finally, many transitioning countries need to develop policies that are appropriate for an aging population, such as lifelong learning, or that take a more flexible approach to retirement age.
- Transforming countries would need to focus on reducing the number of inefficient regulations that may be hindering occupational mobility. This may require updating labor market regulations to adapt to online freelancing and the sharing economy as well as improving data collection efforts to quantify the magnitude of these occupations.

Improving ECA policies with regard to digital skills and more flexible work arrangements will help to leverage ICT for citizens to take full advantage of the Internet and to safeguard against rising inequalities.

Governments

While this book does not devote a specific chapter to the impact of the Internet on government actions, this topic is mentioned in different chapters of the book. In addition, the WDR16 analyzes this topic and draws many policy conclusions that are relevant for ECA countries. These findings are summarized and included here.

The Internet contributes to strengthening government capabilities and to empowering citizens in three main ways:

- Overcoming information barriers and thus fostering the inclusion of all citizens in service provision and elections
- Improving the efficiency of government services through automation of routine activities and more effective monitoring and evaluation
- Lowering communication and transaction costs, including through platforms, and thus fostering citizens' voice and collective action

Nevertheless, in countries where governments lack accountability, there is a danger that digital technologies might lead to greater surveillance and control of citizens rather than their empowerment. Thus, given the importance of government institutions for service delivery, the policy agenda should focus on enhancing transparency while increasing efficiency, for instance, by promoting open government data and citizen feedback mechanisms.

Governments tend to use information technologies more intensively than private sector firms in many ECA countries, as reflected by their relative rankings on the digital adoption index (chapter 1). This use has expanded rapidly to automating core administrative tasks, collecting taxes, improving the delivery of public services, and promoting transparency and accountability. But the return on these investments remains limited and difficult to quantify, often because many e-government services are underused by citizens. This may be because the interfaces are poorly designed or because there is little attempt to market the service, but it also reflects the fact that most citizens use the Internet simply to get information[3] and not necessarily to interact with their government.[4]

But much more could be done, and this generally requires a "must-have" application that can help citizens to get into the habit of using the Internet to interact with their government. In Estonia, for instance, that must-have application is digital identity (ID),which provides a platform to support many other applications (see box O.3 in the overview) and to streamline processes. Azerbaijan and Moldova have developed one-stop shops or service centers that provide services to citizens and businesses (registration, licensing, record, bill payments) via a single portal. The Russian Federation has developed a network of almost 2,000 centers delivering joined-up government services at federal, regional, and municipal levels.

In order to have a positive impact, implementation of e-government systems needs to be accompanied by complementary reforms. For example, in Belarus, e-filing of taxes was part of a broader reform to reduce the compliance costs for citizens (simplified tax code, taxpayer facilitation services, and reaching out to the business community), whereas in Tajikistan, e-filing was not mandatory: its use therefore remained low, and firms continued to submit paper returns.

Some sectors of the economy have adopted digital technologies as part of a more thorough process of reform, such as the Ministry of Justice in Georgia and the Ministry of Health in Estonia. In Georgia, the National Bureau of Enforcement—a legal entity of the Ministry of Justice—has implemented electronic case management since 2010. Thanks to a registration number and a unique code, the parties have online access to enforcement-related materials and follow the process of enforcement proceedings online. In 2008, the Ministry of Health in Estonia introduced an electronic health record system. This nationwide system created a single, common record "owned" by each patient, which can be filled out with data from different health care providers. By logging onto the portal using their electronic ID card, patients can access their doctor visits and prescriptions record. Doctors can easily access their patients' files—including test results and time-critical data that are useful in emergency situations (blood type, allergies, and treatments and medications). As a result, some 84 percent of prescriptions in Estonia are digitized. The Ministry of Health can also make use of these data, in anonymized form, in national statistics to track health trends, monitor epidemics, and ensure that health resources are spent effectively.

These examples show how governments and institutions can successfully leverage digital technologies to respond to the needs and changing requirements of citizens and firms. By adopting an integrated approach, governments can be more efficient, accountable, and transparent in their service delivery and, at the same time, more participatory in the inclusion of citizens and firms in helping to shape service delivery.

Based on these good practices in the area of digital government, the policy priorities for ECA countries can be considered according to the level of existing e-government practices in the country. An action plan may propose which public services should be priorities for automation and emphasize the value of open data. In more ambitious implementations, an action plan may seek to establish a government cloud and an interoperability framework, pursue a data-driven approach, or maximize interoperability between government services in order to minimize the need for citizens to provide documents more than once.

- Emerging countries would need to start developing "must-have" e-government applications in areas such as e-filing of taxes or company registrations or creating a database of driver's licenses.
- Transitioning countries would need to adopt an integrated "whole of government" approach to e-government and accompany this with moves toward systems integration and complementary reforms.
- Transforming countries would need to promote—through a multistakeholder approach—open government data and develop guidelines for the use of big data that are consistent with data privacy and security.

Taken together with the reforms on regulations, skills, and institutions, this will ensure that the benefits of wider Internet use can be converted into digital dividends for economic and social development. But the way in which it is done is almost as important as what is done. The interventions that governments could adopt are best conducted in consultation with the private sector, development partners, user groups, and other interested parties through a multistakeholder approach.

Conclusion

Across the ECA region, the level of digital development varies, based on the economic development of countries as well as differences in their history and geography. This book has proposed a framework for policy formulation that takes into account these disparities and seeks to bridge them. The policies proposed are summarized in table 7.1, which sets out policy and regulatory recommendations for ECA countries, covering both the ICT sector and connectivity issues as well as the analog complements that will help countries to derive maximum benefits from digital technologies.

TABLE 7.1 Proposed Policy Framework for Digital Technologies in ECA Countries

Policy framework	Emerging countries	Transitioning countries	Transforming countries
Connectivity			
Supply of digital technologies	Strengthen market competition in fixed and mobile markets; coordinate at the regional level to promote infrastructure deployment and bring enhanced connectivity to landlocked countries	Develop independent, light-touch regulation of the sector, especially of scarce resources, and national broadband plans through public-private partnership and a multistakeholder approach	Ensure the provision of open-access networks, including local loop unbundling and licensing of mobile virtual network operators, and other measures to limit the degree of market dominance, particularly for digital platforms
Demand for digital services and applications	Use demand stimulation and awareness measures, including selected government services as demonstration projects	Foster regional and global cooperation on cybersecurity and personal privacy (interoperability of regimes and legal systems); consider developing digital identity	Develop a digital ecosystem, including measures to promote open government data and development of rules for the use of big data
Analog complements			
Firms (competition and trade infrastructure)	Reduce or eliminate import barriers; encourage foreign direct investment; facilitate entrepreneurship; improve logistics infrastructure; develop convenient and secure payment systems	Reinforce the competition policy framework; establish a competition office to enforce compliance (if none exists); enhance trade facilitation measures	Adapt the regulatory environment to new issues such as multisided platform markets and emerging cybersecurity risks; reduce trade barriers in partner countries to promote digital trade of goods and services
Citizens (skills and labor markets)	Foster early childhood development; improve the quality of primary and secondary education; remove barriers to Internet adoption; increase new economy skills	Foster professional training; introduce more innovative and creative learning; improve the match between the supply of and demand for skills (at secondary and tertiary levels); reduce barriers to formal work; increase the efficiency of courts and dispute resolution systems; promote lifelong learning and a flexible approach to retirement	Use ICT for more creative learning; address gender gaps in STEM education; foster lifelong learning; reduce barriers to immigration of workers; reduce inefficient occupational regulations; include online freelancing and sharing-economy work in labor market regulations; improve data collection efforts to quantify the freelancing and sharing economy
Governments (institutions)	Focus on overcoming barriers to Internet usage, including supply-side constraints; develop "must have" e-government applications as demonstration projects	Adopt an integrated "whole of government" approach to e-government; accompany implementation of e-government systems with complementary reforms	Promote open government data and develop rules for the use of big data that are consistent with good practice in data privacy and security; develop a multistakeholder approach in consultations regarding policy development

Note: ICT = information and communication technology; STEM = science, technology, engineering, and mathematics.

Building a National Digital Strategy

Implementing this framework in ECA countries needs to be based on a comprehensive digital strategy for each country as well as a regional vision. A national digital strategy that puts forward a suite of guiding principles and actions for ICT can be used to align with the broader socioeconomic goals of the country.

For policy makers in ECA, the main challenges include maintaining "fair" competition—including fostering investment and implementing nonburdensome tax laws—as well as protecting citizens' interests, including their privacy and security.

A national digital strategy[5] may be built around the following axes:

- *Connectivity (quality of access to the Internet):* both international and national connectivity, access to wireless and fixed-line services, and cybersecurity
- *Citizens (skills and labor markets):* training, schools, and education
- *Firms (competition and trade infrastructure):* innovation, ecosystems, and competition policy
- *Government (institutions):* public service platforms, digital ID, interoperability and standards, citizen- and business-centric government, efficient government, and open government.

The strategy should include appropriate performance indicators and performance targets for measuring progress.

Regional and International Coordination of the ECA Digital Market

In addition to national digital strategies, regional and international coordination appears to be necessary, at various levels of the regulatory framework. On the supply side, greater coordination would contribute to opening up landlocked countries by promoting infrastructure deployment. On the demand side, international collaboration would enable interoperability among legal systems with regard to cybersecurity and privacy. On the private sector side, the possibility for transforming countries to negotiate with trading partners would help to reduce trade barriers on goods and services. For these reasons and more, the idea of an extended digital single market in the ECA region deserves examination.

A digital single market is a market in which local, regional, and global barriers and restrictions have been removed, so that individuals and businesses can access and use digital activities and services under conditions of data protection and fair competition, whatever their nationality or country of residence. The potential benefits include fostering investment, promoting economic growth and innovation, and empowering citizens. Implementation of a digital single market involves successfully overcoming some challenges, not least of which are managing the potentially conflicting political and economic interests of individual member states and making the significant investments that may be required.

As noted, digital single market initiatives are already under way within the ECA region. Indeed, creation of a digital single market is one of the 10 political priorities of the European Union.[6] Adopted in May 2015, with a view to delivery by the end of 2016, the digital single market in the EU is expected to generate up to €415 billion (US$45 billion) in additional growth, to create hundreds of thousands of jobs, and to transform public services. It has three main pillars:

- To improve the access of EU consumers and businesses to digital goods and services (by, for example, unlocking the potential of e-commerce and simplifying value added tax arrangements)
- To shape the right environment for the development of digital networks and services (by, for example, fostering access to fast broadband and implementing data protection rules)
- To create a European digital economy with high growth potential (by, for example, using big data and cloud services to realize savings)

The EU digital single market is the most advanced in setting an overall vision, but is not the only project of its kind in the ECA region. In 2016, the Eurasian Economic Commission (2016) started discussing the creation of a digital market within the Eurasian Economic Union (EAEU),[7] underlying the "need to generate common approaches to the development of e-commerce, a single digital infrastructure, the ecosystems of the digital space, and the use of digital technologies in state regulation and control so as to improve the business environment of the EAEU member states."[8] Policy priorities include the development of a regulatory and legal framework (including harmonization of legislation within member states), the creation of a digital single market to boost regional trade by using e-commerce, a more intensive use of ICT to improve cooperation between countries at all levels (government, firms, and individuals), and the implementation of joint programs to support the regional digital transformation.

The EU's regulatory framework is a major driver of regulatory reform in the ECA region—not only in its member states but also in candidate countries seeking to enter the EU (such as Albania, the former Yugoslav Republic of Macedonia, Montenegro, Serbia, and Turkey). Also, in 2014, Georgia, Moldova, and Ukraine signed association agreements with the EU for economic and political cooperation as well as alignment of their regulatory frameworks with European ones, paving the way toward closer regional integration in the ECA digital markets. Policy makers should consider building on these initiatives and start thinking of the whole ECA region as a digital single market, providing an extended space in which best practices could be shared and digital dividends reaped by government, businesses, and citizens.

Notes

1. See http://www.eurasiancommission.org/en/nae/news/Pages/12-07-2016-2.aspx.
2. Intensive in routine cognitive and manual skills.
3. In Europe, only 44 percent of citizens for whom data are available visited a government website at least once to get information in 2015.
4. Only 29 percent of citizens returned a form online in 2015.
5. The OECD (2014) proposes additional guidelines.
6. See http://ec.europa.eu/priorities/digital-single-market_en.
7. The EAEU is an economic union of five member states (Armenia, Belarus, Kazakhstan, the Kyrgyz Republic, and Russia) that came into force in January 2015. Its integrated single market encompasses 183 million inhabitants.
8. See http://www.eurasiancommission.org/en/nae/news/Pages/26-11-2015-2.aspx.

References

Eurasian Economic Commission. 2016. "The Digital Space of the Union Means Fundamentally New Opportunities for the Implementation of the EAEU Objectives." Eurasian Economic Commission, Moscow, July. http://eec.eaeunion.org/en/nae/news/Pages/12-07-2016-2.aspx.

European Commission. 2015. "Digital Single Market: Bringing Down Barriers to Unlock Online Opportunities." http://ec.europa.eu/priorities/digital-single-market_en.

European Commission. 2016. "A Digital Single Market Strategy for Europe." European Commission, Brussels, May 6. http://eur-lex.europa.eu/legal-content/EN/TXT/PDF/?uri=CELEX:52015DC0192&from=EN.

Jeffreys-Jones, R. 2016. "State Surveillance Is More Ethical Than Private Sector Intrusions." *Wired Magazine* [U.K. ed.] (January-February):. 97.

OECD (Organisation for Economic Co-operation and Development). 2014. "OECD Principles for Internet Policy Making." OECD, Paris. https://www.oecd.org/sti/ieconomy/oecd-principles-for-Internet-policy-making.pdf.

World Bank. 2016. *World Development Report 2016: Digital Dividends.* Washington, DC: World Bank. http://www.worldbank.org/en/publication/wdr2016.

ECO-AUDIT

Environmental Benefits Statement

The World Bank Group is committed to reducing its environmental footprint. In support of this commitment, we leverage electronic publishing options and print-on-demand technology, which is located in regional hubs worldwide. Together, these initiatives enable print runs to be lowered and shipping distances decreased, resulting in reduced paper consumption, chemical use, greenhouse gas emissions, and waste.

We follow the recommended standards for paper use set by the Green Press Initiative. The majority of our books are printed on Forest Stewardship Council (FSC)–certified paper, with nearly all containing 50–100 percent recycled content. The recycled fiber in our book paper is either unbleached or bleached using totally chlorine-free (TCF), processed chlorine–free (PCF), or enhanced elemental chlorine–free (EECF) processes.

More information about the Bank's environmental philosophy can be found at http://www.worldbank.org/corporateresponsibility.